ERNST VON KHUON: The Invisible Made Visible

The Invisible Made Visible

The Expansion of Man's Vision of the Universe through Technology

Ernst von Khuon

New York Graphic Society Ltd.
Greenwich, Connecticut

Translated from the German by Paula Arno

International Standard Book Number 0–8212–0536–6
Library of Congress Catalog Card Number 73–80230

Originally published in German under the title
DAS UNSICHTBARE SICHTBAR GEMACHT by Econ Verlag, 1968.

Published in the English language 1973 by New York Graphic Society Ltd.,
Greenwich, Connecticut 06830.

Printed in Germany and in The Netherlands.

TABLE OF CONTENTS

PREFACE

May all who pick up this book enjoy hours of diversion: those just leafing through its pages, those giving it a fast reading, and those reading it with critical appreciation.

The 275 photographs, I admit, have been chosen according to my own taste. One could quarrel with this or that selection on the grounds that certain worthwhile pictures have been overlooked or left out. However, I sifted through the accumulated picture material (no less than 1500 photos at the start) several times, weighed the gold nuggets, or what I considered to be such, yielded to reason, and then parted with various favorites. After all, I was limited by the number of pages in the book. To be sure, I did retain some of the pictures I had grown particularly fond of—despite certain arguments against them—confidently expecting that they would give others as much pleasure as they have given me.

This book does not pretend to be a textbook or a book devoted to any particular branch of the sciences. For the specialist I can hardly offer anything surprisingly new in his field. My hope is, however, that he, together with the layman, will find stimulation and enjoyment in the survey of those fields of science that are complementary to his own.

Since I am only an amateur in the natural sciences, being neither an astronomer nor a biologist, I obviously owe great thanks to the many specialists in these fields who in the course of thirty years have taught and advised me. I had the privilege of meeting in person many of the greats in the natural sciences who are mentioned in this book; of course, not Röntgen, Rutherford, or Madame Curie, but Niels Bohr, Irène Curie and Frédéric Joliot, Max von Laue, Georg von Hevesy, and Otto Hahn. May these few names stand for the fifty or so Nobel prizewinners with whom I have had personal contact. That I was able to make demands on their valuable time I owed to my profession, one giving me opportunity to broadcast their message in word and picture, over radio and television, to a wide audience.

I am especially grateful to author Otto Willi Gail (1896–1956), my teacher in radio reporting and my counselor during the first years of television. He made me fully realize something when I was a young man that I had been dimly aware of but, in view of my recent experiences at school, wasn't quite ready to believe: that the natural sciences when approached in the right way truly offer an adventure, that the world in the drop of water seen under the microscope is more spellbinding than the latest headline, that the moons of Jupiter when seen through a telescope are more fascinating than the latest criminal cases, that research can be more diverting than many a so-called entertainment.

Enjoyment of pictures does not substantiate the widespread complaints about superficiality. "Images are powers," wrote the great physician Viktor von Weizsacker, "they give rise to 'imagination'." Pictures that provide a glimpse of the infinite can only expand the dimensions of our thought.

To anticipate the future, one must carefully observe the receding shores as one is carried forward by the stream of time. To truly understand the present, one must study the past. And always one must distrust "what everybody knows." At all times inquisitive man has tried to orient himself within the world around him. From the first, man has seen himself as placed within the incomprehensibly large vault of the heavens with its lights glittering from great distances; that eternal rules were operating there was recognized thousands of years before the invention of the telescope. Today, it is difficult to imagine astronomy without this instrument; yet only three and a half centuries have gone by since the telescope came into existence. How much astronomy, the oldest of all the sciences, was able to comprehend without this seemingly indispensable tool! The Maya devised a calendar that was calculated for a hundred thousand years and was more precise than ours. Priests of several early civilizations were able to predict accurately solar eclipses. The knowledge enabling them to make such calculations both justified and gave them their position of leadership over their peoples.

The history of astronomy is the history of the endeavor to conquer the unfathomable void that confronts the terrestrial observer. Beyond, at an unreachable distance, he sees the rising and setting of the sun, the shifting of the constellations. It is remarkable that men of learning in many different civilizations succeeded in correctly interpreting observations made with the naked eye, in perceiving certain regularities, in calculating the paths of the heavenly bodies as accurately as if these were controlled by a clockwork infinitely more precise than those time-measuring devices thought up by man. In the revolution of the stars was seen the image of divine power and celestial harmony, the eternal rhythm of a world-clock that God had set in motion and would bring to a stop at the end of time. Plato calls the universe "an image-in-motion of eternity, given form by a creative deity to make the eternal present." This sense of awe that struck man at the sight of the firmament thousands of years ago has not left him to this day in spite of all that has been explained since then. This is how the great Immanuel Kant expressed it: "Two things fill the mind with ever-increasing wonder and awe, the more often and the more intensely the mind of thought is drawn to them: the starry heavens above me and the moral law within me."

Four and a half centuries ago Canon Nicholaus Copernicus called astronomy the pinnacle of all intellectual pursuits and the occupation most befitting a free man. That he recognized the apparent movement of the sun around the earth as an illusion was a magnificent victory of critical reasoning over what was commonly accepted as self-evident. Copernicus died in Frauenburg, East Prussia, in 1543; he never looked through a telescope. There was no telescope until 1609. Of course, the telescope could have provided only circumstantial evidence, not proof, to support the accuracy of his system of the universe: The earth is not the center; it "circles" the sun, as do the other planets. The earth rotates once a day around its axis—only the moon circles the earth.

Dethroned Earth

At all times it has been hazardous to challenge the apparently self-evident, to profess new ideas—especially when by doing so one gives offense to the scientific community. Copernicus hesitated to the very end of his life to present his thoughts to the public. His *De Revolutionibus (On the Revolution of the Heavenly Bodies)* appeared only in the year of his death, with a strange preface by the publisher Osiander that changed the whole meaning into its opposite. Osiander was a Protestant; Luther didn't think much of Copernicus's new theory either. The Catholic Church, on the other hand, left it unchallenged until 1616, two generations after his death. In any case, the whole thing was not taken very seriously; it was not considered a rejection of the "eternal verities" or an attack against the authority of the Church.

Giordano Bruno was to find out how dangerous such an attack could be. A Dominican friar who fled his monastery and then roamed restlessly from one university to the other, he proclaimed his fascinating theory of the universe in this way: "There are innumerable suns, and innumerable earths that all circle their suns in the manner of the seven planets of our system. The countless worlds of the universe are no worse and no less inhabited than our earth. Since they, as well as we, receive life-giving rays from a sun, it is impossible for a rational mind to imagine that they would not be inhabited and would not support beings similar to ourselves or even superior. The innumerable worlds of the cosmos are all of the same shape and order, subject to the same forces and laws.... You can be sure that the time will come when everyone will see things the way I see them."

Giordano Bruno's idea of the universe is stated categorically and ecstatically. What the one-time Dominican friar saw before him represented an unheard-of challenge to human imagination, a challenge also to the Church's authority, a vision that shook the whole edifice of the teachings of the "only beatific Church." With the power of his vision he shattered the fragile structure of the medieval world-concept. When he was seized by the Inquisition, the heretic Bruno couldn't hope for clemency. On February 17, 1600, he was burned alive in Rome, sharing the fate that befell the heretic Huss at Konstanz, the Virgin Joan at Rouen, and the many thousands of witches who had confessed under torture. Giordano Bruno never looked through a telescope; he suffered his violent death eight years before it was invented.

The Telescope and the New Horizon

At the Frankfurt Fair of September 1608, a telescope was offered for sale. At that time it was no more than a toy for fanciers of curiosities. Nevertheless, it seems to have created quite a sensation. Galileo Galilei, then a forty-four-year-old professor at Padua, bears witness to this: "Some ten months ago I heard a rumor that a Netherlander has made an eyeglass that enables one to perceive visible objects that are far away as clearly as if one looked at them from close by. About this truly astonishing invention there circulated a number of eyewitness accounts that were given credence by some but not by others. A few days later I received confirmation from Paris, in a letter from the French nobleman Jacques Badouère. This finally prompted me to devote myself completely to the task of investigating the principle and devising the means by which I might invent a similar instrument. I succeeded shortly thereafter, having first immersed myself in the theory of light refraction: first I prepared a lead tube and then I fitted into its ends two glass lenses, one convex and one concave. With my eye held to the concave lens I saw objects three times closer than with the naked eye. Thereupon I made a more precise instrument that enlarged objects more than sixty times. Finally, I succeeded in building so excellent an instrument that I saw objects about a thousand times enlarged and more than thirty times closer than by looking at them with the unaided eye." This letter, written in Galileo's handwriting, shows that without false modesty he considered himself the inventor, yet did not want to conceal that the inspiration had come from someone else. In fact, Galileo had instantly grasped the principle of the telescope and purposefully assembled an instrument that the Netherlander (most likely the spectacle-maker Hans Lippershey from Middelburg) may have arrived at just by happy accident. The facts of the case should probably not be judged by today's standards anyway; there was no patent office then. Galileo, as he hastened to demonstrate his telescope to the Doge of Venice, possibly kept silent about the earlier achievement of a Dutch lens-grinder. In any case, he did make it clear to the Venetian Council of Ten what it would mean to be able—with the aid of his invention—to ascertain at an earlier moment and at a greater distance the origin and intentions of approaching ships. The Council promptly showed its appreciation. Galileo was assured of lifetime tenure as professor at Padua with triple pay. It should not be held against the learned man that he was so quick to look after his own interest. He was, as we would say today, vastly underpaid, and tried to make ends meet as best he could by private tutoring. When the greatest universal genius of all time, Leonardo da Vinci, applied in 1481 for employment at the court of Ludovico Sforza in Milan, he had to list his skills the way a salesman lists his merchandise: that he knew how to build assault machines, that he knew how to devise theatrical effects for court festivities, and so on. Only as an afterthought came his offer to produce "the best in painting, sculpture, and architecture." Galileo too had a right to make himself noticed by the powerful of his day. The scholar showed himself far from shy; he lost no time, cashed in on his success, and trained the telescope that he had recommended to the captains of the Venetian fleet—on the stars.

In quick succession, before the spring of 1610, he discovered what no one—or hardly anyone—before him had seen. Galileo saw the moon differently: "I saw at first glance that it is not the smooth ball about which the fraternity of philosophers babbles. It has heights and depths, large dark stretches that might be oceans, and many circular mountain ramparts that look like the eyes on peacock feathers. On the moon the sun casts shadows, just as on earth. High peaks shimmer in white light while night still veils their base. From the long shadows it can be concluded that the moon's mountains are higher than those on earth." One wonders just what it is about such a description that makes it so oddly affecting. Galileo's notes contain for us who live twelve generations after him only an enumeration of the obvious, a description devoid of the new or surprising. They are statements made in simple sentences although with forceful imagery. What then is it? Wherever one encounters the early, the very first, descriptions by the discoverers, there is this strange poetry, this morning freshness of a first day which can never recur. In the same way everyone remembers from his childhood one special situation, one word, one fragrance perhaps, something that will never again touch one with the same impact.

Galileo's eye at the telescope searched for the planets and found that "their little disks appear to be round and exactly circular like small moons. The stars, on the other hand, are seen by the naked eye never as circular shapes but always as radiant and constantly twinkling. Through the telescope they appear the same, only somewhat larger. The reason is that they do not present themselves to the eye in their simple—so to speak, naked—outline, but shining with a glittering aureole, as if furry with sparkling rays. The angle of vision is not determined by the body of the star but by the aureole that surrounds it."

His next observation, Galileo related, "concerns the matter of the Milky Way. It can be seen so clearly with the aid of the telescope that the whole quarrel that has plagued the philosophers for centuries becomes meaningless, and there is no need for me to become involved in lengthy discussions. The Milky Way is nothing but an agglomeration of innumerable stars grouped together in clusters. Whatever segment of the Milky Way one observes, one immediately sees an immense multitude of stars. . . . What will be considered even more astonishing is the fact that the so-called *nebulae* (fogs) are in reality extremely dense clusters of small stars."

The Moons of Jupiter

After Galileo reported on his observations of the moon, the stars, and the Milky Way, he turned to what he regarded as his most important discovery: "planets which, from the beginning of time up to our own day, have never yet been seen." He related: "As I looked at the heavenly bodies through the telescope during the first hour of the night following January 7 of the current year, Jupiter entered my field of vision. Since I had fashioned a most excellent instrument for myself, I perceived something that had never before been observed: I saw three small, but bright, starlets next to Jupiter. Although I numbered them at first among the stars, they nevertheless amazed me since they appeared to be situated on a perfectly straight line parallel to the ecliptic and seemed to shine more brightly than the other stars of the same magnitude." Galileo saw two of the small stars on the left and one on the right of Jupiter. By the following evening, all three of them had gathered on the right. The evening after, only two remained, and now they had moved to the left side. The third could only have been hiding in Jupiter's shadow since it suddenly reappeared during the very same night. Finally, a fourth starlet, which acted similarly, made its appearance.

For two months Galileo observed these starlets and very soon concluded from their changing positions that they, as he wrote, "revolve around Jupiter, like Venus and Mercury around the sun.... We observe four stars that circle around Jupiter as the moon around the earth, while all of them together with Jupiter revolve in a wide orbit around the sun during a time span of twelve years."

As early as March of 1610, there was published in Venice the *Sidereus Nuncius*, the *Star Messenger*, Galileo's preliminary report on his "observations with the aid of a new type of eyeglass." He had been in a great hurry to put it on paper, obviously fearing that someone else might get ahead of him. That, above all, Galileo did not want to risk, although he was confident that he would make further discoveries. And he did: by the end of July of 1610, he had already noted the outline of Saturn but with his primitive instrument was unable to see the ring; in November he observed the spots on the sun; in December, the phases of Venus.

The great scholar was extraordinarily touchy in regard to questions about priority. He passionately defended his claim to having been first. He knew that he had to rush: with the invention of the telescope, and circulation of the news about it, further astronomical discoveries had become just a matter of time. In fact, perhaps on the very same day as Galileo, Simon Mayr of Gunzenhausen, called Marius, trained his telescope (one obtained from Holland) on Jupiter and also saw the starlets circling the planet. He published his findings two years later, at which time he wanted to name the Jupiter moons "Brandenburg Stars" in honor of his patron the Margrave of Brandenburg-Ansbach, but Galileo had already named them "Medici Stars" for the greater glory of the Grand Duke of Tuscany. Galileo immediately accused Marius of plagiarism; he knew that Marius was acquainted with *Star Messenger*. "This same fellow, Simon Mayr of Gunzenhausen, who loves to deck himself out in borrowed plumes, has shamelessly published as his own, discoveries that I have made." Galileo was especially incensed over the fact that Simon Mayr claimed priority on the basis of the Protestant, not the Gregorian, calendar, which gave him the advantage by more than a week. Historians have looked into the quarrel. Result: Simon Mayr, independently of Galileo and almost simultaneously, did indeed discover the Jupiter moons. He even observed more accurately and calculated more accurately, but was much too slow in publishing. Ten days later than Galileo and Simon Mayr, and independently of both, Jupiter's satellites were discovered again—this third time by Thomas Harriot of Oxford.

The Splendor and the Sorrow of Galileo Galilei

All this proves that scientific discoveries inevitably and closely follow upon technical advances that increase man's capacity to see. Yet, it should not be assumed that the astronomers of the day all reached enthusiastically for this new tool, the telescope. Far from it. Many learned men refused to look through the telescope; to do so would only confuse the mind, as one of them wrote. Other professors did look through the telescope and claimed they couldn't see the moons of Jupiter. Many others did see something, but explained that it was an optical illusion. Angrily, Galileo scolded his opponents for having barricaded themselves in their studies; they would rather consult their Aristotle than take an unbiased look through the telescope.

It became evident that seeing and perceiving are not the same. The eye "perceives" only that which the mind is ready to take in. The idea of an apparently unshakable "truth" can be so powerful that images appearing on the retina simply are not accepted. Pictures on the retina which will provoke confirmation or rejection from an open mind, will not be registered by others, or will be denied or dismissed as optical illusions. Certainly, there were also more personal reasons: inertia, reluctance to learn something new, disinclination to even consider the possibility of something new, or, even worse, reluctance to give recognition to something that might bring fame to another man.

Among those who did not want to see, or who denied having seen, there may even have been some learned men who expected or at least suspected what was to come. Wasn't this carrousel of stars around Jupiter evidence against the teachings about the privileged position of the earth? Was perhaps the earth really just one planet among many, all circling the sun as Copernicus had claimed? Wasn't Galileo, the man who had become famous overnight throughout Europe, heading into dangerous waters in spite of his good connections with cardinals and even the pope? Wouldn't the Church consider its authority endangered once the true nature of this spiritual challenge was realized? Could it permit henceforth an optical instrument to decide how the structure of the universe was to be conceived? What happened next is common knowledge. The telescope reinforced Galileo's conviction that Copernicus had been right. At sixty-nine years of age he stood before a court of the Inquisition. This would not have been inevitable; envy, intrigues, and jealousy contributed their share, and even blunders by Galileo himself who was crossed not only by his adversaries but again and again by his own temper.

The court proceedings ended with Galileo on his knees, hand on the Bible, swearing: "I deny under oath, condemn, and detest with a sincere heart and true faith my errors and heresies. I swear that in the future I will never again say or assert anything in spoken or written words that could arouse similar suspicions against me. So help me God." Punishment was mild. Galileo was placed under house arrest. He retired to his villa outside Florence, near the convent where his daughter was. He was permitted visitors. Descartes, the philosopher of "clear and distinct" thought, visited him there. The poet Milton who passionately defended the right to free expression came to speak with him. Galileo lived on for another eight years; toward the end he was blind.

It is only legend that Galileo, before the Holy Office, pronounced the defiant words: "But it does move." Who would hold it against the old man that he grasped at the offered straw in the hope of living long enough to see the truth, his truth, triumph? Wasn't it just a question of time for Galileo's conviction to become everyone's truth?

In retrospect it seems narrow-minded that the telescope was not immediately accepted for what it is: a magnificent tool for advancing our knowledge of the universe. Yet, to judge fairly, one has to try to go back in time to the mental outlook of the people of that age. Only then can one truly understand the leap forward they were expected to take. The visible world lay before medieval man like an open book illuminating the glory of God. The invisible was the uncontested domain of religious faith. Above and behind the visible, beyond the spheres, extended the boundless realm of the angels and the blessed, and the chasms of the damned. How many pictures have tried to give substance to the figures of this phantasmagorical world! How could men have guessed then that even of their own world only a small segment was accessible to the senses, that there existed in nature the hidden, the never-yet seen. Galileo, in his *Dialogo (Dialogue on the Two Chief Systems of the World),* wrote: "Since in our time it has pleased God to grant the human mind so marvelous an invention which is capable of multiplying the acuity of our eyesight, innumerable objects, previously invisible because of their distance or their extreme smallness, have become visible with the aid of the telescope." It was just this aspect that was so difficult for his contemporaries to understand: Through glasses, one could see things that from the beginning of time had remained hidden from the normal eye. A reference to spectacles, from which the telescope, "this true gift from heaven," had been derived, didn't simplify matters. Spectacles aiding weak eyes could only make more distinct that which was seen clearly by sharper younger eyes without aid. Just how was one to comprehend the claims that through the telescope one could pull the moon across the dizzying abyss of space to within one's reach; that a glittering dot of light would turn out to be a ball-shaped heavenly body circled by small previously invisible moons? Wasn't it natural then to dismiss what one saw through the telescope as optical illusion or even as the work of the devil?

Theories differ regarding the beginning of the modern age; dates of different historical events, inventions, discoveries, new frontiers are all used to substantiate widely divergent opinions. The invention of fire arms, the discovery of America, the development of printing from movable type count as such milestones. One could date the beginning of modern times with equal, indeed with greater, justification, from this event: Galileo turned his telescope on the stars. That is: 1609. From then on everything changed. Man's curiosity about the hidden was constantly stimulated. Evidence mounted that what was visible so far to the naked eye was only a small segment of the real world, and that this wider reality is of the truly temporal. There was a new realization that the invisible holds unimagined shapes, holds new worlds; that it is given to human perspicacity to render them visible. They are only waiting to be "grasped."

Only ten years after Giordano Bruno's death at the stake, that is, soon after Galileo's time, the great German astronomer Johannes Kepler owned a telescope too. "Whoever holds you in his hand, isn't he king, isn't he lord over God's creation?" he wrote in exultation. In an open letter to Galileo, Kepler looked into the future, into our own time of space travel: "Who would once have believed that a voyage across the immense oceans could be calmer and safer than across the narrow Adriatic, the Baltic, or the English Channel? Just provide the ships, or suitable sails for the air of the heavens, and there surely will be men without fear of the horrible void. Let us, therefore, devise an astronomy for the courageous travelers as if they were already at the door, I, that of the moon, you, Galileo, that of Jupiter."

Riddles of the Sun

Galileo's almost total blindness during the last years of his life, due to what today is called detached retinas, may have been a delayed effect of having incautiously trained his telescope on the sun without blackening the lenses. Galileo felt that he had a right to claim that he had also been the first to observe the mysterious sunspots. This embroiled him in a violent quarrel with the German Jesuit priest Christoph Scheiner of Ingolstadt, who not only had also observed this phenomenon, but had recognized its significance. The very first, though, to report on the sunspots had been the East Frisian Johannes Fabricius. He had watched excitedly how the spots traveled across the disk of the sun, how they disappeared over the western rim, and he had waited impatiently for them to reappear ten days later over the eastern edge.

With the observations of Fabricius, Scheiner, and Galileo, scientific investigation of the sun essentially starts; up to the present it has solved many riddles but for each one solved it has turned up new ones. An astronomer of our own day, the Englishman Fred Hoyle, once said, "No literary genius could have invented a story one-hundredth part as fantastic as the sober facts that have been unearthed by astronomical science." What are those facts? Do we see farther than the wanderer in the medieval woodcut who tries to look behind the gears of the heavens?

Long before 1530, when an unknown master created this naive picture, the sun occupied man's thoughts. In ancient Egypt it was believed that the pyramids offered repose to the sun god. The sun found a place of rest on their once smooth stone mantles. The voyage of the sun god in a barque is represented in the tomb of Tutankhamen. On the backrest of his throne, the divine sun caresses the pharaoh and his queen with tiny hands. Ikhnaton, who reigned before his son-in-law Tutankhamen, wrote in a hymn of praise: "Eternal Sun, thou art beautiful, thou art magnificent, thy rays embrace all thy works. Thou didst create the world according to thy desire."

Three thousand years later, during the conquest of the new world, European men stood before the great Pyramid of the Sun of Teotihuacán on a high plateau in Mexico. According to tradition it was consecrated to the sun. The sun sanctuary of Stonehenge in England, fashioned out of rings of huge stones, has already stood for more than three thousand years.

Goethe once said to his secretary Eckermann: "If I am asked whether it is in my nature to venerate the sun, my answer is: Indeed. The sun is a revelation of the supreme being, it is the mightiest that we mortals are permitted to behold. I worship it as light and as the creative power of God which gives life to us and to all the plants and animals with us." This sounds no less ecstatic than Ikhnaton's "Hymn to the Sun." Schiller, for his part, complained that poetry had yielded to gray science: "As our wise men now describe it:/ Soulless a fireball they see,/ Where once his golden chariot guided/ Helios, in silent majesty." The contemporary German solar researcher Karl Otto Kiepenheuer talks of the "sun of the astronomers" and defines it as a "ball of hydrogen with a constant surface temperature of about 10,000°F, rotating at variable speeds." Only three lines further down he confesses that the sun, in spite of its increasing complexity, has lost none of its worshipful beauty and majesty. Artists have tried to make the incomprehensible entity "sun" more understandable in human terms. No other subject exists in as many representations, pictures and symbols.

The question of size and distance of the sun is a challenge to the mind. And so it was already for the learned men of antiquity. Anaxagoras, who lived at the time of Pericles, ventured to make an estimate which would have cost him his life if he had not fled in time. He claimed that the sun was a fireball as large as the Peloponnesus, the southern peninsula of Greece. This was so unbelievable that it was considered mockery of the gods and of common sense.

Today we know by how much Anaxagoras underestimated the size of the sun. The sun's diameter is 109 times, let us say, roughly 100 times, that of the earth. If we assume that the earth is the ⅛-inch glass head of a stickpin, then the sun is a ball of 1 foot diameter. And if we stay with this scale then the pinhead revolves around the ball at a distance of 90 feet.

It is one of the claims to fame of Thales of Miletus, one of Greece's Seven Wise Men, to have accurately predicted a solar eclipse. Four thousand years ago, two imperial Chinese astronomers paid with their lives for having slept drunkenly through an eclipse.

The Austrian nineteenth-century author Adalbert Stifter wrote about the eclipse of 1842: "There stood the glowing sickle, as if cut out of the darkness with a pen knife. The last sunlight was just melting away like the last spark from a dimming wick. The moon was centered over the sun. All around it a marvelous halo of radiance...."

Men of the twentieth century are no less impressed by a cosmic happening such as the visual encounter of sun and moon. Scientists studying the sun have chartered planes for such occasions prolonging the duration of the eclipse by "flying along." Although a jet plane flies at an average speed of 620 miles per hour, the shadow of the moon is so much faster that the aircraft is quickly and unavoidably overtaken and outdistanced by the moon's shadow. Nevertheless, in this way 100 seconds of eclipse are turned into 144 seconds of observation time. The event is much too rare and, even if prolonged, much too short to lose one single second. Eclipses of the sun have been systematically studied for only about a century. During those 100 years, a total of perhaps 120 minutes have been available for observation. Eclipses of the sun provide unique opportunities to investigate the composition of the solar atmosphere.

Just before the onset of a total eclipse, there can be seen—unfortunately only for a few seconds—flames shooting out from the sun's rim, flares of glowing hot gases that attain speeds of 20 miles a second. They surge outward to heights of 60,000 miles, gigantic jets that erupt at speeds faster than sound; an "inferno" that boils with an "ear-splitting noise," although no listening devices can register it; a horrible froth above the furious ocean of gas with its hot geysers rising to the surface. It has been established that these wild eruptions are directed by the mighty magnetic fields of the sun.

After the steadily narrowing sickle of the sun has completely wasted away and total eclipse has set in, an enchanting phenomenon appears: there glows a halo of rays, a soft down of almost metallic luster that reaches far into space, a magnificent vision about as bright as the full moon. This is the corona, visible only because of the shielding effect of the moon during an eclipse, and normally made invisible by the sun's brilliance. The extremely thin gas around the sun, heated in the corona to about 2.5 million °F, probably reaches, at times, beyond the earth's orbit; our planet then moves in the veils of the corona's outermost margins. They may well be causing the wonderful northern lights. However, these links have not been established definitively.

Many aspects of the relationship of sun to earth, the conditions on the surface and inside the sun, are still riddles. "Our theory about the inside of the sun will always correspond closely to the current views of theoretical and nuclear physics and, like them, is constantly changing" (Kiepenheuer).

Our sun is the only one among the countless suns of the firmament whose surface and shape we can observe. All the others remain dots of light, even when seen through the largest telescopes. The distance from the sun to the earth is small when measured by the speed of light. Light needs only eight minutes to cover the distance. But this is 100 million miles, roughly 4,000 times the earth's circumference. What can be detected on the surface of the sun, across this "abyss" of space, through the telescopes of the solar observatories? The smallest visible details extend in reality over some 300 miles; a spot the size of West Germany would still be visible. Just for comparison: on the moon, details of less than 300 feet would still be visible through a good telescope.

It is possible to observe that, as seen from earth, the gasball of the sun rotates around its axis in about four weeks. It does not do this like a solid body. At the sun's equator it takes twenty-five days to complete a turn; at the poles of the sun, thirty days. The masses of sun gases do not rotate in unison.

One should look at the sun only through strongly light-absorbing black glass; it is even better to project its picture through the eyepiece onto paper. If the picture of the sun is unsteady, this is because of turbulence in the earth's atmosphere. If the air is calm, the surface of the sun, when greatly magnified, appears pebbly, almost as if "paved with cobblestones." This is the so-called granulation. The "cobblestones" dissolve and re-form; they change within minutes. As mentioned before, the surface temperature of the sun was established as roughly 10,000 °F. Redhot iron has a temperature of 1,112 °F, the incandescent filament of a lightbulb up to 3,700 °F. It is understandable that the luminous gases would be in constant violent flux. If one trains a spectroscope on the sun to analyze its chemical composition in detail, one finds about 20,000 of the so-called Fraunhofer lines. Kiepenheuer likens the information contained in the sun spectrum to that given in the telephone directory of a city of 200,000 inhabitants.

Since 1930, there exists a device that permits the observation of eruptions on the rim of the sun, and also of the corona, without having to wait for an eclipse. This is the coronagraph of the Frenchman Bernard Lyot. A black conical shutter covers the disk of the sun, just as the moon does during an eclipse. The instrument is brought out of the dust and haze of the lowlands to the pure dry air of high mountains. The perfectly polished lens is made of especially clear glass, without the smallest blemish, the tiniest scratch or particle of dust, because any of these could cause scattered light. Only a little scattered light is enough to outshine the delicate corona and thereby make it completely invisible. The sunspots as they are registered by the solar observatories look like small freckles but are in reality of enormous size. They may have the diameter of the earth or even five times that. They are described as cooling zones. First there appear dark "pores"; these then coalesce into larger spots that in the course of several days blossom to their full size and then fade away. Only rarely do they outlast several rotations of the sun. Strangely, the sunspot activity increases and diminishes periodically. The increase from minimum to maximum takes five to six years. For months the sun appears nearly free of spots. During the maximal period it is speckled. In pictures taken at that stage, through large telescopes, the sunspots look flowerlike. Within these "flowers," unimaginable forces, gigantic magnetic fields, and electric currents are at work. They light the torches, ignite the gases, they cause the bursts of light, the discharges of gas, the eruptions that send forth uncommonly strong ultraviolet and X-ray radiation. They are able to scatter a shower of atomic particles that causes magnetic needles to dance, paralyzes our radio communications, and produces the magnificent aurora borealis.

Through the coronagraph one can see the so-called prominences, jets of very hot gases that burst forth from the sun. They stand out on its rim, glowing red against the dark background, as they flare out into space for a distance many times that of the earth's diameter.

In this captivating spectacle, the magnificent ferocity of our sun is manifest. And yet, science classifies our sun as a star of only average size and average luminosity. It ranges between the "red gigants," which are one hundred to one thousand times brighter, and the "white dwarfs," which are very small, comparable in size to the earth, but extremely hot. A one centimeter cube of material from a "white dwarf" would weigh one thousand punds.

Julian Huxley, the English biologist and philosopher, has tried to give us an idea of what the unimaginable multitude of these terrifying sums in the universe means: "... Nowhere in all its vast extent is there any trace of purpose, or even of prospective significance. It is impelled from behind by blind physical forces, a gigantic and chaotic jazz dance of particles and radiations...."

Probing the Depths

The world of the stars, the world of the constellations, has always stirred man's imagination, has touched the heart of simpletons and challenged the mind of thinkers. Ovid, the celebrated poet, saw the gods ascend to the palace of Zeus along the Milky Way, the ceremonial road where the immortals dwelled. According to a Greek legend, painted by Tintoretto in 1575, the Milky Way's origin goes back to the moment when Zeus placed Hercules, born to him by Alcmene, at the breast of his sleeping wife Hera. The boy took such a mighty draft of the divine milk that it splattered all over the heavens.

The medieval world believed the stars to be affixed to a spherical shell, the outermost of the transparent spheres. The telescope pierced this shell, and Galileo recognized how much deeper space must be: "Beyond the stars of sixth magnitude you see through the telescope a host of others that are hidden to the naked eye. They are so numerous that it is hard to believe."

Since Galileo's time scientists have repeatedly plumbed the depth of the cosmos. Time and again their calculations were overthrown. Using earthly measurements required unwieldy numbers that the non-mathematician could not even name. Therefore, the astronomers introduced a cosmic unit of measurement: the light-year, the distance that a ray of light travels in a year. They still arrive at a number in the billions, just as impossible to grasp, but at least nameable.

The measuring of the speed of light was undertaken by the Dane Ole Römer, sixty-six years after the discovery of the Jupiter moons, and after careful observation of their orbits. The largest of them circled the planet in 42 hours, 27 minutes, and 33 seconds. It regularly disappeared into the shadow of Jupiter and reemerged from it. The going-out and flashing-on of its light could be timed to a fraction of a second. Ole Römer found that the Jupiter clock was inaccurate. It was progressively slow week after week, and then, after half a year, it speeded up more and more. He found that the Jupiter moon became slower the farther the earth in her year-long travel around the sun receded from the much more distant orbit of Jupiter. Conversely, the moon persistently regained the lost time as soon as the earth drew closer to Jupiter again. This led to the conclusion that the light ray is not simply here instantly, as had been believed. It needs time for its journey. Römer's calculation showed that light travels at fantastic speed. He came very close to the correct figure of 186,000 miles per second! Thus was given the unit of measurement for the universe: in a year, light travels 365 times 24 times 60 times 60 times 186,000 = about 6 trillion miles. A numeral with 12 zeros, shortened to the concept "one light-year." It is a monstrously large measure if one attempts to grasp it. For the school children of the twentieth century it is a commonplace concept.

Based on present-day observation, the universe is described by Walter Baade, astronomer at the Observatory of Mount Palomar, as "a sphere with a 2 billion light-year radius, with billions of star systems scattered within." Nevertheless, the universe appears "almost empty." The idea of "three bumblebees over Europe" illustrates the density of the star systems in the universe. Were one to imagine our sun and the closest other sun, Alpha Centauri, as two pinheads, the distance between them would be about twenty miles.

Albert Einstein undertook to calculate the radius of the universe: 5.8 billion light-years! A ray of sunlight cosmically bent would return to its source after 56 billion earth years.

Einstein thought of the universe as being composed of the three dimensions of space plus a fourth one, time. James Jeans has undertaken the probably impossible task of rendering the unimaginable in the guise of an immense soap bubble: "The universe is not the interior of the soap-bubble but its surface, and we must always remember that, while the surface of the soap bubble has only two dimensions, the universe bubble has four—three dimensions of space and one of time. And the substance out of which this bubble is blown, the soap-film, is empty space welded onto empty time."

The refracting telescope that uses lenses is followed by the reflecting telescope that uses a mirror. A concave mirror ground in the shape of a parabola focuses the starlight on the eyepiece or else on the photographic plate. Some thirty years ago an optician from Hamburg, Germany, Bernhard Schmidt, invented the so-called Schmidt-mirror which combines features of both kinds of telescope. The light passes through a very thin lens before it hits the concave mirror, the shape of which is not parabolic but spherical. The thin lens counteracts the distortions of the concave mirror. Thus the Schmidt-mirror is able to sharply project a wide field of vision.

Through the telescope on California's Mount Palomar one can see into space to a depth of 2 billion light-years. This telescope is one of the present-day "wonders of the world." It is the technological eye through which we penetrate farthest into the cosmos. A concave mirror collects traces of light from the most distant suns. It is 200 inches in diameter and weighs 15 tons. When it was first mounted in 1948, one tiny imperfection on the polished surface was discovered. In one spot, about 20 inches from the rim, a tiny bump of 1/2,000 millimeter was found. So the mirror was removed from its mounting and the tiny unevenness was ground down.

The instrument on Mount Palomar is not a mammoth telescope but rather a precision camera which is able to automatically track a given target and to register on a photographic plate, during exposure of many hours, traces of starlight—light that reaches us after a journey lasting millions of years. What we see is actually the past, the unimaginably distant past. The stars in the remote depths of the cosmos could have been extinguished long before man existed on this earth and yet we would still see them.

It is highly questionable whether the "eye" on Mount Palomar could be optically surpassed. As any photographer knows, it doesn't make sense to enlarge a slightly fuzzy picture beyond a certain size. We just don't get sharp enough pictures of the stars, not even from the top of Mount Palomar under the clear California sky. Even though the gas shroud that veils our planet is very thin, indeed, by cosmic standards, we nevertheless live as on the bottom of an ocean, and what lies outside can only dimly penetrate.

Clearer pictures could not be hoped for before it became possible to leave the earth's atmosphere. But where really are the limits of the atmosphere? At a height of 75 miles or possibly as far out as 125 miles. Seventy miles out, there is already twilight, the silence of space; the veil of air is so thin that it no longer transmits sound. At 125 miles out, the remaining air is so rare that it exerts almost no braking power even on very fast-moving bodies. Then, after leaving behind the last most tenuous veils of the earth's atmosphere, there begins the great adventure: space travel, that extreme effort in the history of technology that will help produce a "clear" view of the universe.

There are only two narrow "windows" in the atmosphere that envelops the earth, which means that of all the rays that reach us from the stars only those confined within certain narrow bands of wavelengths get through. One window admits the visible light with wavelengths of between 4,000 and 7,000 angstrom (angstrom = 250 millionth part of an inch). This window is made of streaky glass, so to speak, since the varying density of the air deflects visible light. Through the second window pass the radio emissions from certain objects in space, on wavelengths from .6 centimeters to 30 meters. Enormous dish antennas receive these signals. These do not compose themselves into pictures but into curves from which lines of similar radiation intensity can be derived. The corona of the sun and the eruptions on its surface are sources of intense radio emissions, as are the sunspots.

Abnormally strong radio signals come from the Crab nebula, the remnants of a supernova whose explosion was observed by Chinese astronomers in the year 1054. However, most "radio sources" can not be connected yet with any visible objects. Radio astronomy grasps by means of its antennas a world of worlds that so far can be made "visible" in no other way.

Wavelengths other than those admitted through the two "windows" are blocked by the earth's atmosphere. They can be examined only "out there."

Space Travel

The development of rocket technology was promoted by military considerations first, and then a desire for political prestige. The men who make the fascinating dream of journeys of space exploration a reality, see the arms race of yesterday and today as a jumping-off point for the future. They count on reason in politics.

In 1949, the upper stage of a two-stage rocket climbed to a height of 250 miles. At this distance, the gas covering around the earth is already as thin as cobwebs. The sky no longer looks blue but pitch-black. The stars, even during the day, appear as steady lights. But then, the rocket was at a height forty-five times that of the Himalayas. Yet, that was only one thousandth of the distance to the moon. With only four times its speed though, the rocket could have already escaped the earth's gravity.

On October 4, 1957, the age of space travel became audible reality for everyone: an artificial satellite sent out its electronic signals as it circled the earth every ninety-five minutes. It was Sputnik 1 (*sputnik* = "fellow traveler"). One month later followed the second Russian artificial moon, Sputnik 2, with a dog on board. Slowed down by the outer fringes of the atmosphere, the Sputniks, after 92 and 162 days respectively, plunged into the denser layers of the atmosphere and burned up like shooting stars.

Meanwhile, on January 31, 1958, the United States had succeeded in placing its first satellite, Explorer 1, into orbit; in March 1958 the American Vanguard followed. Compared with the Russian heavyweights, it is an "orange" but its tiny instruments and transmitters convey valuable scientific data. And this miniature moon of the dawn of the space age will orbit for centuries. Explorer 1, thanks to its sensitive instruments, discovered the existence of the closer of the two Van Allen Belts. These are streams of radiation particles trapped by the earth's magnetic field that spiral about the magnetic lives of force from pole to pole. Today's artificial satellites number in the hundreds. The data radioed back from these forerunners in space yields new information about the magnetic and radiation fields within which our planet moves. We have gained a clearer picture of the forces at play in the outer reaches of the earth's atmosphere.

The so-called Tiros (Television-Infrared-Radiation-Observation-Satellite), Nimbus, and several other United States as well as Russian weather satellites have televised hundreds of thousands of pictures to earth. The gaseous mantle of the earth, whose lower layers are the cauldron where weather brews, is now being checked from space. Television cameras watch the formation and progress of cold and warm fronts and hurricanes. These satellites can be quizzed from the ground and they transmit their pictures to the weather bureaus. The cloud cover of the earth, which previously could not be surveyed in its entirety because of the vastness of the oceans, is now being photographed at regular short intervals. Thus, meteorological interconnections become recognizable. Infrared detectors probe the earth's surface through the atmosphere. The difference between the incoming radiation and that which is returned to space is the amount of energy that determines weather activity in our atmosphere.

At a speed of from five to six miles per second, satellites are still bound to earth in a more or less elliptical orbit. Only at a speed of about seven miles per second, the so-called escape velocity, can they break away from the gravitational pull of the earth. On January 2, 1959, the Russian Lunik 1 (a combination of "*luna*" and "*sputnik*"), carrying a 770-pound instrument package, sped by the moon at a distance of 4,660 miles. It was possible to photograph the yellow glowing sodium cloud that was trailing behind it. Today, this rocket is an artificial planetoid, in a fifteen-month solar orbit, closer to Mars than to the earth at its farthest point from the sun. Two months later, the United States' Pioneer 4 also missed the moon and became the second artificial planetoid. On September 13, 1959, the Russians succeeded in hitting the moon.

The second anniversary of Sputnik 1 was celebrated by the Russians with the launching of Lunik 3 which sent back pictures from the dark side of the moon. They did not show much detail yet, but enough to map the far side of the moon for the first time. This map was perfected in 1963 with information from the outstanding pictures taken by the Soviet Zond 3. Thus began a new phase of "television," extraterrestrial television, which sends pictures of neighboring heavenly bodies. The next United States' lunar effort aimed at the transmittal of pictures of the moon at close and closest range, even during the last minutes and seconds before the probe hits the surface of the moon at a speed of 5,800 miles per hour. Ranger 7 in July 1964, Ranger 8 in February 1965, Ranger 9 in March 1965, have transmitted views of the moon's surface such as no telescope on earth had ever attained. The accuracy of the aim was astonishing: Ranger 9 hit only four miles from its target in the crater Alphonsus. In order to properly appreciate the marksmanship one has to consider the factors that enter into the calculations of a moonshot: the rotating earth, the enormous angle of lead for circling moon, the engine shutoff of the rocket which then proceeds without propulsion, the effect of the gravitational pulls of the earth, the moon, the sun, and the neighboring planets.

Less than eight years after placing the first artificial moon in orbit, American technicians sent in the same way a space probe to the immediate vicinity of Venus. Mariner 2 left the launching pad in August 1962. On December 14, it passed Venus within 21,600 miles, a distance of less than three earth diameters. The detectors of Mariner 2 were switched on by radio over a distance of 35 million miles. Mariner 2 began to transmit its findings: "hothouse" Venus was, as expected, shrouded in thick clouds. The infrared-sensors measured a surface temperature of 800 °F. Is Venus unapproachable for human beings, a goal only for robots impervious to heat?

On October 18, 1967, the Russian space probe Venera 4 made a soft landing on the surface of the earth's neighboring planet; before its radio went dead, it reported a temperature of 520 °F. The following day, the United States' probe Mariner 5 passed within 2,500 miles of Venus and, in near agreement with Venera 4, recorded a temperature of 500 °F.

The vicinity of Mars was reached by Mariner 4 on June 14, 1965, after a flight of 240 days. At the launching on November 28, 1964, the targeted point of closest approach, as seen from earth, was "behind the sun." When the spacecraft met the red planet–it passed within only 6,118 miles!–the earth had greatly outdistanced Mars and the probe. The precision of this flight overshadowed everything so far achieved. On the seventh day, Mariner 4's course was corrected. A sensor, a sort of photoelectric cell, oriented itself by the light of the star Canopus. Just imagine: a technical instrument, equipped with sensitive robot eyes is capable of picking out a lodestar–granted that it is a very very bright one–in the ocean of stars. Across a distance of 135 million miles radio signals reported the information that Mariner 4 had been able to gather during its flyby of Mars: the martian atmosphere is extremely thin, making a parachute landing as unlikely as on the moon. The atmosphere on Mars consists largely of carbon dioxide with traces of water vapor. The white polar caps on Mars are probably just a thin layer of hoarfrost.

Twenty-one televised pictures, each made up of 200 rows of 200 dots that ranged in 65 gradations from black to white, showed the surface of Mars. Contrary to the prevailing expectations, it does resemble the surface of the moon. There are ringshaped craters–mountains rise, from some of the craters, just as in the lunar landscape. Transmittal of the twenty-one television pictures took ten days. The arriving radio signals were reduced in strength to an unimaginably weak .000,000,000,000,000,002 watts. Even if one or the other signal was lost through static or became "illegible" because of atmospheric disturbances, the pictures received are amazingly complete. They are irrefutable doccumentation obtained from so far away that to get there, light would have to travel almost a quarter of an hour; an express train, day and night for 300 years. At the time it was a long-distance record for television.

The first soft landing on the moon was made on February 4, 1966, by the Russian spacecraft Luna 9 on the eastern rim of the so-called Ocean of Storms. Luna 9 was an automated space-probe equipped with a television camera that had a range of about a mile. From the pictures transmitted back to earth, scientists concluded that the vehicle had landed on a surface of black basaltic lava densely strewn with rock fragments that were either of volcanic origin or parts of the meteorites that had crashed into the moon over millions of years ago. Extreme temperature changes, and radiation untempered by any atmosphere, have broken up the material without thereby creating the expected deep layer of dust. Immediately in front of the camera, details of one sixteenth of an inch, and on the horizon, fragments only three feet high were clearly visible. The sun, low in the sky, cast long shadows. The sky was pitch black. A component thrown clear of the space vehicle, proved that the landing was none too soft.

Four months later the first United States probe, Surveyor 1, landed on the burning-hot moon surface 6.8 miles from its target in the Ocean of Storms. After the retrorockets had slowed down the craft, the landing was surprisingly soft. When it first touched the surface, it bounced back 2.5 inches, coming to rest after some slight swaying. A few seconds later the probe reported itself operational. A radio command, transmitted by a huge antenna in California, called in the first series of pictures. More than 10,000 photographs of the moon were received in the course of twelve days. In the meantime, the surface temperature of the moon had risen to 250 °F; then the sun set. Fifty-two hours after the onset of the fourteen-day night, the temperature of the solar cells had dropped to −300 °F. There was little hope that the batteries would survive this drop in temperature. When day returned to the moon it became evident that they were in part able to recharge themselves. The moon probe transmitted another 800 pictures. The eye of the camera was directed almost vertically upward, toward a mirror with a 360-degree sweep. Color screens produced color selections; they showed that the moon surface is a neutral gray. The pictures are of outstanding quality, of razor-sharp definition as far as the horizon. They show a harsh desert, obviously totally devoid of life. Since the moon is "naked," without any covering atmosphere, there is also no haze to blur the contours of the mountains in the background. It is a stony world, scarred by the hail of meteorites that for eons has kept up a bombardment from space. This then is the surface of the moon; this, the landing place that a human vanguard was soon to reach. The eye looks in vain for the out-of-the-ordinary. Nothing. Only the location itself is strange.

Now the question is, what are the limits of the adventure of space travel? How far out can man go? It is conceivable, even likely, that men of the waning twentieth century will head for the neighboring worlds in our solar system, and that with some luck, or maybe already with some certainty, they will return. What are likely goals? The innermost planet, Mercury, because of its closeness to the sun is an "inferno." The outermost planets are worlds of frozen rigor. Between Mars and Jupiter a numerous tribe of minor planets (asteroids) orbits around the sun. They will have cooled off more than Mars, be easier to survey, and will have less gravitational pull.

Does another moon perhaps offer the ideal landing place? At present, thirty-two moons are known in our solar system; the last one, discovered (on December 15, 1966) by the Parisian astronomer Audouin Dollfus, was a tenth moon of Saturn, probably covered with ice as are the other nine. The second largest Jupiter moon, Jo (one of twelve), may possibly have an atmosphere. It emerges from Jupiter's shadow covered with hoarfrost that melts in the light of the sun. The two Mars moons, Phobos (37 miles diameter) and Deimos

(7.5 miles diameter), could be curious goals. A space traveler on Deimos would be all but weightless. A soccer ball, forcefully kicked into space, would soon be out of sight.

A speed of ten miles per second, double that of the artificial satellites, would be sufficient to leave the solar system. The likely planets of the nearest star are at a distance of four light-years from earth. Return from there after ten years would be possible if in the more distant future speeds approaching that of light are to become attainable. Photon-propulsion might make this feasible; it supposedly is capable of accumulating the seemingly minimal pressure of light. Amazingly, it follows from Einstein's theory of relativity that for men inside a spaceship traveling at nearly the speed of light, time would elapse more slowly than on earth. It is difficult to make these unimaginably daring concepts of Einstein's "wonderland" comprehensible for non-scientists. This much is certain—no perils will deter man from venturing into deep space, from wanting to see the reachable with his own eyes. And traveling with the speed of light, the first picture documentation will have arrived on earth long before the spacemen have come home.

"No literary genius could have invented a story one-hundredth part as fantastic as the sober facts that have been unearthed by astronomical science." (Fred Hoyle)

The first great target of space travel: sub-orb Luna.

In September 1608, a telescope was offered for sale—as a curiosity—at the Frankfurt Fair. Galileo Galilei, mathematics professor at Padua, heard rumors about it. Soon after, he was able to re-invent the telescope. He immediately offered it to the Council of Venice for use by the captains of the fleet. Rewarded with a raise in salary, he returned to his study to make the telescope what it is today: a scientific instrument. Galileo trained it first on the moon; he saw ring-mountains and dark, flat areas which he took for oceans. To this day a walk on the moon by means of a telescope is an unforgettable experience. The photo on the right was taken with the Zeiss-Refractor of the Deutsches Museum at Munich. The moon is in its last quarter. Exposure time, 1/25 second.

Ing. Ernst Klee, Munich

A robot takes a look.

Pressagency Novosti (APN), Moscow

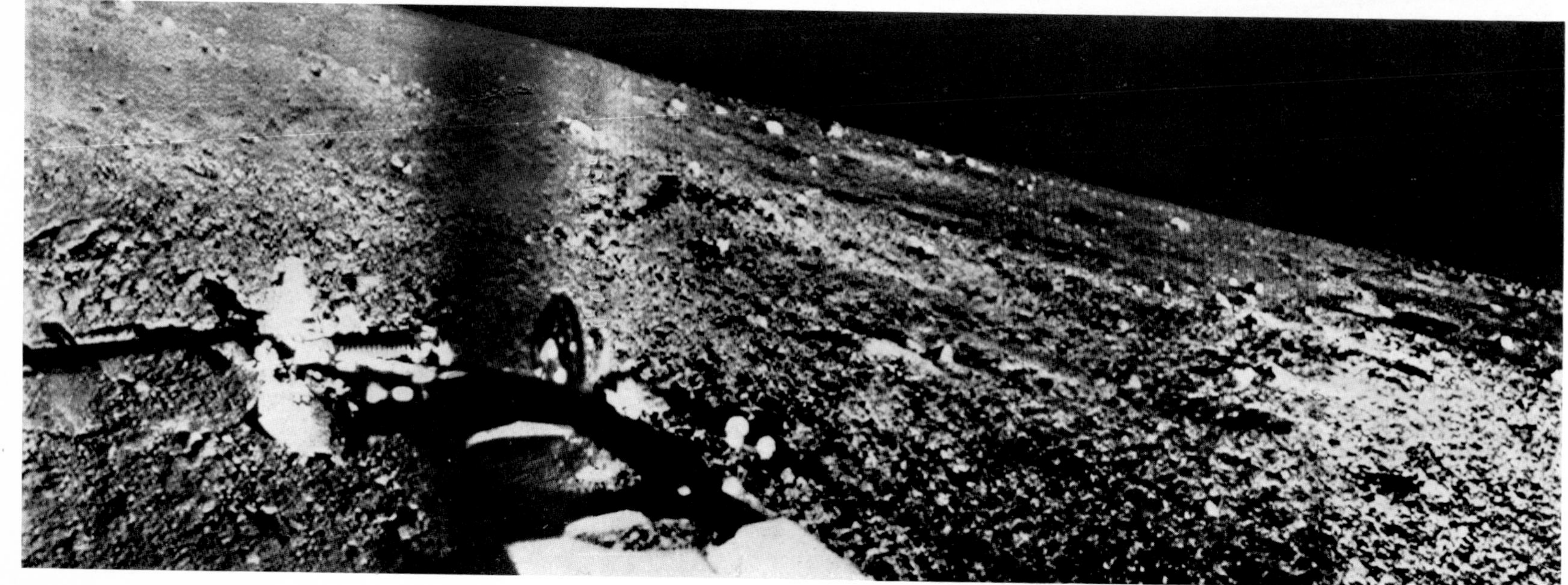

On February 3rd, 1966, the Russian space probe Luna 9 became the first vehicle from earth to make a soft landing on the lunar surface. The course of the trajectory had been corrected from earth. On December 24, 1966, Lunar 13, the second USSR landing craft, reached the lunar surface in the Ocean of Storms, the same area where Luna 9 had touched down. The photos were transmitted by radio waves. Scientists concluded that the surface shown consists of basaltic lava strewn with rock fragments. Parts of the spacecraft intrude in one of the pictures.

Pressagency Novosti (APN), Moscow

The moon from thirty miles up:
a sub-planet without an atmosphere—
a lifeless desert.

Four months after Luna 9, Surveyor 1, the first American probe, landed on the immensely hot moon surface that is unprotected by any atmosphere. On November 7, 1966, the American Lunar Orbiter 2 was propelled into an elliptical orbit around the moon. This picture was transmitted to earth on November 28. It shows the crater Copernicus from an altitude of thirty miles. In the background are the 3,000-foot-high Gay-Lussac foothills of the Moon Carpathians, barely 200 miles distant. The horizontal lines are the seams between individual picture strips televised to earth.

USIS

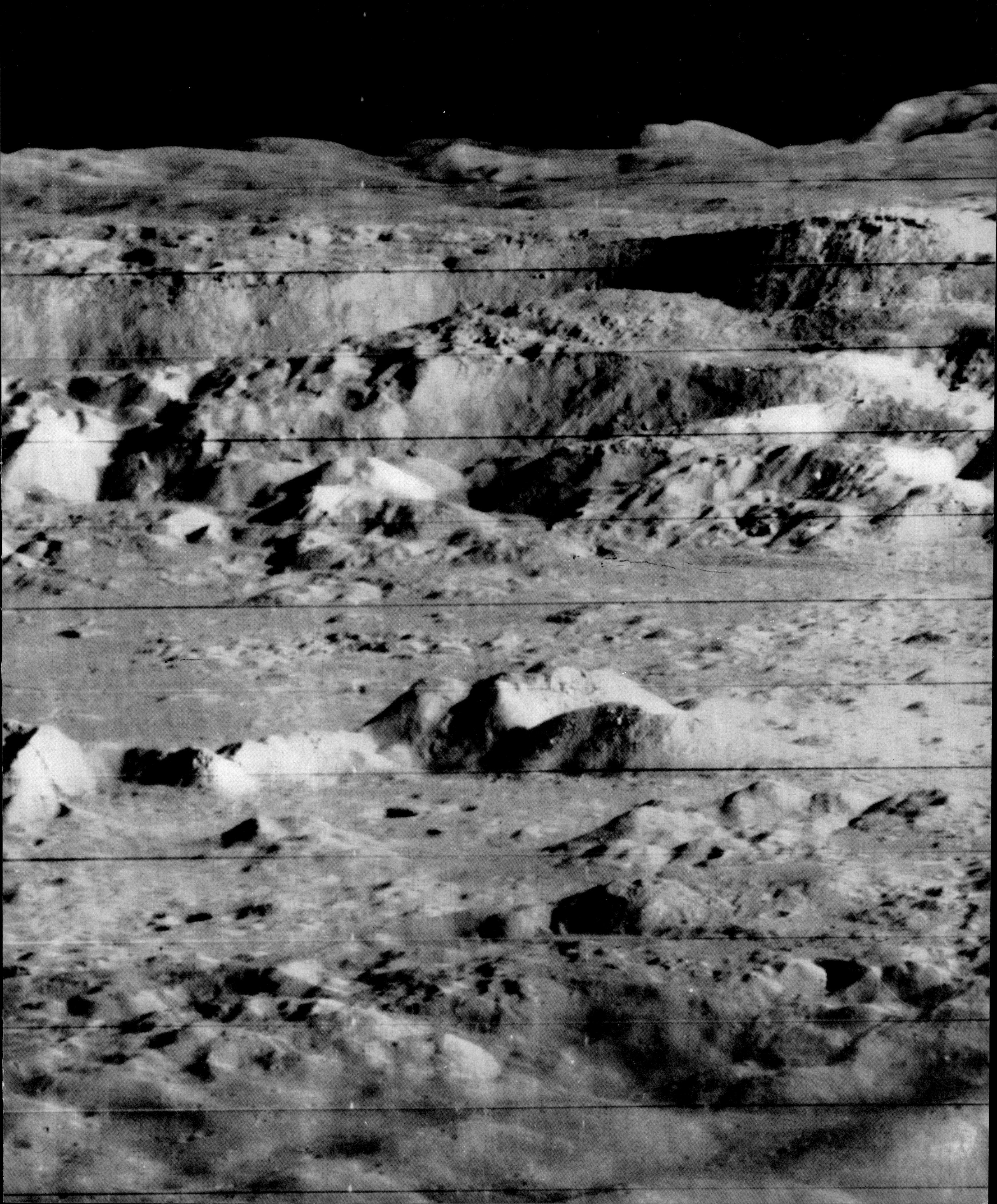

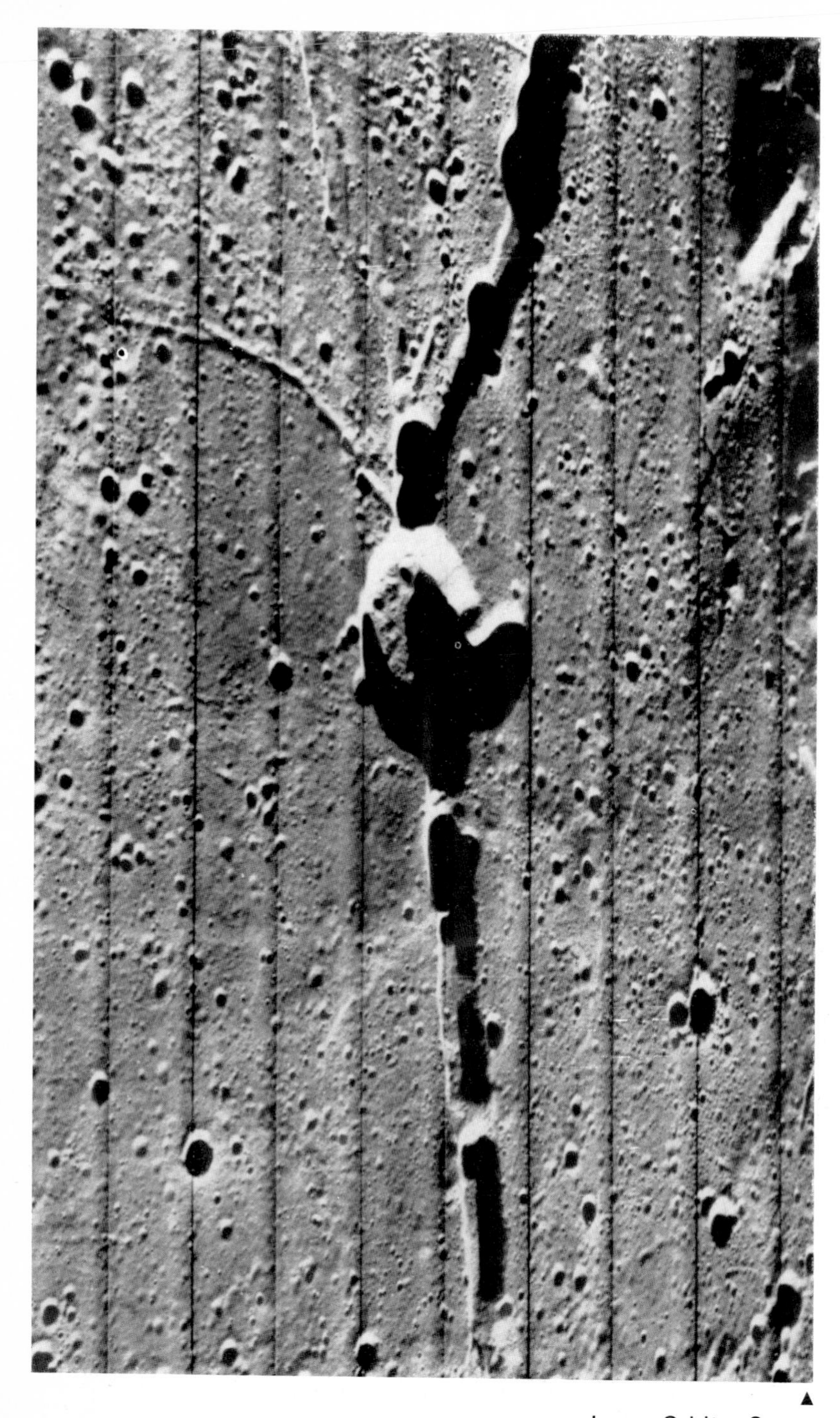

▲
Lunar Orbiter 3 was launched from Cape Kennedy on February 4, 1967, and sent into orbit around the moon. It transmitted close-ups of the so-called Hyginus Rill. The area was considered unsuited for the landing of a manned spacecraft.

A wide-angle shot taken ▸ by Lunar Orbiter 3: The crater Kepler (top center) in the so-called Ocean of Storms measures twenty miles across; it is a little over a mile deep.

USIS

Ing. Ernst Klee, Munich

The moon's surface as seen from earth. Enlarged detail of a photo taken by the twelve-inch Zeiss-Refractor of the Deutsches Museum at Munich.

USIS

This picture was made with a considerably larger telescope: the crater Copernicus, as photographed by the 120-inch reflector at the Lick Observatory, Mount Hamilton, California.

USIS

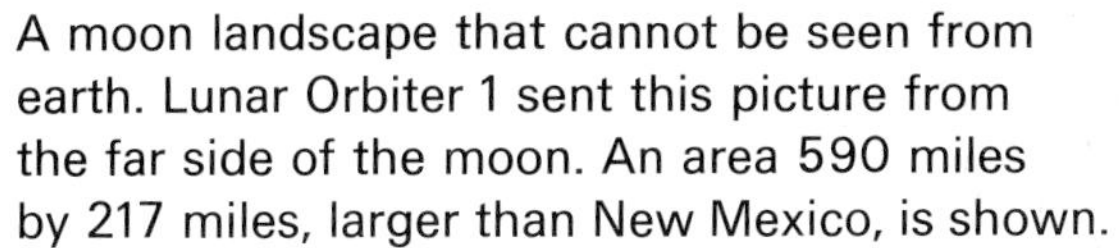

A moon landscape that cannot be seen from earth. Lunar Orbiter 1 sent this picture from the far side of the moon. An area 590 miles by 217 miles, larger than New Mexico, is shown.

USIS

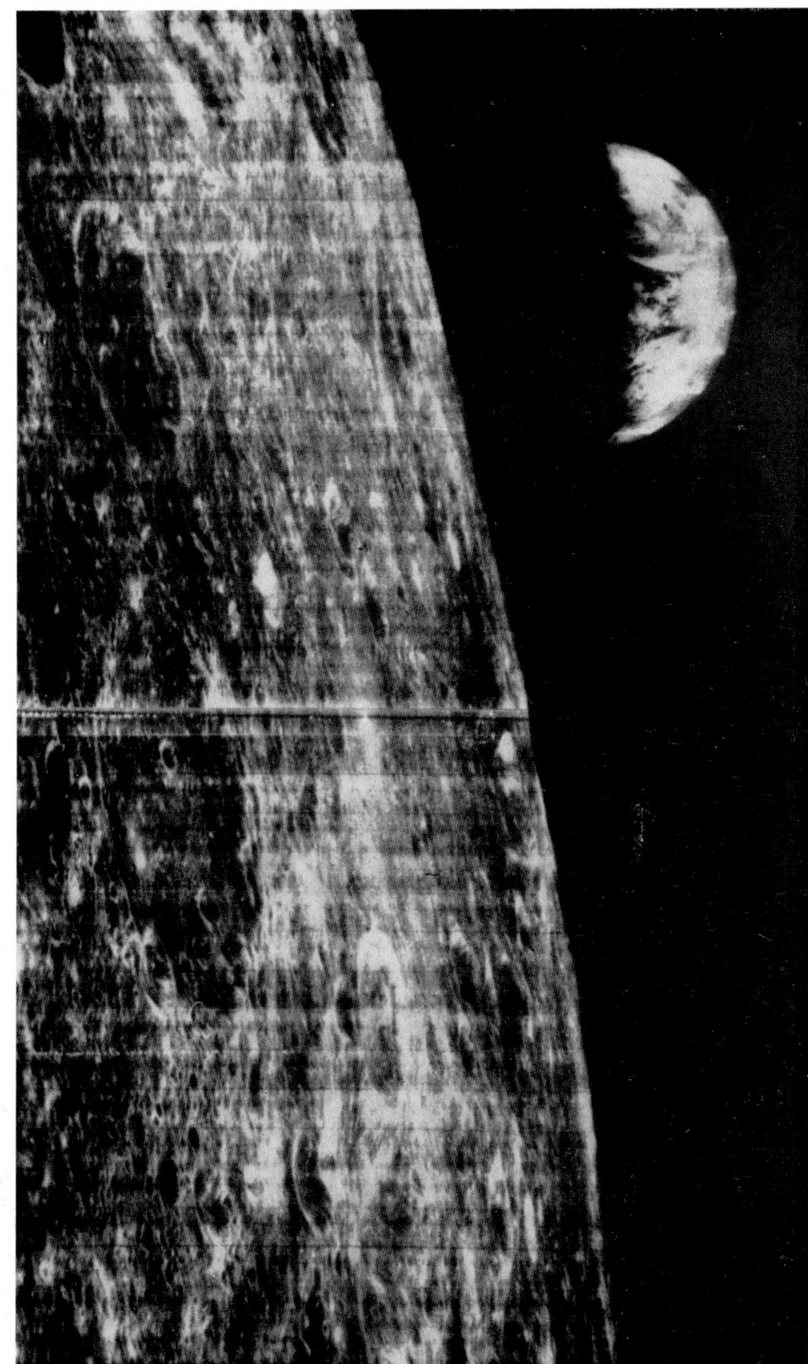

This unusual view was also taken by Lunar Orbiter 1 during its trip around the moon: the first picture of the earth as seen from the moon. The satellite travels at a distance of 745 miles from the moon at a speed of 2,200 miles per hour.

Deutsches Museum, Munich

J. Focas, H. C. Camichel, A. Dollfus

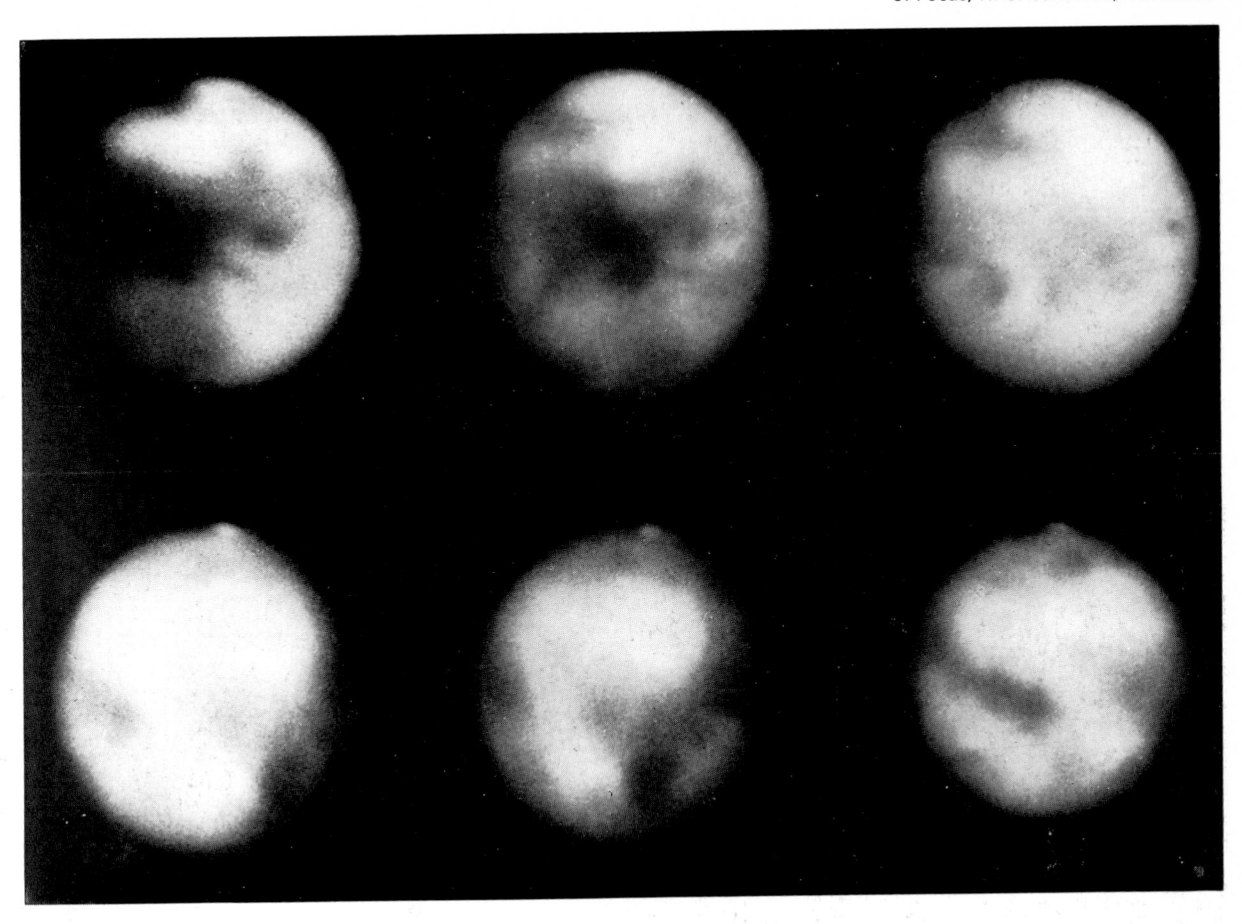

USIS

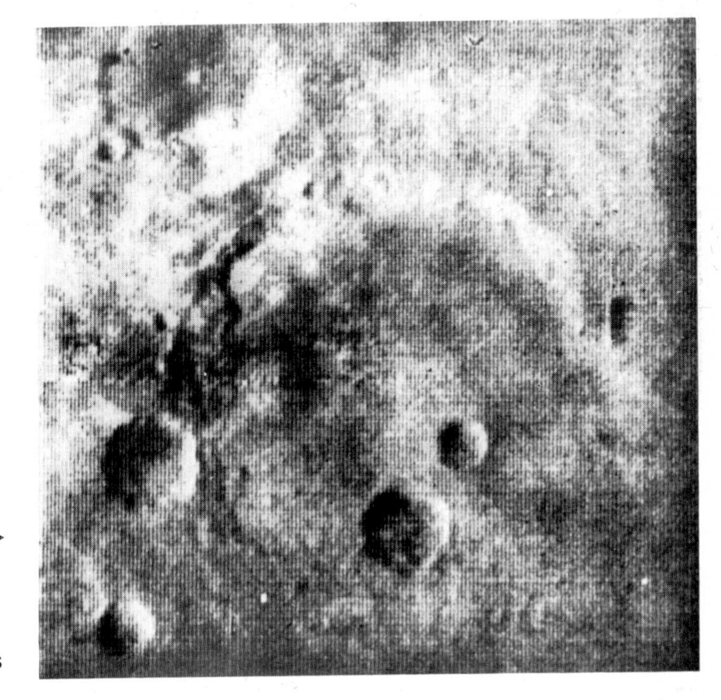

The six photos of Mars were taken through a telescope on the Pic du Midi in the Pyrenees. They show the polar caps that appear in the Martian fall and disappear in the Martian spring. The caps most likely consist of frozen carbon dioxide dry ice and possibly some water.

◄ The tail of a comet, no matter what the comet's trajectory, always points away from the sun; its gas molecules are probably being pushed away by pressure from the sun's radiation. One of the most impressive ones was the Morehouse comet of 1908.

On July 14, 1965, the space probe ► Mariner 4, after a trip of 228 days, passed the planet Mars. It flew by at a distance of about 6,000 miles, a distance less than the earth's diameter. The televised photo shows that crater-studded Mars looks somewhat more like our moon than like the earth.

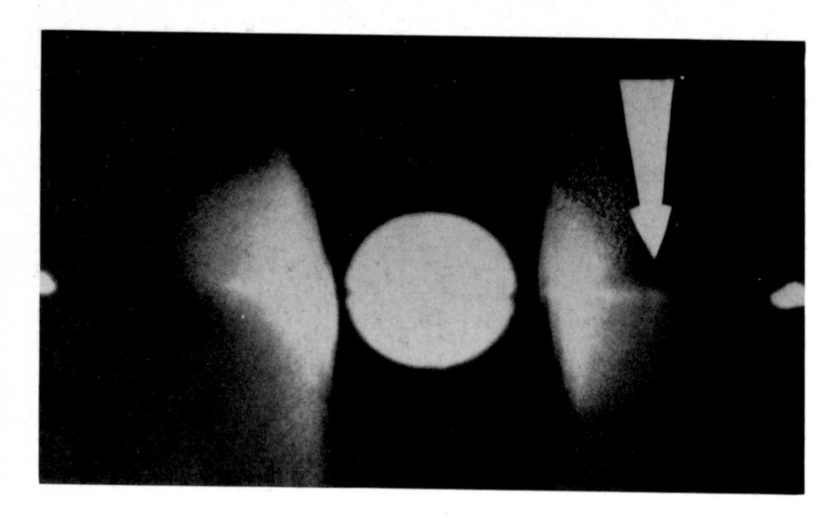

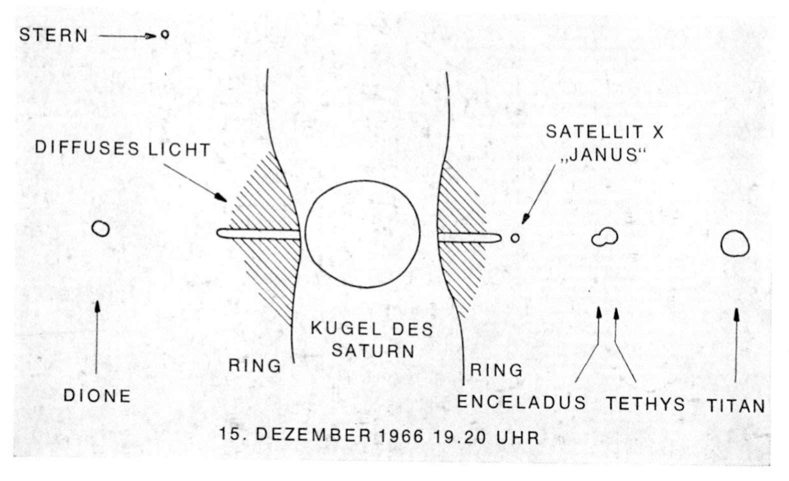

The most beautiful planet is Saturn. Depending on the position of the earth, the rings appear in varying perspective. The picture on top was taken in 1942, the second in 1950. The rings consist of rock fragments and cosmic dust. The third picture documents a discovery: on December 15, 1966, the French astronomer Audouin Dollfus detected a tenth moon of Saturn, Janus. A photograph by the discoverer shows the new moon as a pale speck of light just barely visible against the brightness reflected by the planet and its rings. The arrow points to the newly detected moon.

Jupiter, the largest planet of our sun, with Ganymede, one of its twelve moons. The horizontal stripes are cloud bands. The nature of the large oval spot, "the Red Spot," is still an enigma ▸

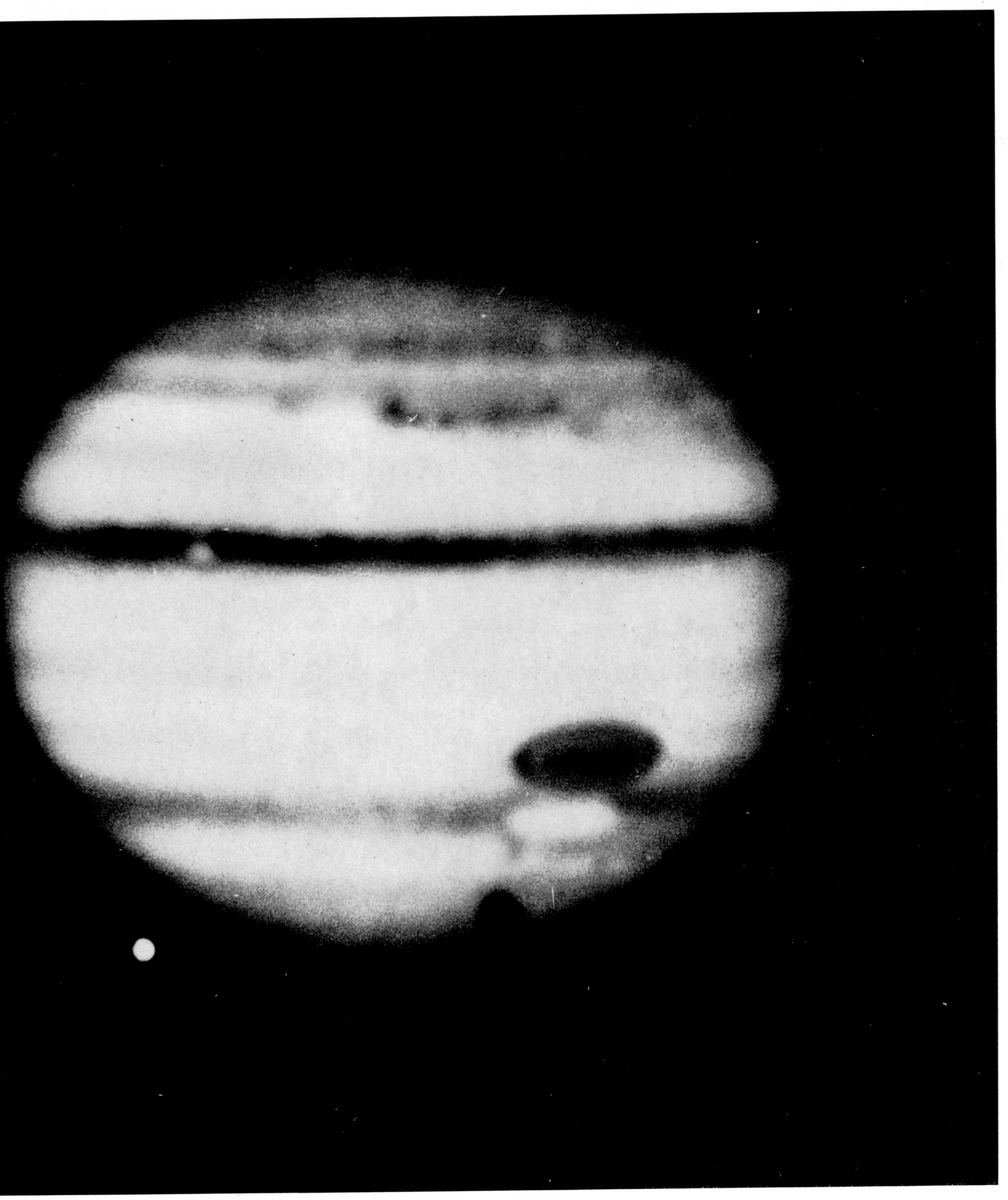
USIS

USIS

Our sun is of only average size among the suns, yet 300,000 times the mass of the earth! During a total eclipse the "corona" shows its radiance, a halo as of metallic down, consisting of solar plasma that reaches far out into space.

USIS

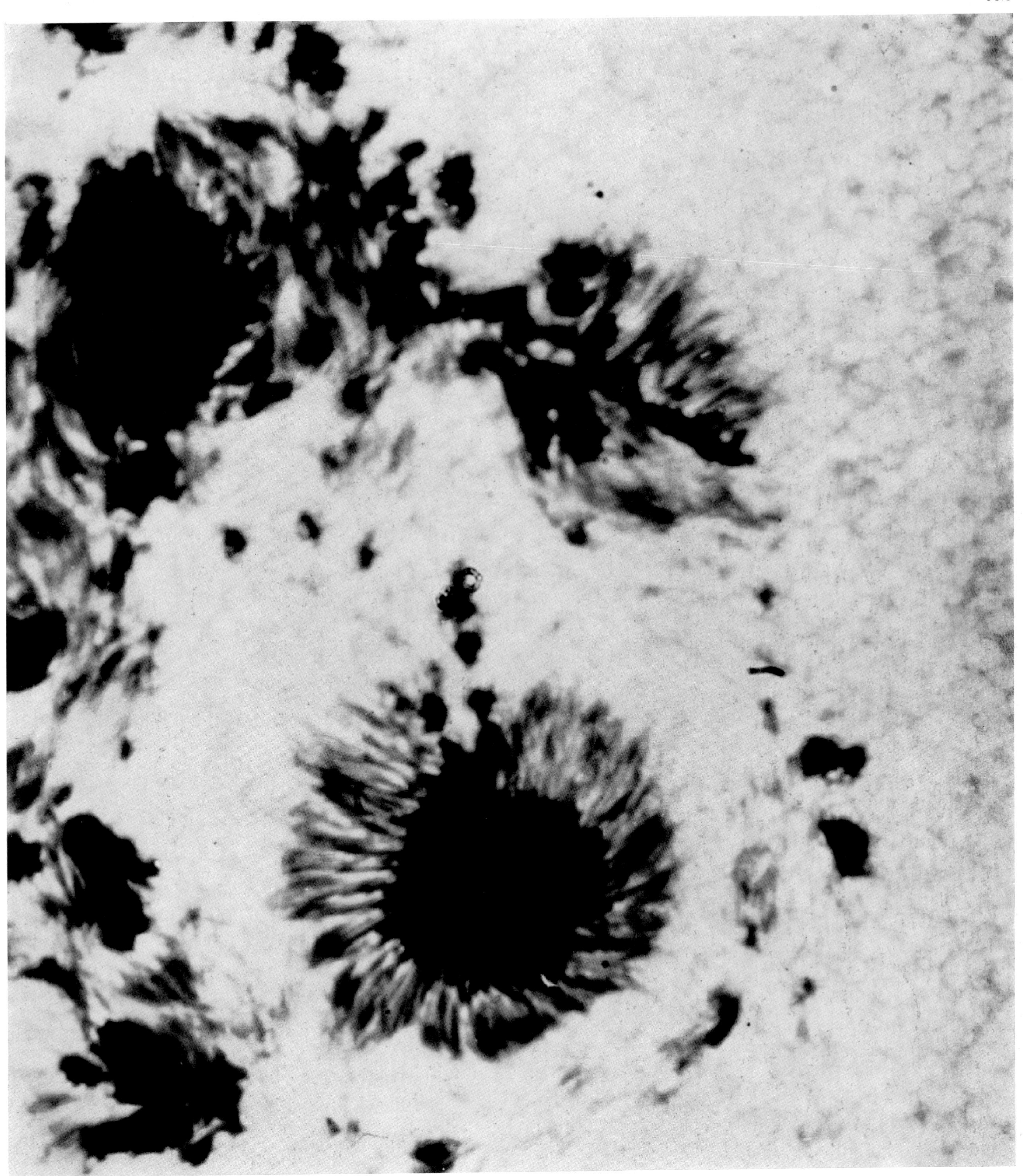

So far the origin of the sunspots has not been fully explained. This exceptionally sharp photo was made in 1959 with a telescope that had been carried by a balloon to an altitude of fifteen miles.

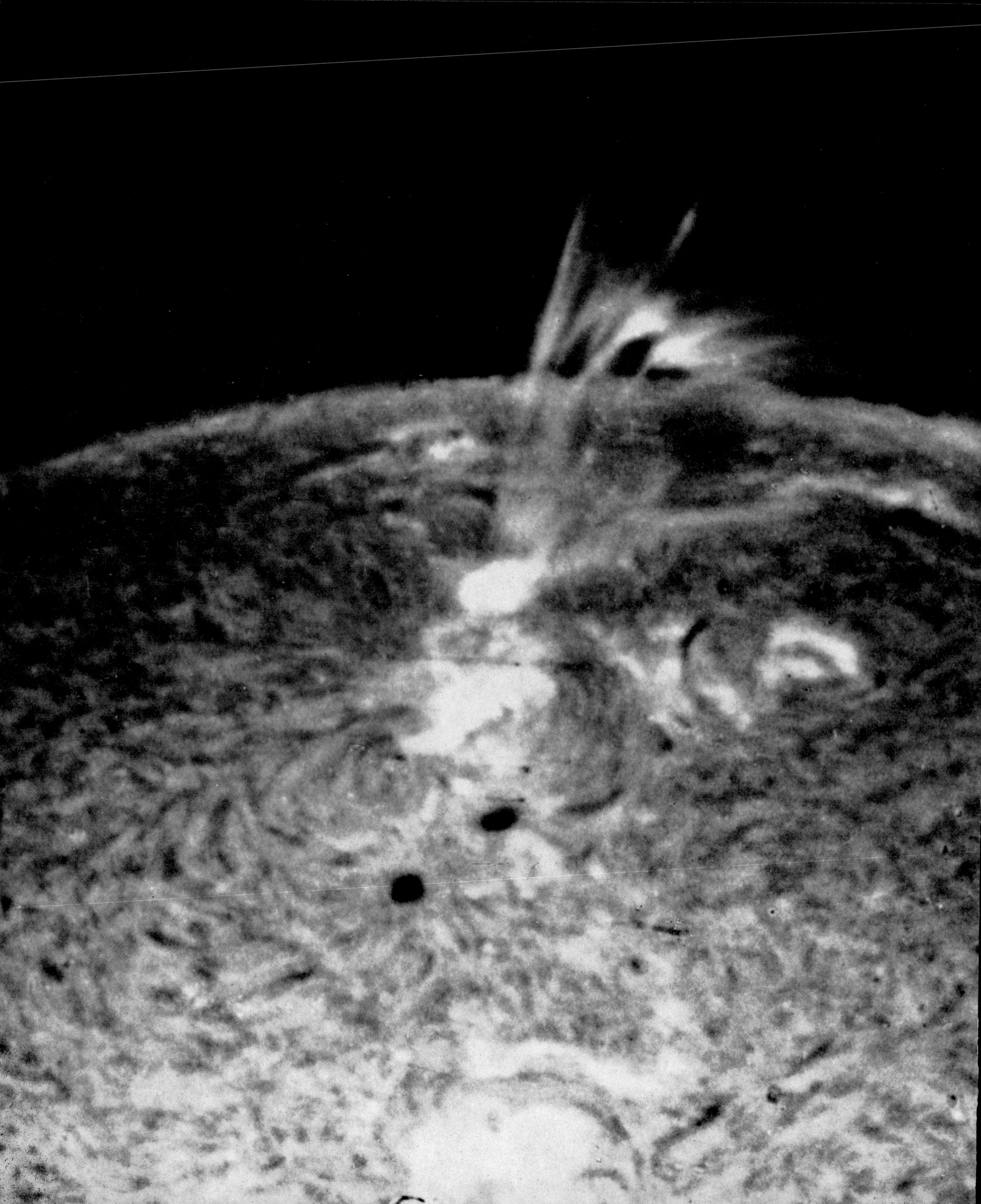

Violent, enigmatic sun.

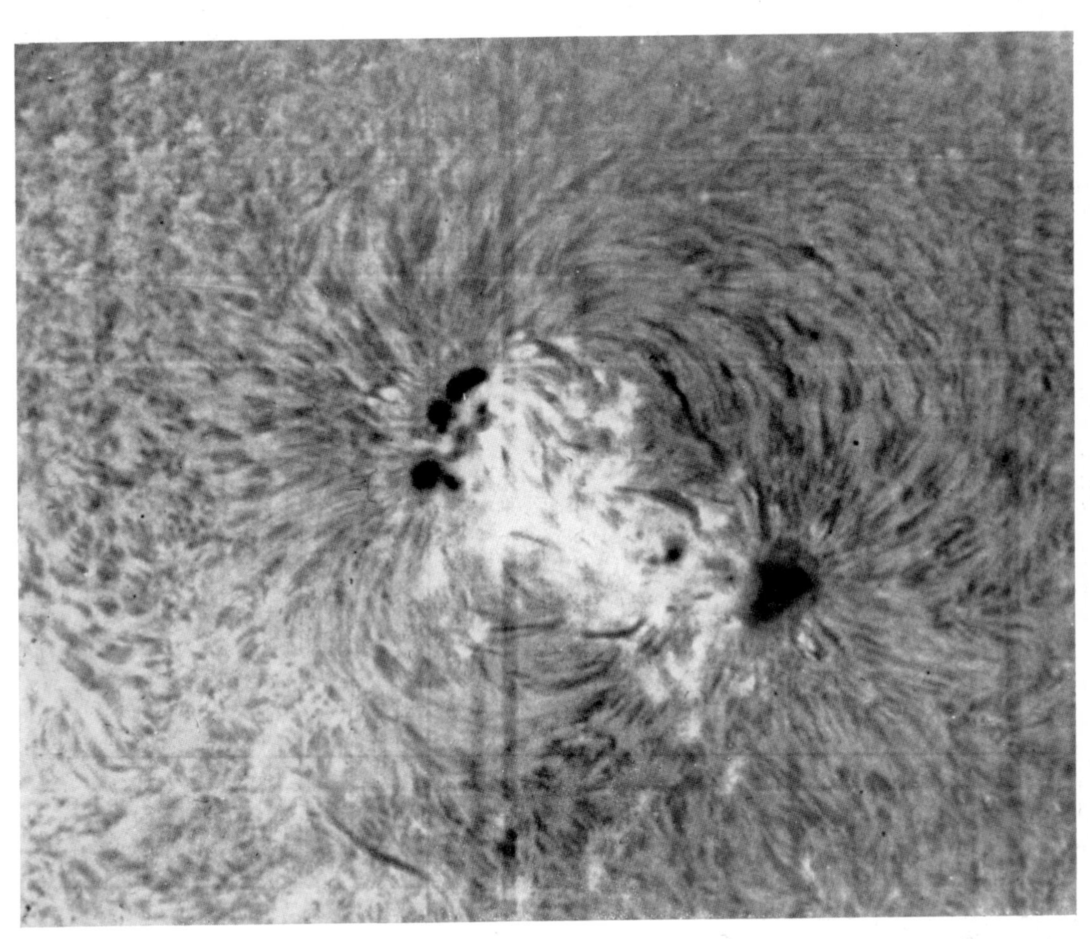

Mount Wilson Observatory

Professor K. O. Kiepenheuer, Fraunhofer Institute, Anacapri Observatory

▲ A magnetic whirl around a pair of sunspots; electromagnetic currents flow to and fro.

◄ The energy that wells up from the sun's core to the surface erupts with ever recurring violence. The sun may be thought of as an ocean of boiling gases, out of whose depth rise fiery turbulences that flare out as giant gas jets. The supersonic speed of these events must be accompanied by a tremendous roar. But no visitor will ever hear it; this hell is unapproachable.

USIS

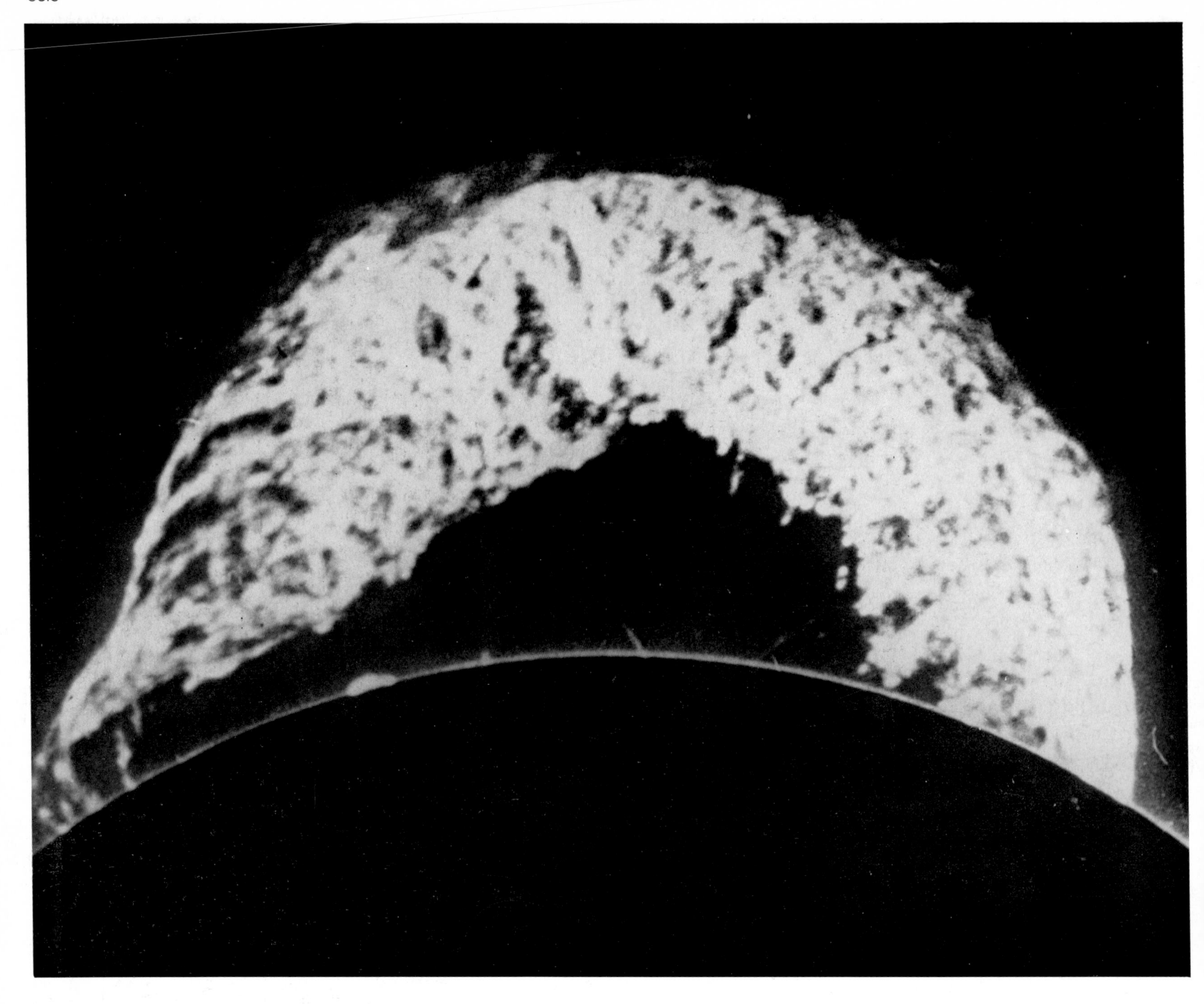

▲
In the coronagraph, invented by the French astronomer Bernard Lyot, a conical device masks the disk of the sun; the result is an artificial eclipse! At the rim, an enchanting phenomenon makes its appearance. The prominences stand out in fiery red against a dark background. Jets of glowing gases flare far out into space. Some prominences slowly rise from the sun's surface in the shape of mighty fiery arches. Some look like giant torches. Others flow back to the sun's surface along magnetic lines of force.

USIS

This picture has been called a "portrait of creation." For astronomers it is the Horsehead Nebula in Orion. A dense, dark cloud of cosmic dust swallows up the light of the nebula in the background. The picture was taken by the Mount Palomar telescope. The film records starlight that has traveled for millions of years. Will it ever become possible to decode the light messages we receive from galaxies hurtling outward at inconceivable speeds? Will we ever solve the riddles of this infinity?

USIS

Deutsches Museum, Munich

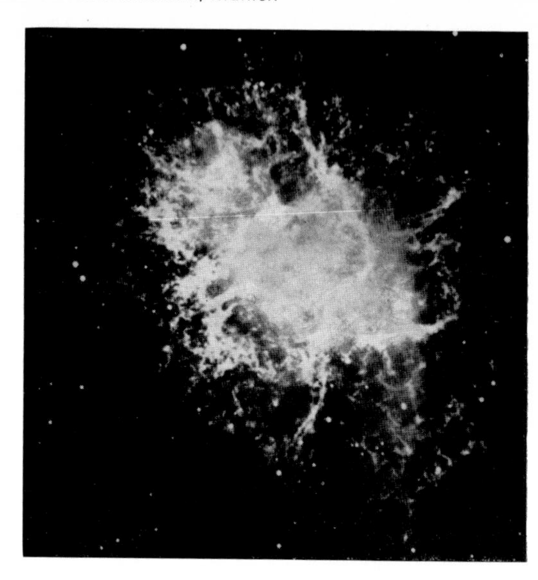

The nearest "island universe in space," a neighboring galaxy of billions of stars, more than two million light-years away, is the Andromeda galaxy. In shape and size it resembles the Milky Way of which we are a part. The inhabitants of a planet orbiting around a sun in the Andromeda galaxy would see our Milky Way system through their telescopes looking much like the picture below.
▼

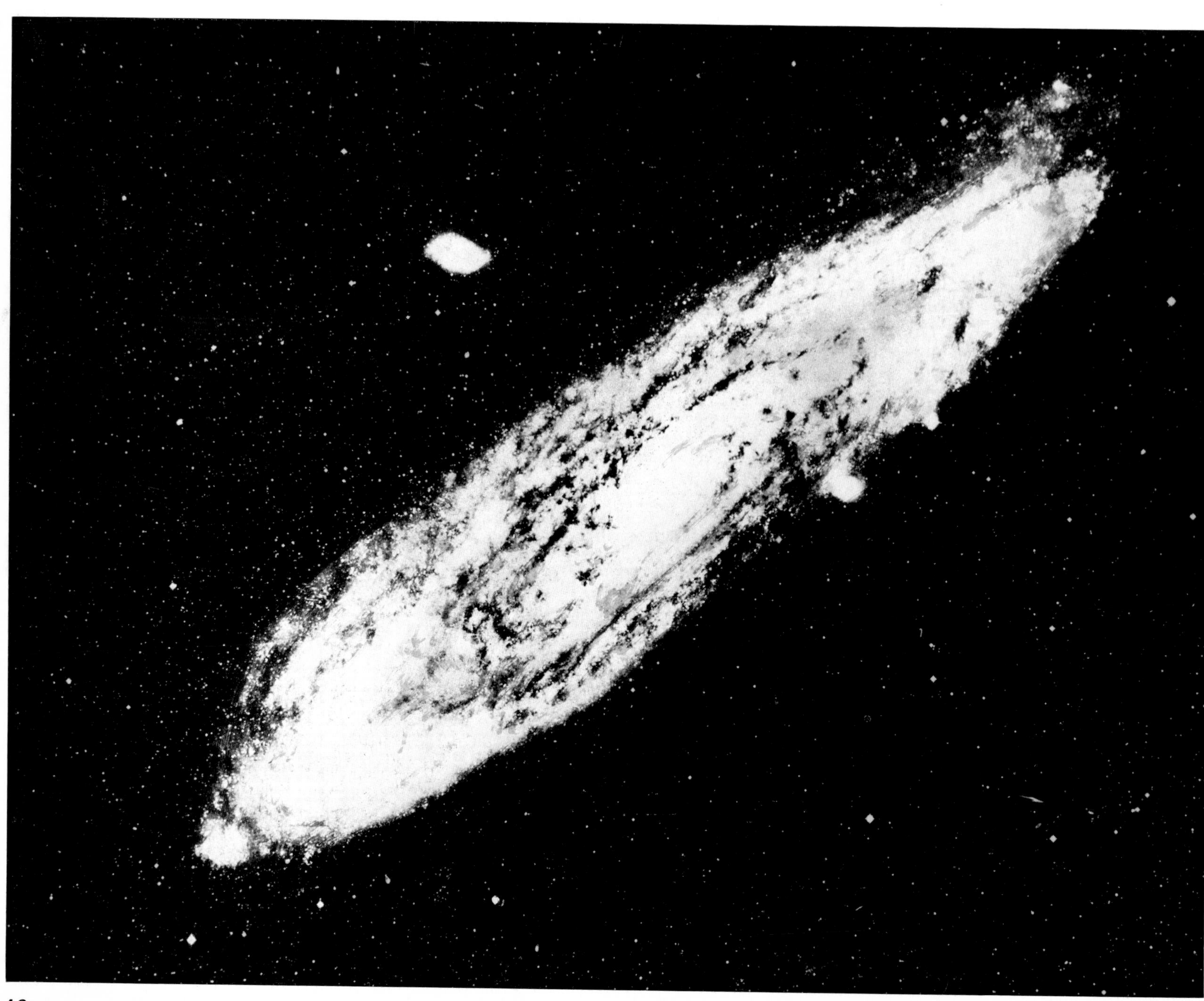

◂ The Crab Nebula is the remnant of a supernova, a star explosion, that Chinese astronomers observed in 1054.

The spiral M81 in Ursa Major, photographed at the Mount Palomar Observatory. It is seven million light-years from the earth. Spiral nebulae are not gas clouds but, as our picture shows, consist of an inconceivably large number of stars.
▾

USIS

Rhode & Schwarz, Munich

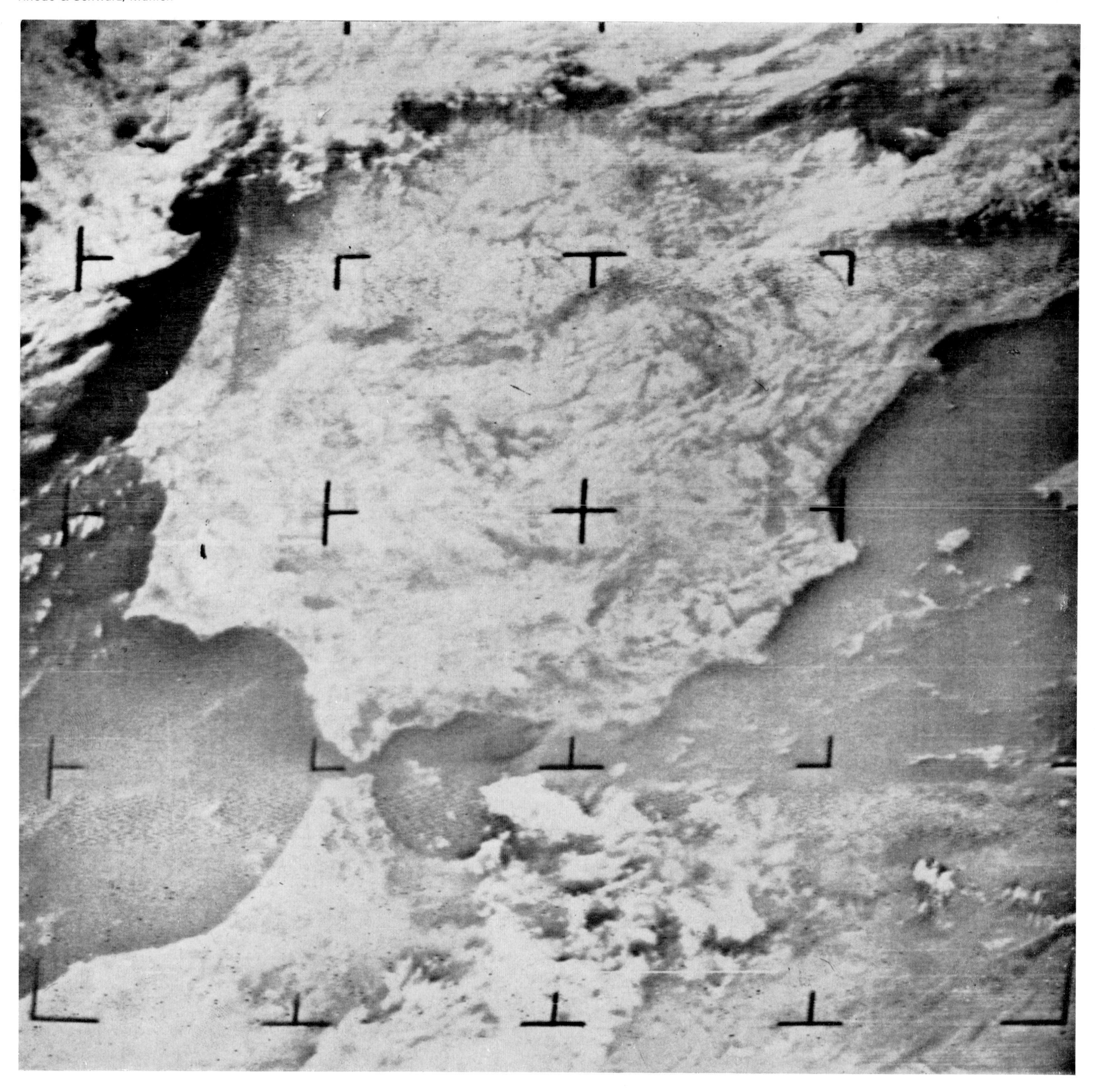

Photos, televised by satellites, confirm what cartographers have charted. The picture of the Iberian peninsula on a sunny day consists of 800 picture lines. It was sent by a U.S. Nimbus weather satellite from an altitude of 300 miles and was received in Munich. Such photographs are of great value for long-range weather forecasting. The distance between each pair of markings is 180 miles. Surveys of the globe become increasingly more accurate.

Electronic eyes in space confirm our maps.

USIS

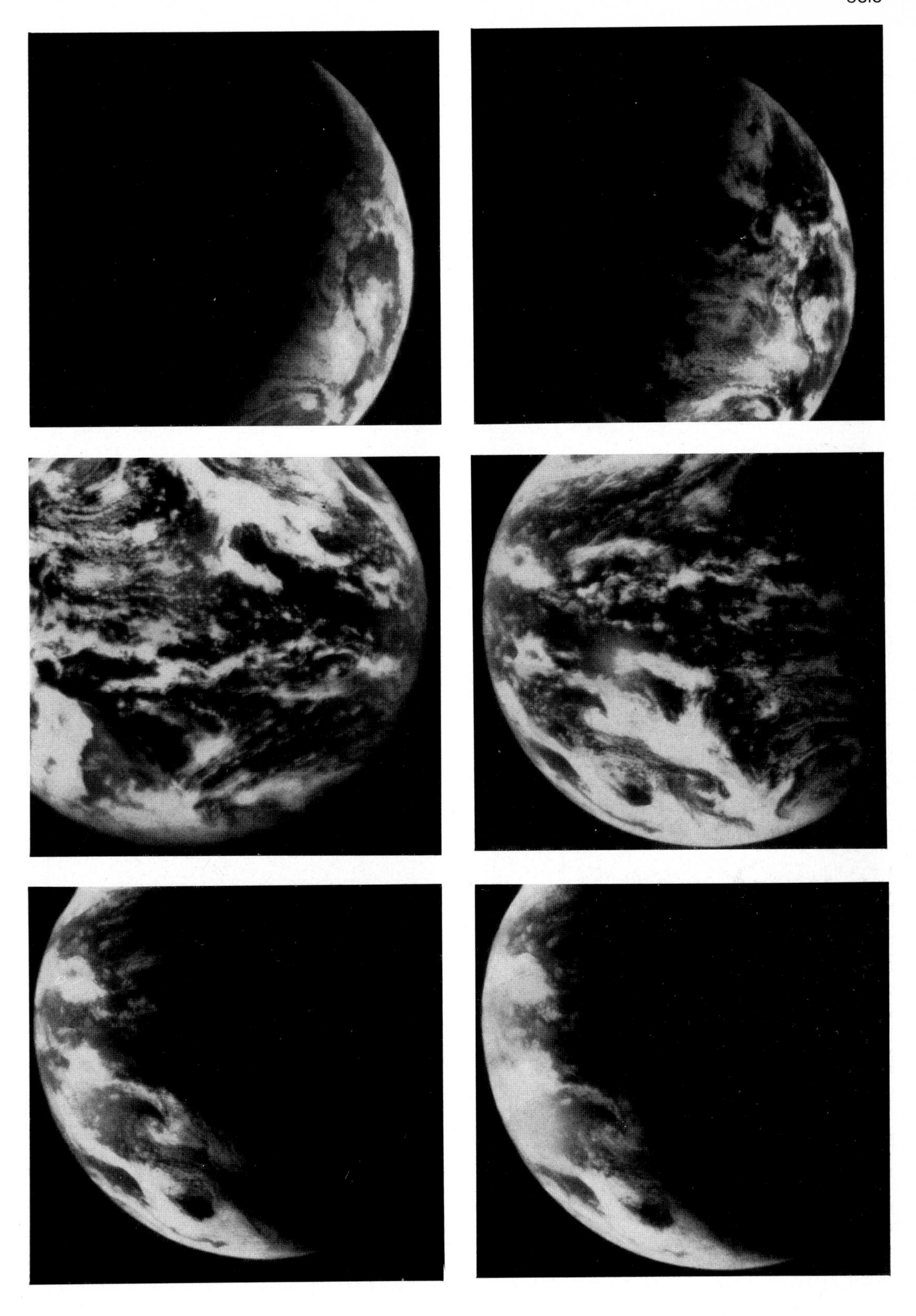

The first photos of our entire globe: not the distribution of oceans and continents, but the cloud cover characterizes the face of the earth. The pictures taken by the American ATS satellite from an altitude of 23,000 miles impressively demonstrate this fact. The satellite was propelled into a synchronous orbit in the beginning of December of 1966. It seems to stand still above the Christmas Islands in the Pacific because it turns together with the earth. The pictures show six phases between sunrise and sunset.

USIS

The Indian subcontinent and Ceylon as they appear from an altitude of 530 miles. This photograph was made by Richard Gordon during his space walk, with a hand-held camera. The orbit of Gemini 11, with astronauts Charles Conrad and Richard Gordon aboard, ranged to a maximum altitude of 850 miles (September 12–15, 1966). Visible in the foreground is the radar antenna of the space ship.

The ultimate goal of space travel will not be the moon, not Mars, not a planet in another solar system—it will be the earth! Out of space travel will grow a "planetary consciousness," an awareness that our planet is the precious property of all mankind, an oasis in the unfathomable desert of space.

It is the enigmatic medium, light, that forms a bridge between universe and man, between man and microcosm.

We perceive our environment via electrical signal patterns on the postage-stamp–sized retina.

The photograph, showing the back of the eye, was taken with a special camera. To the extremely bright light of the electronic flash corresponds a very short exposure time. In spite of the eye's movements, the picture is sharp. At the bright spot, the so-called papilla, the blood vessels that nourish the eye converge and join the nerve fibers that carry the signals to the brain. The papilla is completely insensitive to light and therefore called the "blind spot." Within the dark area is a tiny spot, the fovea, which is the seat of sharpest vision.

When we want to see an object as clearly as possible, we focus our eyes in such a way that the image received through the lense falls directly onto the fovea.

Carl Zeiss, Oberkochen

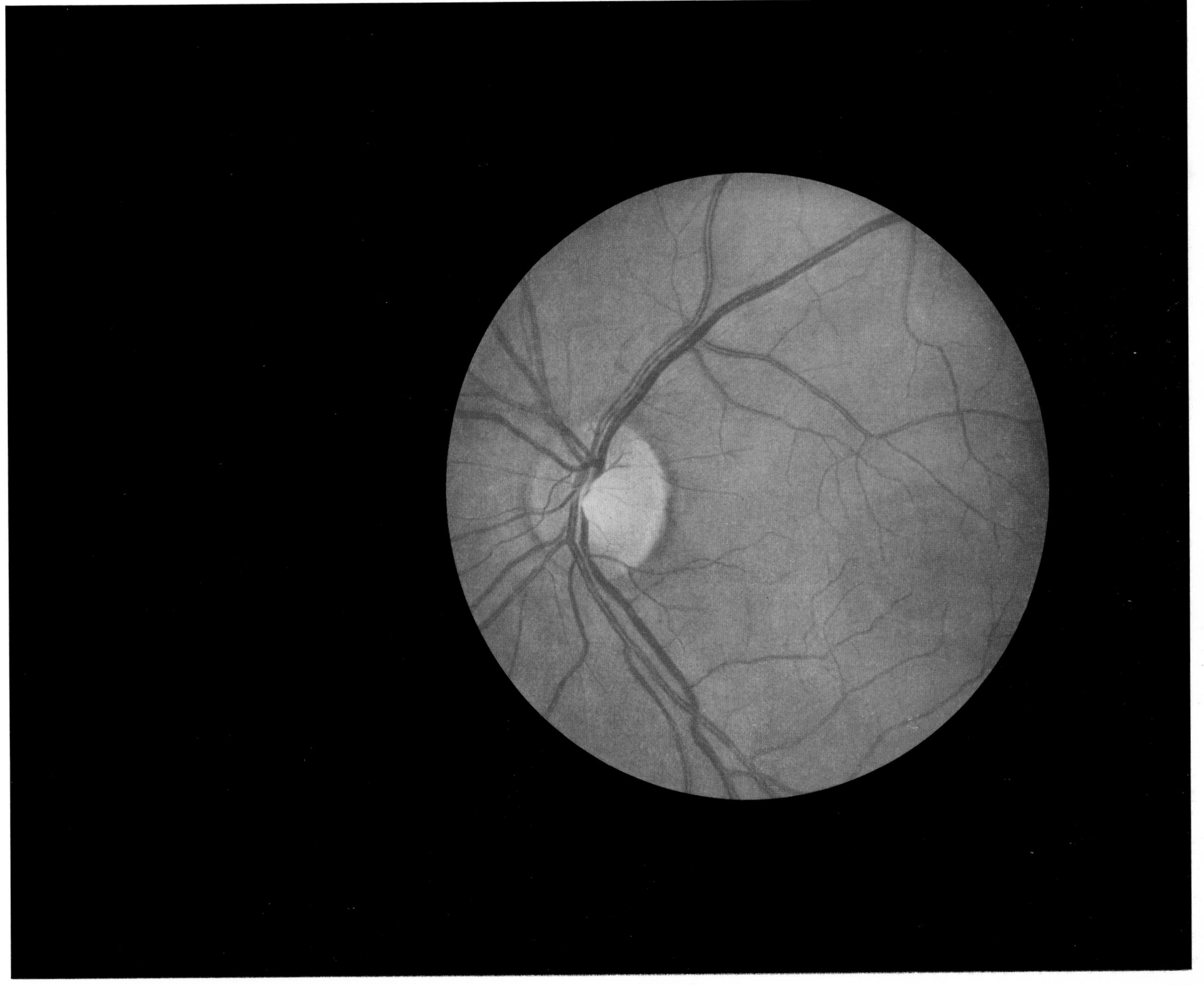

Photomicrography.

The development of light microscopy culminated in instruments that offer a variety of means for observation and are easily manipulated. The aim is to leave the researcher free to concentrate exclusively on his scientific project. ▼

The large fixed eyes of this ▶ horsefly enable it to see almost the entire area surrounding it. For this picture no microscope was needed. It is an extreme close-up made with a special objective in front of a bellows extension. Original magnification: 6.5:1.

Carl Zeiss, Oberkochen

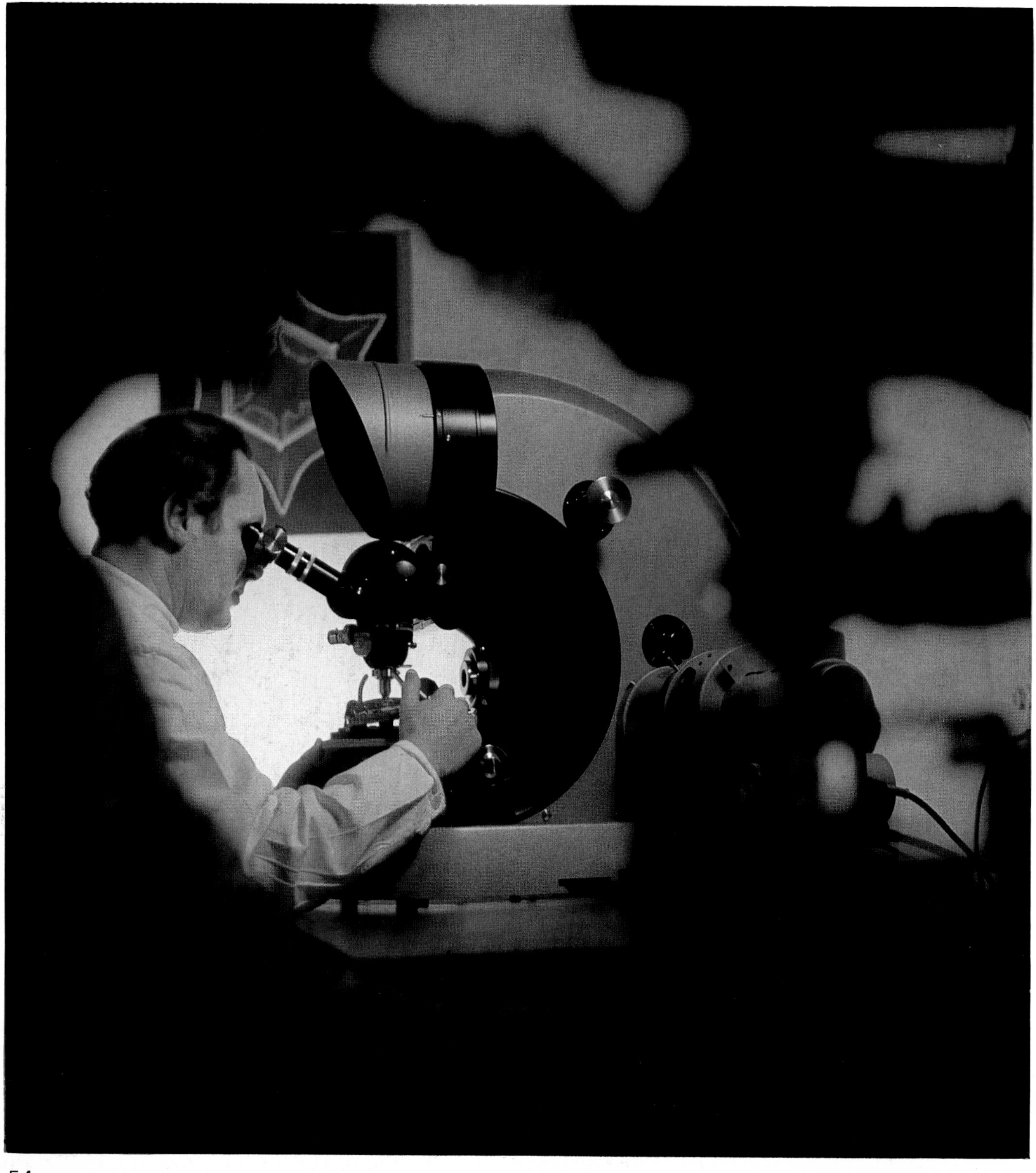

Carl Zeiss, Oberkochen

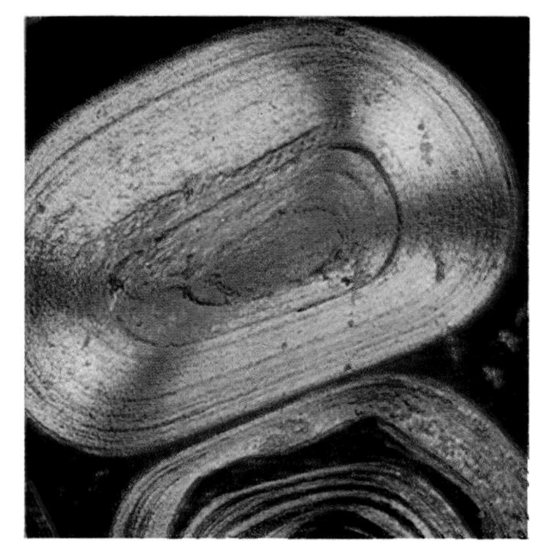

Through the polarizing microscope: a thermal tuff from Karlsbad magnified 30 times.

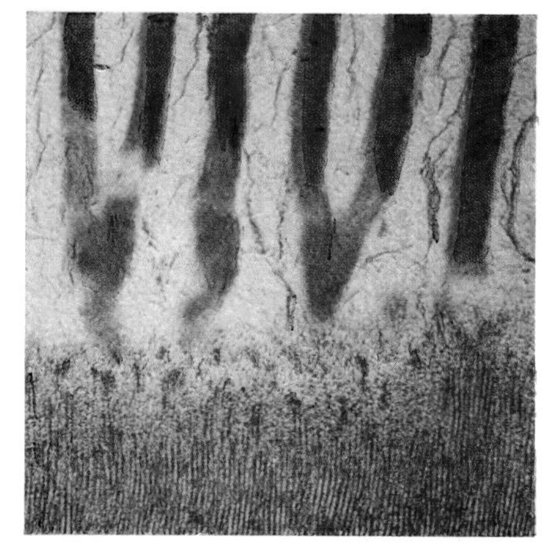

A fossil from the Tertiary period in 420-fold magnification.

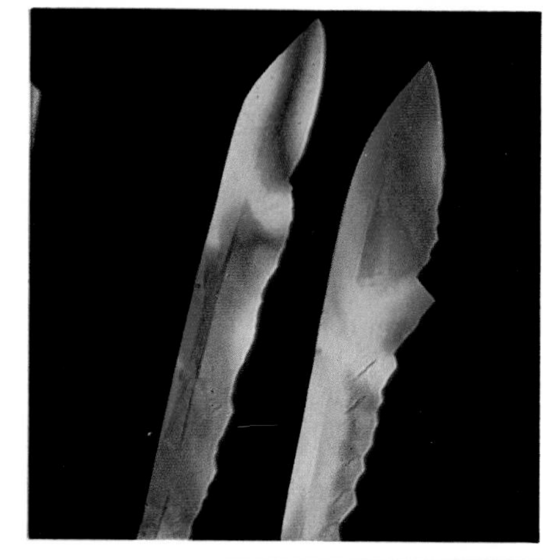

Sulfur crystals grown from a melt. Scale 100:1.

Radially grown caffeine crystals under polarized light, magnified 65 times.

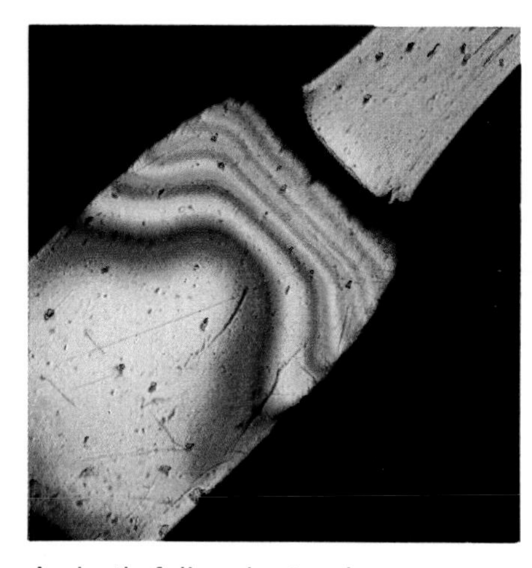

A plastic foil under tension. Polarized light shows the stresses. 30:1.

These caffeine needles, magnified 65 times, crystallized from a solution.

Caffeine from a melt. The ramified pattern is caused by enclosures.

Here caffeine grew from a melt into a steplike configuration. Enlarged 65 times.

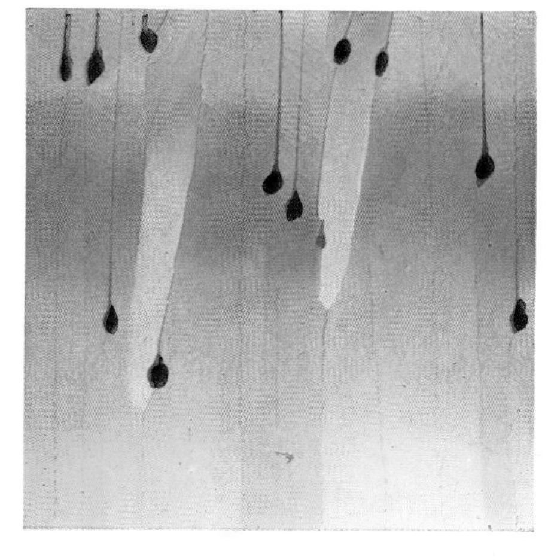

These too are caffeine crystals, with scattered enclosures. 100:1.

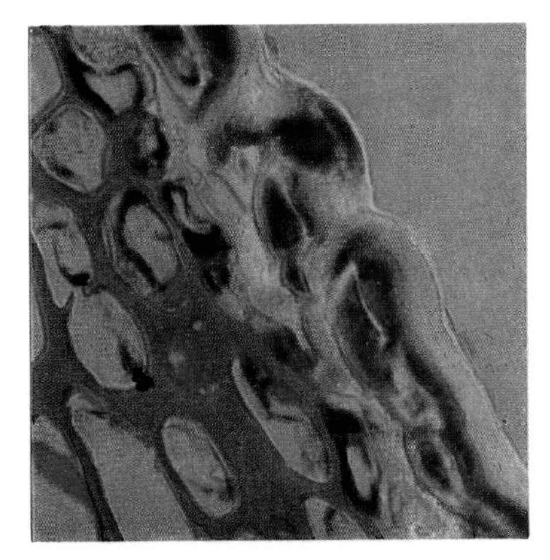

Thin section through the peel of an apple, 400 times magnified. Interference caused the color contrasts.

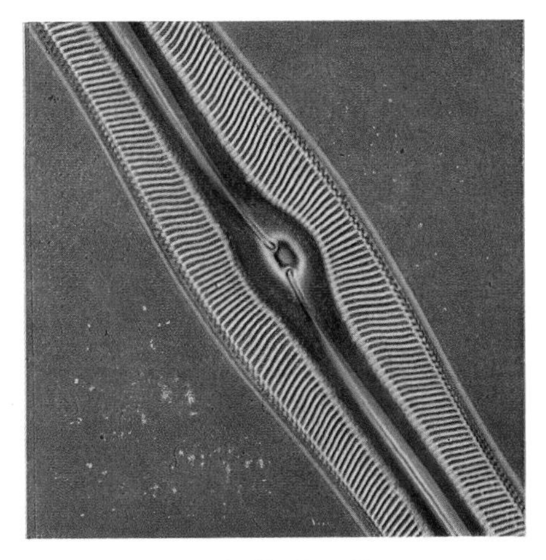

The armored shell of a diatom, a microscopic alga. Scale 420:1.

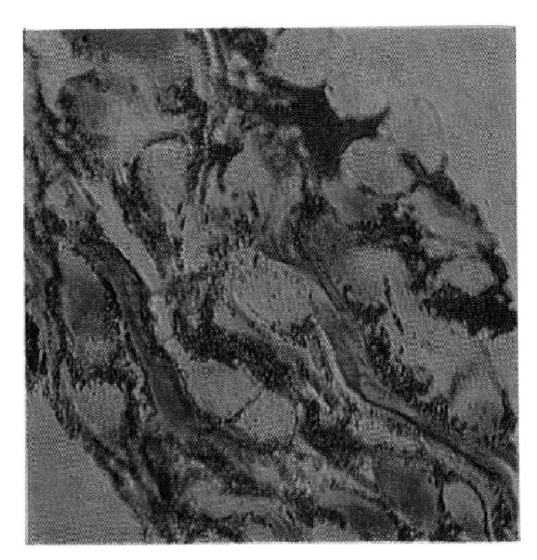

Nylon-reinforced paper magnified at same scale. Contrasts achieved by interference optics.

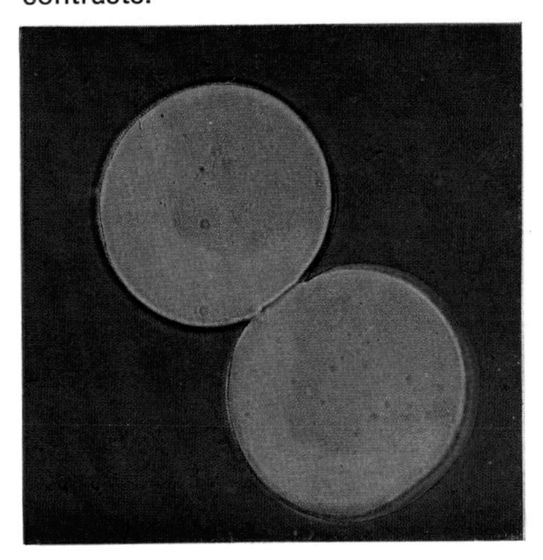

Crosscut of nylon fibers. 420:1. Interference optics brings out differences in density.

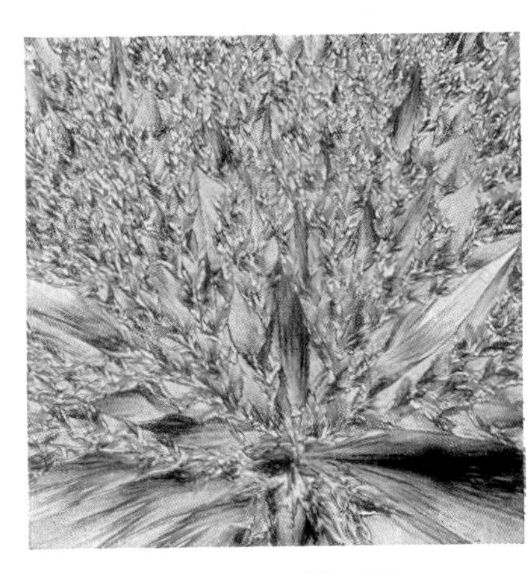

Hippuric acid, crystallized in starshapes from a melt. 100-fold magnification.

This too is hippuric acid from a melt, here in 420-fold magnification.

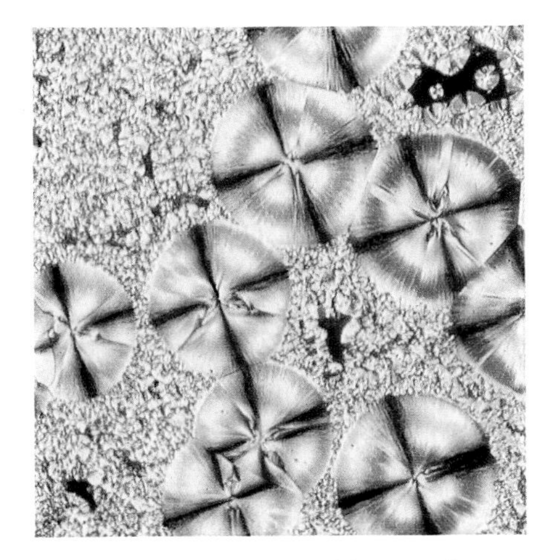

Pictures of hippuric acid crystals, in spite of all their differences, are unmistakable.

Seen from above, the crystal cones appear as little wheels. 100:1.

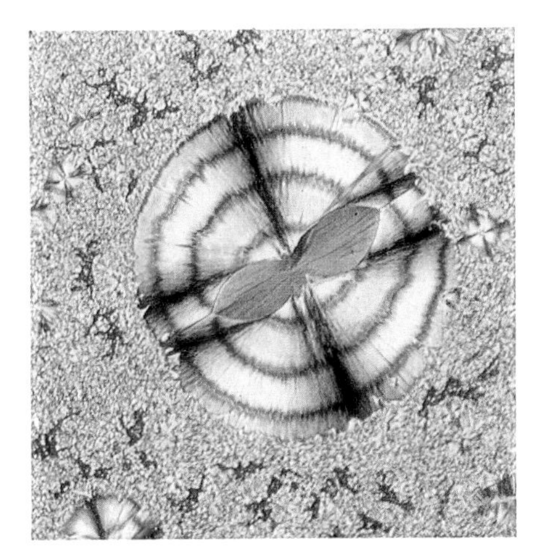

Polarized, "sifted" light defines even further what has been made visible.

Carl Zeiss, Oberkochen

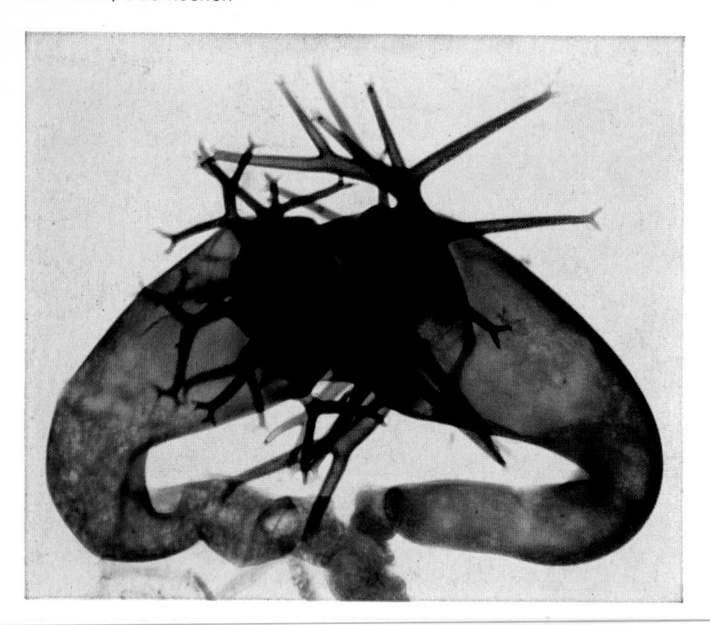

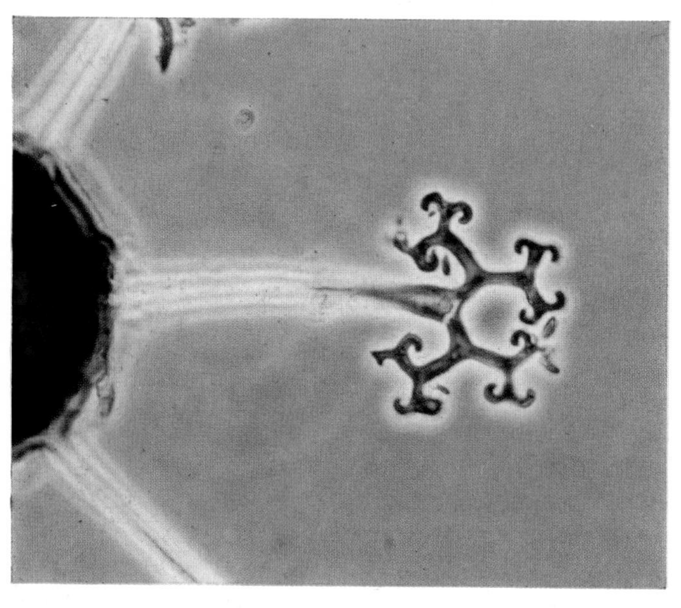

There are close to 100,000 varieties of fungi. Not only chanterelles, truffles, and morels, but also molds, yeast, and mildew belong to this group. Most fungi are microscopically small. The photos show the reproductive organs of such a fungus; after two of the hoselike arms have touched, two spores merge; the "thorny antlers" protect the ripening germ cells from small foraging animals.

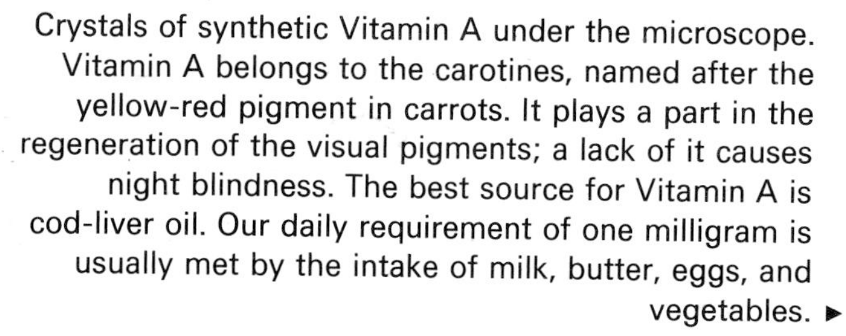

Crystals of synthetic Vitamin A under the microscope. Vitamin A belongs to the carotines, named after the yellow-red pigment in carrots. It plays a part in the regeneration of the visual pigments; a lack of it causes night blindness. The best source for Vitamin A is cod-liver oil. Our daily requirement of one milligram is usually met by the intake of milk, butter, eggs, and vegetables. ►

Deutsche Hoffmann-LaRoche AG

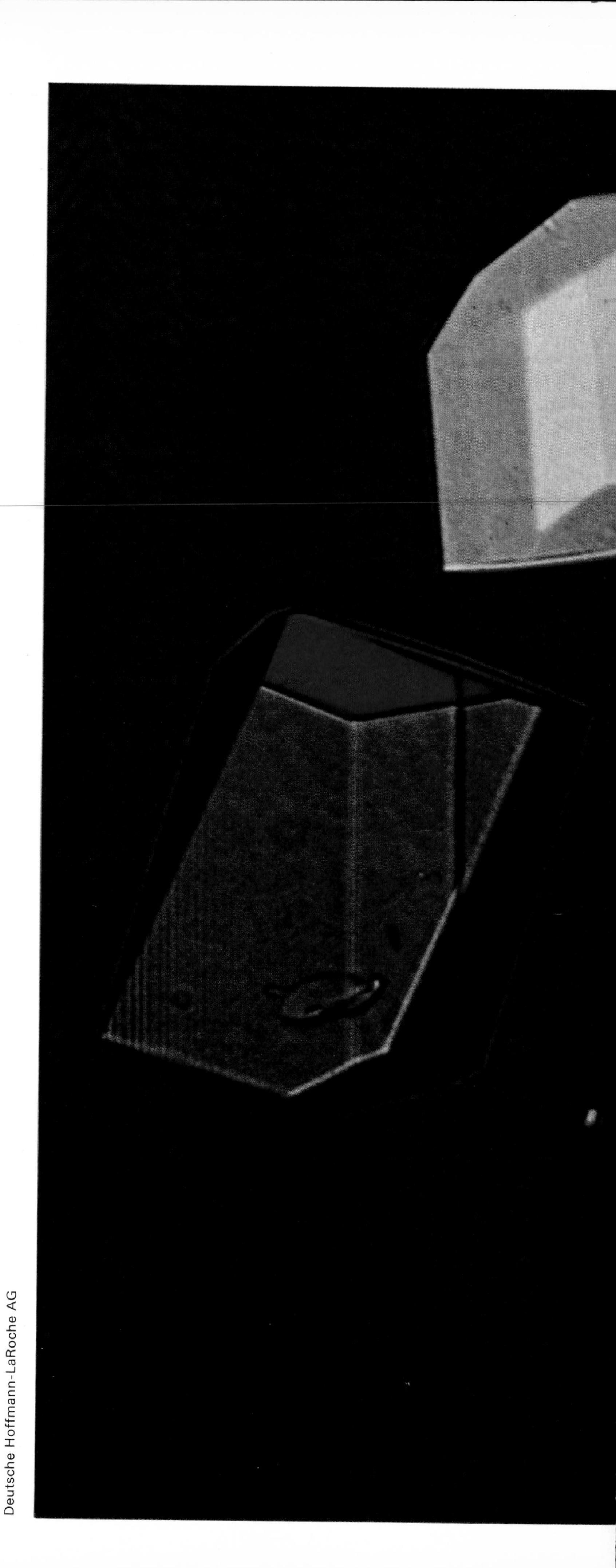

Canada balsam with bubbles, as seen under an interference microscope. Scale 800:1. Canada balsam is the resin of the Canadian balsam fir dissolved in turpentine. Canada balsam hardens to a clear, transparent film and because of this property is used as an adhesive for optical lenses and as a coating for microscopic specimens.

Carl Zeiss, Oberkochen

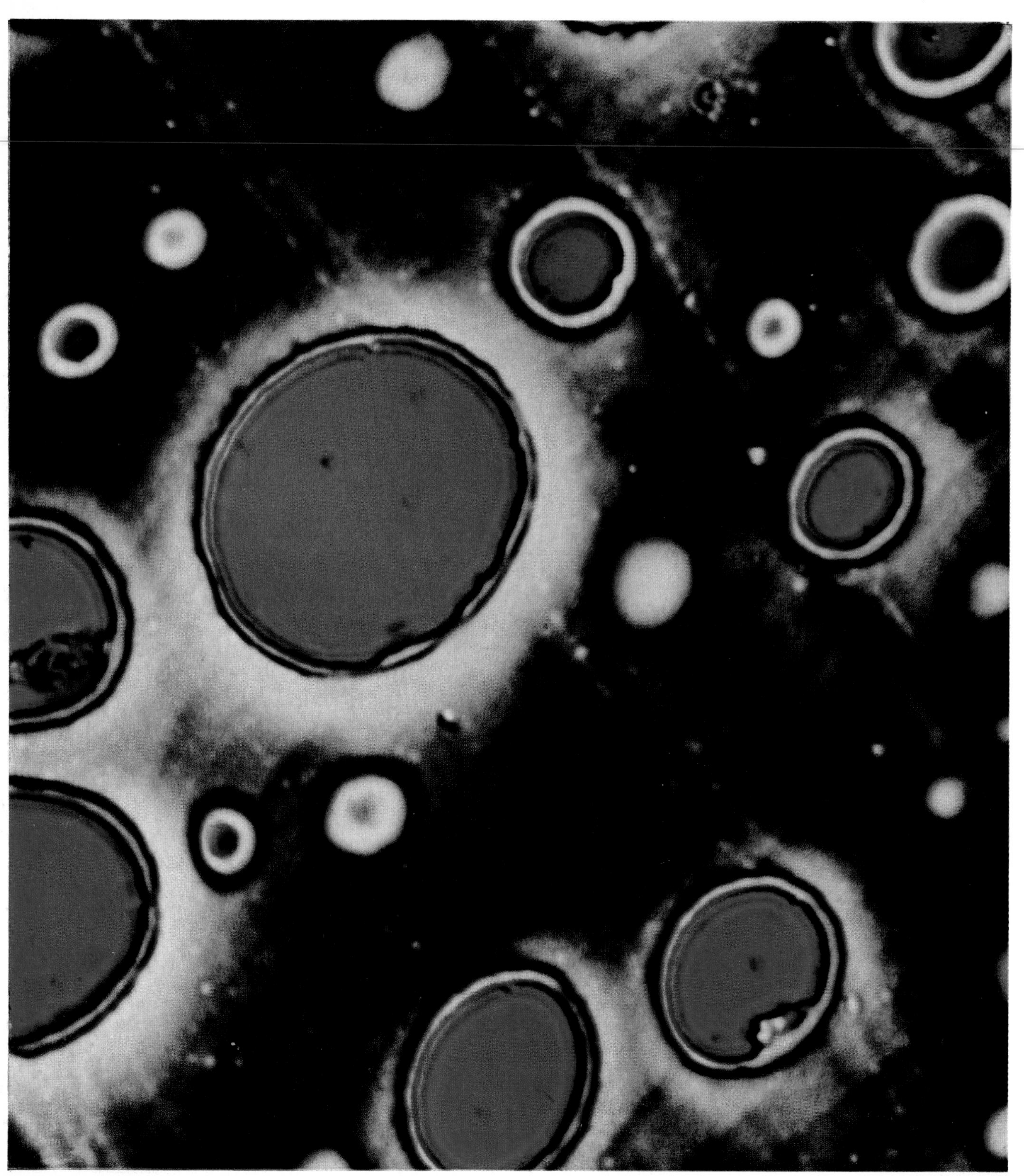

Magnesium nitrate, component of some fertilizers, consists of rhombic columns and needles. Crystallized from a thin layer of solution with an additive, fan-shaped figures emerge that resemble the ice-ferns on winter windows. The detail shown here gives an impression of flaming domes.

Carl Zeiss, Oberkochen

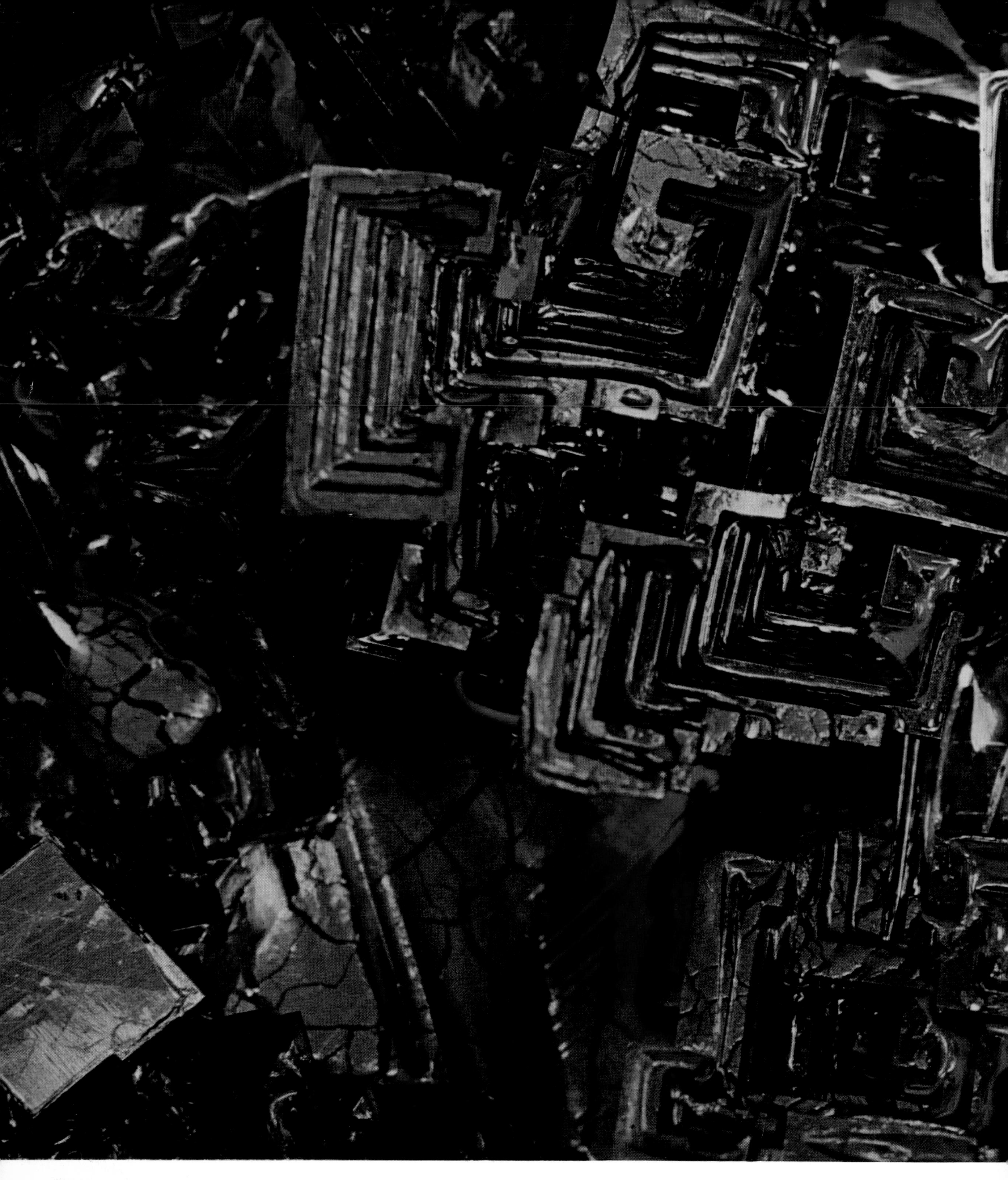

In a bowl the size of a hand, a melt of bismuth slowly crystallizes. Bismuth, considered a metal, is a rare element.

Manfred Kage, Winnenden

A penicillin-potassium drug. Penicillin, orginally produced from a mold, is manufactured synthetically today. Scale 280:1.

Fritz Brill, Hofgeismar

Menthol, also called mint camphor, in the process of crystallization from an alcohol solution. Scale 90:1.

Crystallized copper. (Original photo measured diagonally 3/8 inch.) From a watery-blue copper sulfate solution, metallic copper, reduced by zinc, precipitates. The process corresponds to galvanizing, but it was slowed by the addition of a synthetic resin. In this way tree shapes formed. The red coloring comes from the finely distributed copper. Its effect is comparable to that of gold added to ruby glass.

The process of crystallization is one of the most marvelous spectacles in nature.

Manfred Kage, Winnenden

Manfred Kage, Winnenden

"Nature is capable of creating not only *one* form, it is experienced in all of geometry."

(Johannes Kepler, 1611)

The opal is the most enigmatic of the gemstones; the black opal, the rarest of all the opals. The play of its colors rises out of its depth. Light refraction by tiny globular enclosures causes rainbow effects. The intensity of the opal's colors is comparable to that of iridescent tropical butterflies. The macrophoto (picture measured diagonally ¾ inch) might be the creation of a modern painter.

Our earth is a mere speck of dust in the universe; the tiny, insignificant companion of a sun among billions of suns. And yet, for us the earth has remained what it once was in the eyes of pre-Copernican astronomers: the center of the universe. The earth is at the center of the universe in still another, very special sense: the macrocosm is balanced by the microcosm and man stands between the largest and the smallest.

The Microscopists' Visual Pleasures

Inventive spectacle-makers in Holland devised the telescope and the microscope at about the same time. To this day, it has remained impossible to prove that a telescope was built before Hans Lippershey of Middelburg built his, in 1608. Credit for the invention of the microscope, probably as early as 1590, goes to yet another Middelburg lens-grinder, Jan Janszoon (sometimes spelled Hans Janssen) or to his son Zacharias Janszoon, or to the two of them together. The story goes that the son perfected what the father had put together, and then saw to it that news about the magic glass got around. The true story of the invention may forever remain unknown. A letter, mentioning father and son as the inventors, dates from the year 1665; it can only have been based on hearsay.

This uncertainty may be, in part, the reason why the names of the probable inventors have been all but forgotten. Moreover, the invention of the microscope stirred up considerably less excitement than that of the telescope which in the hands of a Galileo became the sensation of the learned world. The microscope retained its low standing as a mere curiosity for a longer time and was relegated to the sidelines of the stormy developments of its day. It quickly became a sophisticated toy for imaginative dilettantes who met socially in scientific societies. They told each other what they had wonderingly observed, but beyond a little understandable vanity, they never considered questions of priority as of prime importance.

While the telescope pierced the heavenly spheres of the Middle Ages, and challenged, in the view of concerned cardinals, the authority of the Church, the microscope initially provided only visual delights for its devotees. It became a charming and absorbing hobby, a means for the ecstatic admiration of the ever new, ever surprising forms through which creation manifests itself. It was a "delight for the soul and for the eye," a poetic undertaking, though at times it sent shivers down the spine of the microscopist: perhaps when he believed himself close to lifting the veil of the ultimate secrets of life, somewhat like a criminologist nearing the solution of a case, or perhaps when he realized into what new, uncharted depths he was descending. Of course, no matter how lively the initial activity, interest began to flag as soon as the novelty wore off, as will happen with hobbies that are mere pastimes.

To be sure, great men of learning also took notice of this new and bewitching instrument. So did Galileo, though his real passion was directed to the stars. Only marginally did he seem to have realized that the microscope, just as the telescope, opened new vistas, gave a new dimension to vision and comprehension. Galileo, too, stressed diversion in a letter to Duke Cesi, the patron of the Accademia dei Lincei, founded by a group of Roman amateur-scientists: "I am sending you herewith an "*occhialino*" for the study of the smallest objects at close range. I hope that it will be of as much use and delight to you as it was to me. I have studied with endless amazement many of the smallest creatures: the most grisly among them is the flea, the most beautiful are the ant and the moth. It delighted me to be able to observe how flies and other tiny beasts go about running head down on the surface of the mirror." Federico Cesi took up this technological novelty with enthusiasm. He employed an artist to draw what appeared under the microscope: the tiny seeds of ferns, "large as peppercorns," the anatomy of bees. It did not dampen his enthusiasm that the microscopes of that time were still rather primitive. There were no others. The field of vision was small and relatively dark. The colored borders did not annoy him, they were accepted as a given fact.

Preoccupation with the diversionary aspects of the microscope lasted for a generation: thirty-five years after Cesi's death a book came out by an exceedingly versatile and ingenious man, the then twenty-nine-year-old Englishman Robert Hooke (1635–1703), *Micrographia*. Despite the weak light sources of his instruments, the sixty plates show his critical power of observation, his unexcelled care and precision in drawing. Hooke worked by daylight or by the glow of an oil lamp. A water-filled globe and a convex lens concentrated and focused the light. Tiny pests like a flea or a louse show up as armor-plated, astoundingly equipped creatures. A head-louse, twisting its legs around a human hair, looks like a knight with many arms, brandishing a spear.

Hooke concentrated at first on all the things that might interest a child: after the flea, the louse and the silverfish, the wondrous transformation of a mosquito, the eye of a fly. His pencil records how facets of the eye reflect the window of his workroom. He draws the eye of the dragonfly, the needle of the nettle, the sting of the bee, the hollow spaces in cork, which he called "cellulae," cells. That was merely a name at the time, and only later, in the nineteenth century, when the "cell" is recognized as the basic unit of organisms, does it become a meaningful concept. Unfortunately, Hooke occupied himself with biology only for a short time, two years in all. He had so many other interests. He was an astronomer, and discovered the rotation of Jupiter; he was physicist, chemist, inventor, architect. He was a man of many ideas, yet the idea to ask for money for his scientific work did not occur to him. He enjoyed conversation at his club, and was a hypochondriac who made notes of what he ate and drank and how it did or did not agree with him. The fact that he fell out with Newton does not necessarily speak against him. Newton was, no doubt, a genius, but not easy to take. Unfortunately there exists no portrait of Hooke, and we do not know what he looked like. We do not even know where he was buried.

Early Scientific Microscopy

In Hooke's day, there lived in Delft a linen merchant named Anton van Leeuwenhoek (1632–1723), who practiced microscopy as a sort of compensatory mental exercice. He built his own microscopes and it apparently never occurred to him to acquaint the scientific community with what he saw. It was due to the good offices of the anatomist Régnier de Graaf, six years his junior and the discoverer of the vesicles or follicles of the ovaries that bear his name, that in 1673 an exchange of letters between Leeuwenhoek and the London Royal Society was initiated. It was then, and is still today, considered the most prestigious scientific society in the world.

From then on, this merchant and dedicated dilettante scientist sent in reports about his discoveries at regular intervals. They were eagerly awaited. Leeuwenhoek saw the blood circulate in the tail of a tadpole. He saw the hydra reproduce itself asexually through budding. He discovered little creatures of extremely fidgety demeanor in an infusion of hay. "They stop, they stand still in one spot as it were, then they whirl around with the speed of a top and the area they move in is no bigger than a grain of sand." Leeuwenhoek was the first to describe single-celled organisms living in water. He gives their size vividly—and accurately. "They are a thousand times smaller than the eye of an adult louse."

Leeuwenhoek reported to the Royal Society in 1683, that, even though he regularly cleaned his teeth, he had found on tooth deposits he had mixed with rainwater, "many small living animalcules in the said matter, which moved very prettily." He then described them as: "short, straight little rods, balls, curved rods, and spiral-shaped things like tiny corkscrews. Small, living animals that move in the most graceful manner. The largest of their kind uses rapid forceful movements and shoots through the water like an arrow." Today we still speak of these "rods"; "bakterion" in Greek means "the little rod" and "bacillus" means the same in Latin. The first man on earth to have seen bacteria was Leeuwenhoek; no one else in his day succeeded in doing so. In fact, two centuries went by before the rediscovery of what the linen merchant of Delft had seen with his magnifying glasses.

Leeuwenhoek discovered "plate-like bodies" swimming about in blood plasma, "a chain of one hundred of these would correspond roughly to the diameter of a coarse grain of sand."

In 1676, Leeuwenhoek announced to the Royal Society that he had found little animals that "look like tadpoles" in human sperm. Their number, he suggested, might well surpass that of the human population of the earth, which he estimated at thirteen billion. A student of medicine had directed Leeuwenhoek's attention to spermatozoa. The student had found small threads that moved, in the sperm of a man afflicted with venereal disease, and had connected them with the sickness. Leeuwenhoek apologized in his letter to the secretary of the Royal Society for bringing up so precarious a subject. He had secured the material for the study of this subject "from the surfeit of martial conjoining without the commission of a sin." Shortly thereafter, he found that similar little animals were present in the milt of fishes, and in the sperm of horses and other animals. These observations aroused much interest. The learned men saw themselves confronted by one of the great mysteries of life. What kind of function might these tadpole-like little animals have? Could there already be, within this little head propelled by a lashing cord, the future being, a tiny embryo? It took almost two hundred years before these questions could be answered. During all that time the processes of reproduction remained unexplained.

How Leeuwenhoek made his discoveries remains his secret. He never disclosed his methods. One thing is certain—he did not have a compound microscope. It can actually only be called a magnifying glass, an almost spherical lens on a stand, with the object for viewing fixed to a movable pin. Nevertheless, in this way he achieved a better than two-hundred times magnification. Leeuwenhoek left more than four hundred lenses, among them some made of rock crystal. His glass lenses were ground rather than cast.

Leeuwenhoek received visitors with reticence if not with suspicion. He refused to show his best microscopes. It is doubtful whether even Czar Peter the Great, who visited Leeuwenhoek in 1698, was persuited to see them. Hooke, who used drops of glass-paste for lenses, thought he might be able to uncover the secret of the reticent Dutchman. Hooke had found that a drop of water in a thin glass tube gave additional magnifying power. Leeuwenhoek most likely discovered that the effectiveness of the microscope is enhanced by using slanting light and a dark background; in this same way, dust particles in a ray of sunlight are made visible.

In spite of all his work at the microscope, Leeuwenhoek remained the amateur scientist. He apparently proceeded systematically only when trying to invent new techniques of observation. The objects for his microscopical studies were selected according to his whim. The joy of seeing something surprising, even the pleasure of repeating the same experiment over and over again in always more perfect form, was more important to him than the actual scientific investigation. In order to show how the blood circulates in the tailfin of an eel, he rigged up a special mount for his microscope. "A spectacle that enraptured me more than anything I had seen so far. Not only did I see the blood flow through the finest capillaries from the center of the tail to its extreme parts, but also how each vessel makes a bend and returns the blood to the center of the tail so that it may flow back to the heart."

In 1671, the Royal Society received a work entitled *Anatomy of Vegetables Begun*, by Nehemiah Grew, a physician from Coventry (1642–1712). Under the microscope, Grew had skillfully dissected leaves, stems, roots, flowers, fruits, and seeds. His drawings, while systematically factual, are of poetic charm. Later, physician Grew returned to his original calling. He wrote *Comparative Anatomy of the Stomach and the Intestines.* The title sounds modern and so was the conception of the work.

At about the same time that Grew sent in his paper, Marcello Malpighi (1628–1694), professor of anatomy at Bologna, sent his *Anatomia plantarum (Anatomy of the Plants)* to the Royal Society. Malpighi had a many-faceted imagination. He saw in the riches of nature a thousand riddles asking to be solved. In the gall-nut of the oak he discovered the wasp's egg; the growth, therefore, turned out to be neither witchery nor spontaneous excrescence, but a sort of larder for the hatching maggot. Malpighi investigated the relationship between mistletoe and host tree. He noticed on the underside of the oleander leaf tiny channels through which—as was discovered only much later—the plant breathes. He drew the nodules at the roots of bean plants, long before their role in supplying nitrogen to leguminous plants was understood. In the silkworm Malpighi found the tracheae, the respiratory organ of insects, and those vessels of excretion that have been named after him. Malpighi followed the astounding genesis of the lepidoptera from egg to caterpillar to butterfly. He scrutinized the organs of man through his microscope; in the kidney he saw the fine channels; in the spleen, the Malpighi corpuscles; in the cortex, the pyramidal cells. In 1660, he detected the capillaries in the lung, and thereby found the until-then mysterious connection linking arteries and veins. (Only this contribution made William Harvey's discovery of blood circulation complete.) Malpighi investigated the development of the chicken embryo. He described the structure of the central nervous system and of the bony socket of the eye.

It was an exceptional man who sat there bent over his microscope. His unprejudiced way of observing remains admirable even in those instances where his conclusions later turned out to have been off the mark. The university professors held Malpighi in low esteem. His observations at the microscope were dismissed as useless, ridiculous trivialities, especially so since he lacked a gift for lucid explanation of his findings.

Malpighi invented new techniques. He was superbly skilled in preparing specimens, and no one surpassed him except Jan Swammerdam of Amsterdam (1637–1680) who succeeded in preparing specimens of the nervous system of the honeybee and the larva of the ephemeral fly. Swammerdam attacked his work with suicidal frenzy and was driven to religious mania in his late thirties. His magnificently illustrated *Biblia naturae* was published during 1737–1738 in honor of his hundredth birthday. "In the anatomy of a common and repulsive insect, the louse, I have encountered wonder upon wonder. I was amazed to find even here, summed up in this tiny speck, the wisdom of God."

Oddly enough, the microscope then disappeared for quite some time as a tool of scientific inquiry. Too many scientists felt that it had reached the limits of its possibilities and regarded it as a toy for innocents in search of edification. "Serious" scientists held it against the amateurs that the visual joys of microscopy had become almost a fad, an entertainment accessible to anyone with some pretensions to erudition.

The Microscope's Performance

The era during which the microscope was to open up new challenging fields for science did not dawn until the second half of the nineteenth century. From then on the microscope became the characteristic symbol of scientific research. As an aside, it could be mentioned that there had been very early attempts to achieve greater magnifying power with materials other than glass. Pierre Borel, the personal physician to King Louis XIV of France, proposed casting optical lenses from fishglue.

The "compound" microscope, a combination of two or more lenses, was adopted only in the eighteenth century. The great Swiss mathematician Leonhard Euler was able to calculate, as early as 1764, the gain in magnifying power of such doubling of lenses, but the actual assembling of microscopes remained for a long time to come a matter of trial and error. Appropriate lenses were selected and the most effective distance between them was tested. "Testers" were still employed during the early years of the Carl Zeiss Works in Jena.

The era of mathematical optics actually started with Joseph von Fraunhofer, the Munich professor who made research into the properties of glass his lifework. The first microscope built on a strictly scientific basis goes back to Ernst Abbe, who was chosen by Carl Zeiss as his associate. Ernst Abbe created the so-called apochromats, objective-lenses that almost completely correct aberrations and distortions. Until the 1930's optical systems were calculated with the aid of logarithmic tables. Today, the task is performed by electronic computers. Creative ideas and constructive reasoning pose the problem. Modern data processing combined with the experience of decades result in the precision of today's scientific microscopes.

Ernst Abbe found that the theoretical limit to the performance of a microscope is given by the wavelength of visible light. Accordingly, one could at best count on a resolution (the smallest distance at which two points can still be discerned separately) of one hundred thousandth of an inch. Anything smaller than that seemed beyond the reach of a microscope, and would probably remain forever hidden, or so the reasoning went. But the splendid advances in light microscopes had not come to an end. By using certain refined methods it became possible to achieve the seemingly impossible—namely, to go beyond the limit mathematically established for the light microscope. By means of the "dark-field illumination," a method perhaps already known to Leeuwenhoek, considerably smaller objects can be made visible. Obviously, neither the outline nor the structure of things that thus become barely visible can be determined with complete certainty.

In 1953, the Dutchman Frits Zernicke received the Nobel prize for the invention of the "phase-contrast microscope." In this microscope, a phase-shifting platelet is placed in the path of the light. This little device changes the timing of the wave motion of part of the light so that it interferes with light passing through a transparent specimen. Contrasts are reinforced in this way, and what formerly looked vague and hazy emerges with surprising sharpness and depth. Once, before his death in 1966, I had occasion to talk to Zernicke about his invention. He said with a smile: "there are two ways for a scientist to become known, either he is ahead of his time and does something extraordinary, or he finally solves a problem that was close to solution long before. That's how it was in my case."

Wonders in a Drop of Water

Today's microscopes permit magnification beyond the thousandfold. The light microscope magnifies up to about 1,500 to 2,000 times. Under such magnification, a drop of lakewater holds the most astounding surprises: There floats an amoeba, a protozoan, a little lump of slime. The phase-contrast picture shows how the amoeba—a protozoan without limbs, without organs, in size perhaps five thousandth of an inch, yet a living being that moves, eats, digests, and reproduces—bulges out and how the content of the cell flows into the bulge, the outstretched "false foot." Next to it is another single-celled animal, a sun animalcule, a heliozoan, with long plasmatic rays projecting from its body. There move some flagellates with lashing tails. Like plants, they manufacture sugar and starch from water and carbon dioxide. A ciliate swirls food into its cell-mouth. The coordinated movement of its cilia (hairs) drives it through the water as if it were propelled by oars. A paramecium swims into sight. Within its body flower-like shapes form and dissolve: with these pulsating hollow spaces within its body it can squeeze out water and thereby equalize pressure. The cell colony of a spherical alga, a mobile plant colony, drifts by like a glassy blackberry. Slowly a volvox rotates like a dotted transparent ball. Inside it float smaller balls, daughter colonies under protection. The beginnings of "parental care" can be recognized. There are already differentiated or specialized cells; some are responsible for reproduction, others for nutrition and locomotion. A hydra, a freshwater polyp, moves its arms. A simple tube is the primitive mouth. As with the Hydra of the classical myth, severed parts will re-grow the complete animal. The freshwater jellyfish, a tiny umbrella, contracts into the shape of a cup and thereby expells the water from its inside. The recoil propels it forward. The freshwater medusa is the tiny relative of the jellyfish that float in the oceans and defend themselves against a swimmer's touch with stinging nettle filaments. Part of the world in a drop of water is the multitude of tiny crustaceans that strain the water through the bristles on their feet in order to catch minuscule plants. The glassy crustaceans are as transparent as demonstration models in a school exhibit, one can observe how the large eye moves, how the cardiac valve palpitates. The digestive tract is visible and next to it the oil drops that facilitate floating. A growing young moves its legs jerkily inside the brood space of a crustacean as if inside a corked bottle.

The clearer and cleaner the water, the fewer such animals it contains. Water that is tempting to bathe in is in most cases uninteresting for the biologist. It might be called a "desert" because it harbors hardly any life.

Nature's capacity for inventing ever new forms within the microworld seems inexhaustible. The most minute creatures build from the silica in the seawater delicate trestlework, filigree structures that are light enough to float yet strong enough to withstand the deep-sea water pressure. They form bizarre, fantastic casings, as beautiful as they are utilitarian. The abundance of forms and constructions invites comparison with the many creations of artists and engineers. Within the ocean plankton, the "wandering," the "floating," there are myriads of diatoms in such a multiplicity of forms that it will most likely never be possible to catalogue them all. Nature has an infinity of variations to offer. Improvements are tested in the course of millions of years, and competitive forms may appear one day.

It was once believed that living things that built complex skeletons had to be highly developed animals. That single-celled beings should be able to do so, seemed simply unimaginable. Even the famous German zoologist Christian Gottfried Ehrenberg, who admired these skeletons in the mudsamples he had collected, was overwhelmed and expressed his feelings in these worlds: "The smallest in the world is wonderful and great, and of the smallest, the worlds are made." Some two thousand years earlier, the Roman naturalist Pliny the Elder expressed the same thought more dryly: "It is precisely in its smallest and simplest structures that nature shows itself most perfect and accomplished."

In the year 1876, the English corvette *Challenger* returned to Portsmouth. It was a war vessel whose cannons had been removed to make room for a scientific laboratory. The *Challenger* had been away for more than three years, roughly seven hundred days of it spent on the high seas.

The results of this voyage, which was undertaken to explore the secrets of the deep sea, were turned over to a team of scientists for evaluation. Ernst Haeckel, who had discovered radiolaria (an order of marine protozoans) in the Gulf of Messina, undertook an examination of the mud samples from the ocean bottom. It took him eleven years to complete volume eighteen of the *Challenger* report's fifty fat volumes. Volume eighteen contains more than 2,000 pages of text and 140 drawings done by Haeckel. By 1887, Ernst Haeckel had discovered no fewer than 4,318 different kinds of radiolaria, three times as many as there are of mammals. He subdivided them into more than 700 species, 85 families, 20 orders, and 4 "legions." Just imagine: There are untold billions of these beings, the largest of which is .04 inch in diameter, floating in the oceans, each kind at its own characteristic depth. Haeckel later published a selection from his scientific picture atlas of the radiolaria to instruct the layman and to give pleasure to a wider audience. *Kunstformen der Natur (Artforms in Nature)* was the title of this book that, regrettably, is almost forgotten today.

Haeckel drew at the microscope even though microphotography already existed and its worth as documentation could not be denied. But these radiolaria were three-dimensional structures, often trellis-globes, most difficult to photograph. Under the light microscope only one plane is in focus, while when drawing, one can optically scan such a delicate trellis-globe by moving the micrometer screw on the specimen mount.

Only with the modern scanning electron microscope can objects be scanned line for line by the electron ray. Thus, unusually well-defined three-dimensional-looking pictures result. With magnification from 20 times to 30,000, even 50,000 times, this instrument ranges between the light and the transmission electron microscopes (see chapter three).

Haeckel and other drawing microscopists have been accused of idealizing their objects in their drawings, of modifying them according to their preconceived notions, even of falsifying them. Commented Wilhelm Bölsche in 1923: "anyone who considers these drawings exaggerated should be convinced by the original specimens that no reproduction, not even the most artful one, comes anywhere near the great natural beauty of these unique objects."

One thing should always be kept in mind when looking at radiolaria: these silicon structures under the microscope are, so to speak, "specks of dust" from the muddy bottom of the seas. With the death of what once animated them, they have abandoned their state of floating and have slowly sunk down. Myriads of such filigree skeletons have settled in the course of millennia to form the sediment of a living world that was unknown a century ago. Within each mud sample brought up to the surface by a research vessel, a condensed history of life in the sea is recorded.

In the meantime it has become possible to study living radiolaria through photomicrography. The zoologist Karl G. Grell of Tübingen has presented some exceptionally beautiful and impressive documentaries in collaboration with the Station Zoologique Villefranche sur Mer and the Göttingen Institute for Scientific Cinematography. They were shown in 1963 on German television, with the same aim that Ernst Haeckel had in mind when he published his picture portfolio, *Artforms in Nature*—namely, to inform a wider public.

Three generations ago it still was believed that life could not possibly exist in the deeper strata of the oceans. Now, we can see under the microscope the most delicate, most fragile structures, living beings brought up from abysmal darkness, animals that have withstood the deep-sea pressure, artful spherical or trestle casings through which life radiates fine needles in all directions to facilitate floating. Even at mountainous depths where not a glimmer of sunlight reaches, there exists beauty. For thousands, for millions of years, long before man could conceive of beauty, long before artists and engineers could give it form, all this already existed in the world of creation! In mud, in the sediments of the oceans, there is a treasure trove of shapes whose existence was undreamed of before the microscope revealed it.

Dr. Horst Reumuth, although confronted daily with problems he has to solve for industry in his Karlsruhe Institute for Applied Microscopy, has remained faithful during his rare leisure hours to the passion of his youth: the "observation of biological structures in microscopic scales." He has taken hundreds of microphotos of diatoms, those tiny siliceous algae that form part of the plankton of the oceans, that live in salt water as well as in fresh water. Reumuth writes: "micronature is inexhaustible. It builds with art, experience, and care... it varies forms, structures, construction materials, it creates functional designs, it applies the rules of economy in material and weight, in the building of its light, floating, flying, or swimming structures...." With regard to "floating," nature has shown itself especially inventive. It uses materials sparingly and achieves strength through special reinforcements and braces. Millions of years before man conceived of technological construction, nature tested shapes that use the principle underlying the shape of egg cartons where the high-and-low embossing gives the paper-pulp containers enough strength to carry the weight of a man. The finest tubules of our physicians do not measure up to the hollow needles of the microworld. Their diameter is eight microns; the thickness of their walls, one micron, a few hundred thousandth of an inch. The absorbency of silicious marl, the diatomaceous earth, that is mined where ancient sea bottoms are now dry land, is due to those capillary tubules.

Within the microworld of the water droplet, Horst Reumuth found again and again "prototypes of man's artistic and utilitarian objects: exquisite decorative boxes looking as if made of rock crystal, round, oval, or openwork boxes, with deeply cut designs or with raised ribs, but also stars resembling decorations, chains, brooches, and bracelets, then again swords, halberds, crowns, goblets, bowls, pots, and tankards! Only when hundreds of these are joined together do they become visible to the naked eye, as one speck of dust."

Many diatoms have amiable names that express the admiration of their discoverers: *navicula alongata* = the elongated little ship; *amphora ovalis* = the egg-shaped winejug. The honeycomb and trestle-work of diatom casings serve as touchstone for the performance of a microscope. Diatoms are often encased in silica shells that fit together like the top and bottom of a jewel box. But this comparison still does not do justice to the graceful daintiness of these structures that protect and sustain the tenuous life within.

The photographic camera has been able to show how such a being reproduces itself within its box. When the cell body is ripe for division, the shells part. The plasmatic lump within divides, both shell halves become lids, and a new receptacle forms under each lid. The inevitable diminution in size of the descendants stops at a certain point. Then, two individuals emerge from their casings, combine to form a new being, which then builds a new box, this time receptacle and lid, around itself.

These microconstructions obey mysterious guiding forces, forces that man's technical experience has not been able to reveal. It remains a secret in what way nature builds these works of art, what means life's microtechnique employs. How do these characteristic needlesharp spears develop, how are the geometric pores and holes not punched out, but built around? How does the infinitesimal creature, building simultaneously on many sides, achieve the symmetry of the whole? How does nature supervise the execution of its blueprints? What is the role of the living protoplasm in guiding and shaping the silicic acid? Shapeless lumps of slime have the gift to produce silica skeletons in thousands of different designs in ever new modifications, like a muscial composition with unending variations. This microworld offers probably greater adventures, stranger apparitions, than a visit to Venus or Mars. A clouded drop of water bears comparison with the Milky Way. Just as the telescope revealed the Milky Way as a dizzying multitude of stars and thus made it comprehensible, so the microscope revealed the clouded drop of water as a fantastic world of countless shapes. Here, perhaps, it is shown most clearly how much the microscope has expanded the world that can be experienced by our senses, how deep the spaces are that our vision can now penetrate, and how we have gained thereby a truly new dimension. The beauty of the virtually infinite variety of forms in this living world is unsurpassed. Outside the organic world, such a wealth of forms is encountered only among the crystals, in that vast realm between the diamond and the snowflake.

Snow Crystal Filigrees

The oldest representation of snowflakes appears in a book published in Rome in 1555, written by a learned ecclesiastic from Uppsala, named Olaus Magnus. In his book a woodcut meant to depict drifting snowflakes looks somewhat more like a rockslide. In Descartes' *Specimia philosophiae,* which came out in 1637, snow crystals are shown as stars, but, even here, the illustrator did not yet know anything about their typical six-rayed pattern. The great Johannes Kepler gave some attention to this consistent regularity. In 1611, he wrote in a humorous and poetic New Year's letter to a friend: "As I was crossing the bridge, deep in thought and worried and chagrined by the miserable circumstance of coming to you without a New Year's gift, chance had it that because of the intense cold the water vapor condensed into snow, and isolated little flakes fell on my coat, all hexagonal, with feathery rays. Hey, by Hercules, this is quite something, smaller than a drop of water, yet of symmetrical beauty. Hey, this is a most welcome New Year's gift for a friend of the naught! And just as fitting a gift from a mathematician who has nothing and gets nothing, except such as falls down from heaven and resembles the stars." It seems that Kepler had repeatedly looked at the perishable crystals under the reading lens. "Why is the snow pattern six-cornered, why not just for once five-cornered or else seven-cornered?" he asks in his letter.

The first accurately observed snow crystals were drawn in 1655 by Robert Hooke for his *Micrographia,* that is exactly a hundred years after the naive little picture in the book by Olaus Magnus. In our own time, another three hundred years later, there appeared two collections of photographs devoted to the astonishing variety of snow crystals: one in the United States, by W. A. Bentley and W. J. Humphreys, *Snow Crystals,* that contains 2,400 pictures; one in Japan, by Nakaya, with 1,500. Even so splendid a sampling as the one we owe to these great dedicated enthusiasts in their field, can give no more than a handful out of the immense number of variations with which nature toys.

Today, the structure of snow crystals in their hexagonal symmetry is being studied in temperature-controlled chambers where snowflakes can be artificially produced. This process may never fully explain what happens to a crystal nucleus in the atmosphere, or which circumstances produce which shapes. The slow fall through the clouds, the long drift through layers of varying temperatures, the repeated ascent and descent within the currents of a storm determine whether snowflakes, or sleet, or many-layered hailstones large as pigeon eggs will come down to earth. When there is little water vapor in the air and it is very cold, small, relatively simple snow crystals form. In fog, the crystals are sparse. With rising temperatures and more water vapor, richer "skeletons" grow around the dust particles which provide the crystal nuclei, and feathery rays sprout out of the hexagonal platelets. Nature chooses among the possibilities at her disposal to continue her constructions. The snow crystal sinks through the unevenly warm, unevenly moist layers of the atmosphere; it may fall for a long time and over long distances. Finally, an hour, even three hours, may have passed. In most cases several crystals stick together, in a small snowflake perhaps a hundred, in a large one as many as three thousand. It is a stroke of good fortune if a single crystal alights on the sleeve of one's overcoat, a crystal that has remained separate and undamaged by collisions.

Anyone who wants to try to place these wondrous structures under a microscope in order to record them in a microphotograph, has to make all sorts of preparations. It is best to take up position in the open air, perhaps on a covered balcony. How does one prevent the delicate crystals from dissolving too quickly? Walter Seyfarth, a well-known German amateur photographer, recommends wearing leather gloves; otherwise the warmth, transmitted from the fingers to the microscope, spreads too fast. Furthermore, he recommends covering one's mouth and nose with a shawl. Even without the effects of heat, there is danger that evaporation may obliterate form and sbustance; in that case, the crystal disappears without dissolving into water. Seyfarth suggests letting the snowflakes fall on a thoroughly cleaned deep-cooled slide. Unfortunately, the crystals are often already injured; or they do not settle down horizontally, which inevitably results in fuzzy pictures. When dealing with snowflakes we are dealing with ice. We perceive, therefore, no inner structure, only surface structure, of which we can show a certain aspect through special lighting. With vertical bright-field transillumination, snowflakes look like fine ink drawings. Not much more than their outlines show. By slightly tilting the mirror of the microscope and thus using slanting light, and by substituting a dark background, one gets a more three-dimensional picture; through dark-field illumination, the surface structures stand out. This method may cause, though, an unevenly lighted field of vision. A 20- to 50-fold magnification has proved to be best suited. Exposure time must be tried out in test pictures; it ought to be between one and three seconds.

Urs Beyeler, who has photographed snowflakes on the Jungfrau-Joch in the Swiss Alps, positioned a heat-screen between the low-voltage lamp and the microscope mirror, cooled his instruments with dry ice, placed dry ice and instruments on a block of ice, and covered the dry ice with a woolen rag to prevent the microscope from fogging up. For the gathering of snow crystals, he recommends, as does Seyfarth, a piece of black velvet stretched over cardboard, and for the handling of the delicate little stars, two fine-hair brushes. Perhaps these hints will inspire some readers to make a try themselves, and will prevent failure of the attractive undertaking through lack of experience.

The Magic World of the Crystal

Mikroskopische Augen- und Gemüths-Ergötzung (Microscopic Amusements for Eyes and Mind) is the title of a book by Martin Frobenius Ledermüller from the year 1763. Ledermüller was a man of many interests, a baroque character who enjoyed prestige as a scientist. The name "infusoria" for microorganisms found in infusions was coined by him. He went over some of the ground that Hooke had covered. The copperplates in his book illustrate the many things he had shown under the microscope to the Margrave of Bayreuth, whose collection of natural curiosities he directed: fish scales, mosquito wings, flower pollen, lace from Brabant.

Figure 58 is dedicated to the "configuration" of alum: a drop of solution evaporates. Crystal points shoot into the picture, shapes materialize like frost flowers. Stars and comets blossom. Crystallization of ammonium chloride is particularly recommended to the reader as a spectacle. "Since this salt tends more than any other to evaporate quickly and 'configurate,' amateurs could hardly choose a better one for their entertainment."

A mysterious trait manifests itself in crystals: the playful urge of forces that in spite of all their freedom of choice still must obey certain laws of nature. Nature acts according to a structural imperative, a mathematical plan that is the foundation of our world. Crystals growing inside a mountain seem beyond geological explanation, an almost magical happening, even though nature has been copied in the laboratory, in the production of artificial rubies and sapphires and in the synthesis of small industrial diamonds. Gems are among the most mysterious treasures on earth; our fascination with them cannot be explained by their money value alone. In the hardness, brilliance, transparency, and brightness of precious stones we find realized the wish-dream of imperishable beauty.

Crystallization from a solution is always a captivating spectacle: under the microscope one can follow step by step in every detail how the crystal needles grow, how the designs abruptly change! Polarizing filters may enhance the experience even more. Ordinary light vibrates in all directions at right angles to its path. Polarizing filters allow it to vibrate in only one direction. If one places one polarizing screen across the lightbeam under the condenser, and another one under the eye-piece of the microscope, crystals appear in glowing colors. If one turns the eye-piece screen, the colors change into their complementary colors. Potassium nitrate, potassium chlorate, ammonium oxalate, barium chloride, thiocarbonate—all substances available in drugstores—afford phantasmagorical spectacles, as does magnesium sulfate, commonly used as a laxative under the name of Epsom salts. The colorless crystals are dissolved in water and a drop of the solution is placed under the polarzing microscope. A heated covering glass shortens evaporation time. Crystallization begins normally at the rim of the drop. Needles grow inward as if directed by a concave mirror: luminous lances, sprouting bundles of colored needles! The growth of crystals is a fascinating process. It looks "purposeful." Yet, at each new stage, nature seems bent on surprising the spectator with some "new idea."

Manfred Kage (born 1935) is one of several younger people who try to influence crystallization under the polarizing screen, in order to bring about the formation of esthetically pleasing shapes and colors. He calls this "polarchromatic variations." During the process of crystallization he adds moistures, alcohol, and resins to the evaporating solution. He cools the blends at varying rates and thereby produces "mysterious eyes, oases, rivers mountain ranges, rhythmic waves, trees, deep-seascapes." He chooses the substances, the crystallization conditions, and with the help of a so-called polychromatizer, he selects the color tones and color harmonies. "After having crystallized many hundreds of times a preparation of the same substance, one begins to penetrate its nature. After several hours of work, which is very much like meditation, one becomes aware of strange things. When one identifies so thoroughly with the character of a substance it becomes hard to tell whether one creates a picture or whether one just witnesses its emergence."

Manipulated Light: Polarization and Interference

By far the greatest part of matter is composed of crystals: rocks, coal, minerals, building materials, ceramics. To examine them in transillumination, one has to grind down the rock specimens as thin as twenty-five thousandths of a millimeter. Under the polarizing microscope one gets vividly colored pictures that define the rock composition.

Photoelastic stress analysis has the faculty to reveal, under polarizing light, internal stresses in transparent glasses and synthetics. Stress points in glass may cause ruptures. Heatproof household glassware is given an apposite pre-stressing to insure against cracking under high temperatures. Heatproof glass, when transilluminated with polarized light in conjunction with a small plaster-of-Paris plate which produces a red interference color, shows the desired pre-stresses in blue or yellow. Through the Polaroid lenses of sunglasses, the windshields of some cars look strangely paterned; the lenses reveal the built-in stresses, designed to make the glass crumble in case of an accident.

In the examination of thinly ground rock specimens and in photoelastic stress analysis, it is interference that brings out the colors and thereby reveals the hidden. The opalescent colors of a soap bubble, of an oil slick, of Newton's rings are due to the superimposition of light waves. In certain spots these waves reinforce each other, in others they cancel each other out.

In the interferometer a lightbeam is divided, the separate beams are guided along different paths and then reunited. This is done by means of optical systems incorporating mirrors. In this way one can make visible and measure changes of air density, air pressure, and temperature. The invisible appears as if by magic; commonplace processes acquire supernatural beauty. The flame of a candle, seen through the interferometer, is framed by bands of color. Ether, escaping from a bottle, looks like evening clouds at sunset. The carbon dioxide, exhaled by a person, appears in color. Temperatures and pressures show themselves through the interplay of colorful interference bands. The interference microscope produces strange relief pictures. They make possible detection and measurement of surface imperfections in the order of magnitude of light wavelengths, that is, in fractions of one thousandth of a millimeter. Two overlapping lightbeams "draw" minimal surface unevenness as if it were the topography of a landscape.

The Function of the Infinitely Small

One fifth of all microscopes are used in technology; four fifths serve medicine and biology. One would assume that new knowledge, intelligently presented, would be quickly and widely accepted; that it would develop like a growing tree spreading its branches, or a river swelling with affluent brooks. But no, progress often takes a circuitous route, at times comparable to those strange currents of water that disappear underground and resurface only after a great distance. This is what happened with the observations of a Leeuwenhoek, Hooke, and Malpighi. Around 1800, at the famed Sorbonne, the microscope was "in low esteem as a toy for amateurs, and almost shelved" (Erwin E. Ackerknecht). In Germany, on the other hand, it continued to enjoy an almost romantic devotion. In Berlin, Johannes Müller, a biologist and universal intellect, pursued the microscopic study of the living body. One of his pupils, Theodor Schwann, determined in 1838 that all animals and plants are ultimately composed of cells; another Müller-pupil, Rudolf Virchow, based the whole of pathology, the science of disease, on the study of sick cells. Medicine was thereby given a new foundation, the one it is still based on today. It was the microscope that made this step forward possible. From then on scientific research could penetrate further into the vast enigmatic area of physiology and pathology. Biology and related natural sciences also conquered new territory.

Robert Koch and Louis Pasteur ("the function of the infinitely small appears to be infinitely big") established bacteriology as a science. What is obvious today was not always so. Let us see then what Pasteur set forth before the Paris Academy of Medicine: "This water, this sponge, this cloth used for washing or covering a wound, contain germs which are able to multiply rapidly within tissue, and which will inevitably and in a short time cause the patient's death unless the life forces of the body counteract them. Regrettably, natural resistance is insufficient in all too many cases. The patient's constitution, his debility, his psychological makeup, inadequate dressings often cause a weakening of the defenses against invasion by these minute organisms, inadvertently carried to the injured part of the body. Considering the risks the patient incurs because of the microbes present on the surface of all objects, especially in hospitals, I would–if I had the honor of being a surgeon–use not only absolutely clean instruments, but I would also clean my hands with the utmost care. I would hold them briefly into a flame, a procedure that would not cause greater discomfort than that experienced by a smoker changing a piece of glowing coal from one hand to the other, and I would use only cloth, dressings, and tampons that have been exposed to a temperature of 130–150°C! [266°–302°F]." Bacteriology soon led as a matter of course to antisepsis and asepsis in surgery.

When Robert Koch, in 1876, was able to demonstrate that a bacillus caused anthrax, a living microorganism was for the first time made "resonsible" for an infectious disease. In 1882, Koch identified the tuberculosis bacillus; in 1883, by means of a special tinting technique, the microorganism that causes cholera. Making something visible under the microscope was in each case the point of departure. Recognizing the enemy was followed by the battle against him, the search for a suitable remedy whose effectiveness could again be studied under the microscope. In such a way Louis Pasteur developed the rabies vaccine.

"The bacteriological discoveries made in France and Germany between 1875 and 1905 were the most important contributions to medicine ever.... They made possible the great advances in surgery, the somewhat later ones in drug therapy, and most importantly those in preventive medicine that led to the practical disappearance of a whole number of diseases, and thereby to today's much higher life expectancy as compared to 1900" (Erwin E. Ackerknecht).

How does the world of bacteria, that Leeuwenhoek was the first to observe, look under today's mechanically and optically perfected microscopes? Streptococcus bacteria, among them those causing pus to develop, look like strings of pearls; staphylococcus bacteria that cause blood poisoning and are also present in pus look like grapes; diphtheria-producing bacteria resemble small cudgels. The tetanus bacilli look like drumsticks; the producer of cholera, like a crosier.

The spirochete pallida, the microorganism causing syphilis, discovered by Fritz Schaudinn in 1905, may be compared to a pale corkscrew. It is just barely visible under a light microscope.

The pathogens are only part of a whole world of microorganisms. The rest is useful to man, even supports our lives. Cheese and vinegar are the products of bacteria. Bacteria in the soil collect the nitrogen in the air and make it available to the plants. Bacteria of decay convert the remains of living things into fertile humus; without these bacteria life on our planet would long ago have suffocated under a layer of corpses. "Without microbes, life would come to a standstill," says the eminent American microbiologist René Dubos.

Only the microscope made it possible to recognize that blood is a "fluid tissue," an organ composed of cells. For closer study, blood cells have to be stained. But stained blood cells undergo changes and die. Previously unobservable details of living blood cells can now be studied with the phase-contrast method invented by Frits Zernicke. A look into the realm of the blood is one of the strangest experiences. The realization that what one sees is simultaneously repeated a million times in one's own blood stream, causes the same feeling of awe as that experienced when looking into the depths of the Milky Way. The secret of life excitingly reveals itself. The cells that look like dented platelets are the red corpuscles; Leeuwenhoek already observed them. Five million red corpuscles in a cubic millimeter of blood! They deliver the blood's oxygen, then carry away the waste carbon dioxide. Here and there, within the blood cells, light and dark bodies can be observed. One sees them floating back and forth in the transparent plasma. What does this mean? Red corpuscles live from three to four months; our blood renews itself constantly. When red corpuscles disintegrate small particles show up that arrange themselves into chains and take on the shape of bacteria. Such waste products have repeatedly been mistaken for carcinogenic material. Under the microscope, a white corpuscle wriggling through a fibrin network can be observed. Fibrin is a fibrous protein instrumental in blood clotting. Of white cells we have "only" a few thousand per cubic millimeter; for every thousand red blood cells there is only one white one! In the opinion of prominent blood researchers, science, through the observation of living blood cells, stands on the threshold of new insights.

The life sciences are confronted by an ocean of mysteries. They try to plumb its depth. Continuously biology dredges up new facts that are answers to yesterday's questions, and at the same time brings up new riddles that become the questions for tomorrow. Biology may well change our lives to such an extent that it will inaugurate a new era and give it its name.

In retrospect, the utilization of nuclear energy will perhaps be considered of less importance than these developments now in the making. The thought that one day it may become possible to redesign man himself, is eerie and haunting. The realization of the ancient wish-dream of creating a homunculus, of creating life in a test-tube appears to be only a question of time. Charles C. Price said in 1965, when he was retiring as president of the American Chemical Society: "We may be no further today from at least partial synthesis of living systems than we were in the 1920's from the release of nuclear energy, or in the 1940's from a man in space."

What sort of mechanisms govern life, the preservation of life, heredity? No doubt, the ocean of life's secrets is deepest where these riddles lie. Can the light microscope, with its approximately thousandfold magnification probe that deeply? As a matter of fact, it was the light microscope that brought about the near solution of just such a key question.

Adolf Butenandt, who in 1939 was awarded the Nobel prize for his hormone research, said in a lecture: "Biochemistry tries to analyze the fundamental units of life that are shared by all organisms. Depending on the question posed, and the method employed, organisms from the kingdoms of the animals, the plants, or the microbes can be used in trying to find an answer."

The classical laws of heredity, applying to all living creatures were to a large extent discovered in the insects. The American biologist and Nobel prizewinner Thomas Hunt Morgan (1866–1945) found in the fruitfly, drosophila, the ideal experimental animal for the geneticist. Within a single year, twenty-four generations of fruitflies are produced! Mutations are easily ascertained. Their connection with the mysterious chromosomes, the "ribbons" in the cell's nucleus that are the carriers of the hereditary units, can be better demonstrated in the fruitfly than elsewhere. The cell nuclei of the drosophila's salivary glands contain giant chromosomes one hundred times larger than those of most other living beings. They, as well as other chromosomes, can easily be stained—therefore their name. Under the microscope, the giant chromosomes look like strings of platelets. Each of these platelets represents a possible location for genes; these contain the "information" for what we call heredity.

Man has twenty-three pairs of chromosomes with a multitude of genes. They see to it that human beings produce other human beings, that the species is preserved. Dependent on the chromosomes are the physical appearance and characteristics of the individual as well as his reaction to the stimulation and demands of his environment. Also predisposition for certain diseases depends on the genes.

More recently geneticists, like Wolfgang Beermann at Tübingen, have been studying the giant chromosomes of the midge *Chironomus tentans*. The biologists are especially interested in one strange irregularity. Generally, the crosscut slices look dense and thin. But now puffed-up areas have been observed where the chromosomes appear to be thicker. It has been concluded that these odd "puffs" must consist of activated genes that are in the course of conveying instructions for various chemical activities to the rest of the cell. The microscope alone is hardly able to let us find out what happens here and how it happens. Many approaches are necesary, above all biochemical ones devised by scientific researchers to penetrate the weirdly complicated processes that take place in the chromosomes and in the cell.

"I believe a leaf of grass is no less than the journey-work of the stars," wrote the poet Walt Whitman. The road that leads the scientist in his quest for knowledge toward the most distant galatic islands in the cosmos may well turn out to be shorter than the distance that separates us from an understanding of life here on earth.

"This, then, is that godly science of optics that brings what was concealed in utter darkness to startling light."
(Athanasius Kircher, 1646)

In 1876, the English corvette *Challenger* returned from a three-year voyage of exploration. Ernst Haeckel who had discovered the so-called radiolaria in the Gulf of Messina undertook the examination of the collected mud samples from the ocean bottom. He published a picture atlas of radiolaria and followed it with a popular selection entitled *Artforms in Nature.* Our picture shows one of the radiolarium skeletons, the size of a speck of dust, that Haeckel drew under the microscope. It has been redrawn here as if under dark-field illumination.

Drawing by Ernst Haeckel, redrawn by Dr. Ing. Horst Reumuth, Karlsruhe

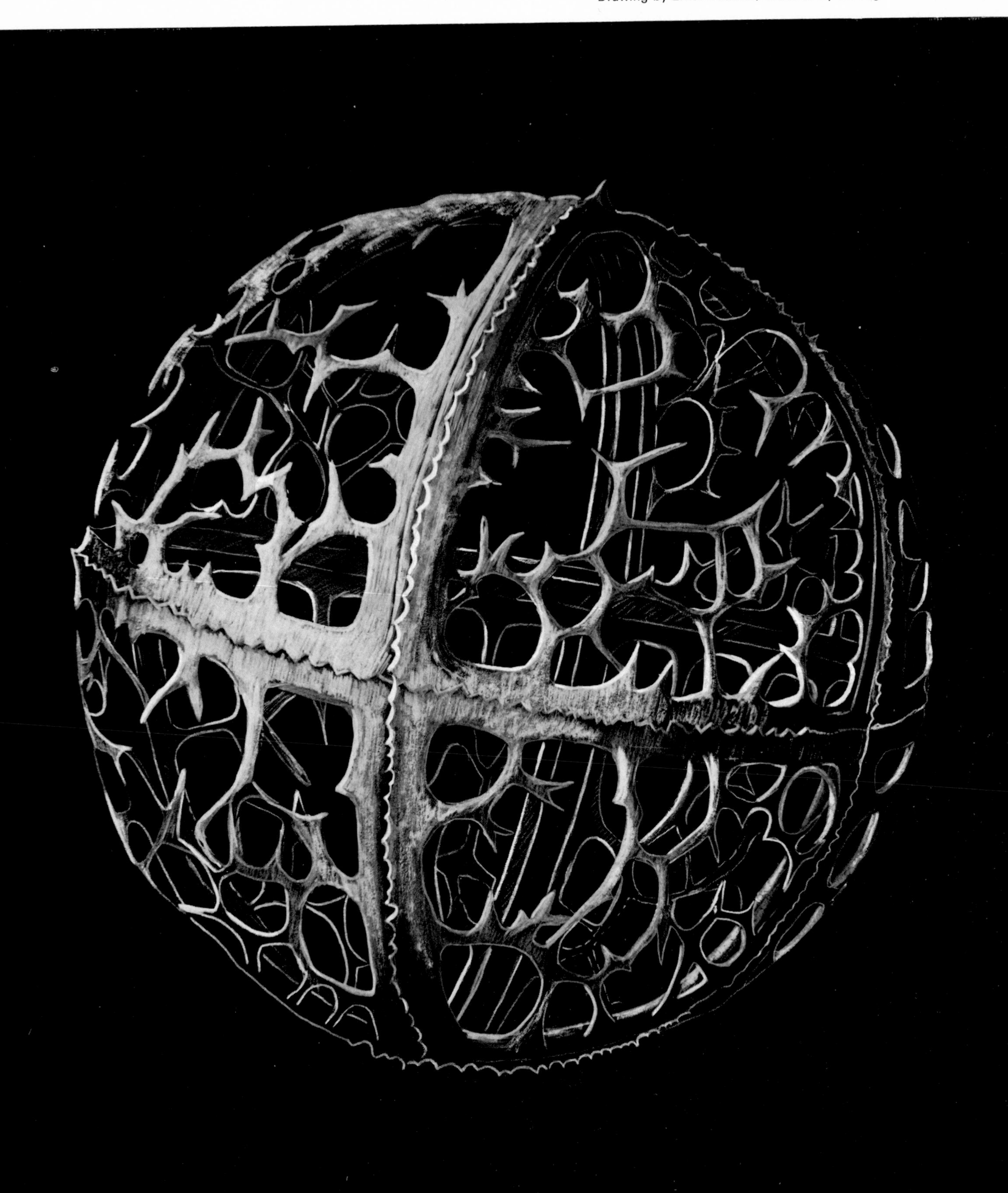

H. J. Huber, Institute for Applied Microscopy, Karlsruhe

These radiolaria skeletons were photographed with the aid of an electron beam that scanned the radiolaria line for line. The results are unusually sharp pictures in three-dimensional detail. With a magnifying power of from 20 to 50,000 times, the scanning electron microscope, SEM, bridges the gap between the light microscope and the transmission electron microscope.

H. J. Huber, Institute for Applied Microscopy, Karlsruhe

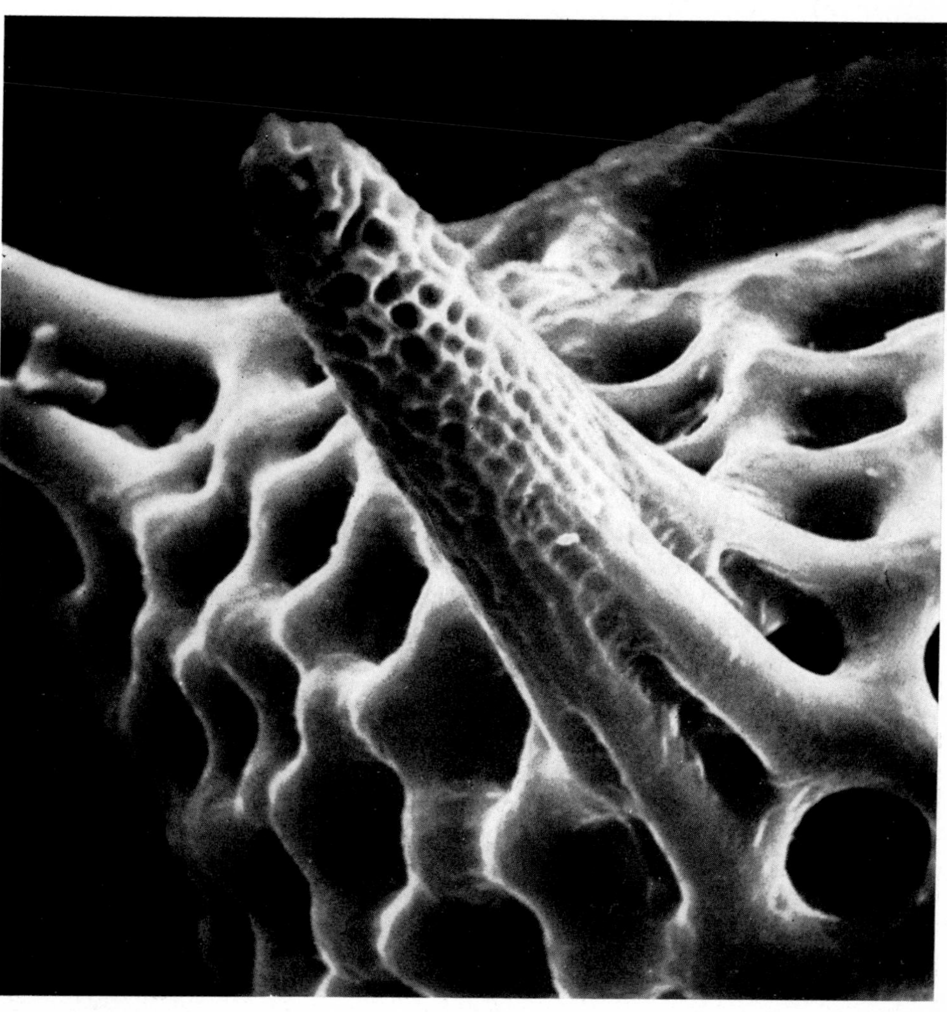

Dr. Ing. Horst Reumuth, Institute for Applied Microscopy, Karlsruhe

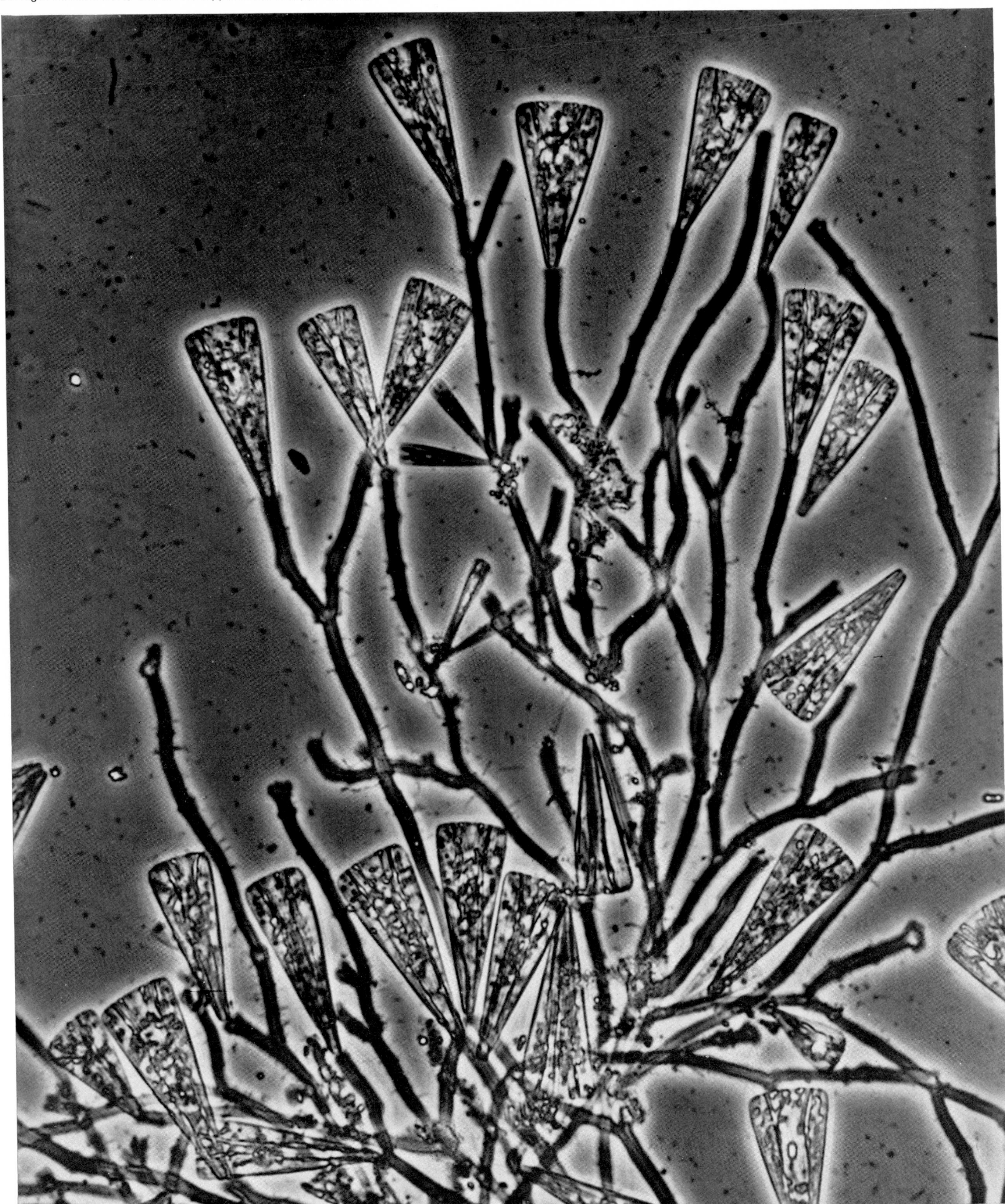

Dr. Ing. Horst Reumuth, Institute for Applied Microscopy, Karlsruhe

The casing of a diatom ▶ measuring .004 inch is enlarged here 350 times. Nature has invented an especially suitable light construction for these floating creatures. These filigree trellis-structures are as beautiful as they are utilitarian; they withstand enormous pressures. Shown here is the same diatom in three microphotographic techniques: dark-field, bright-field, and phase-contrast.

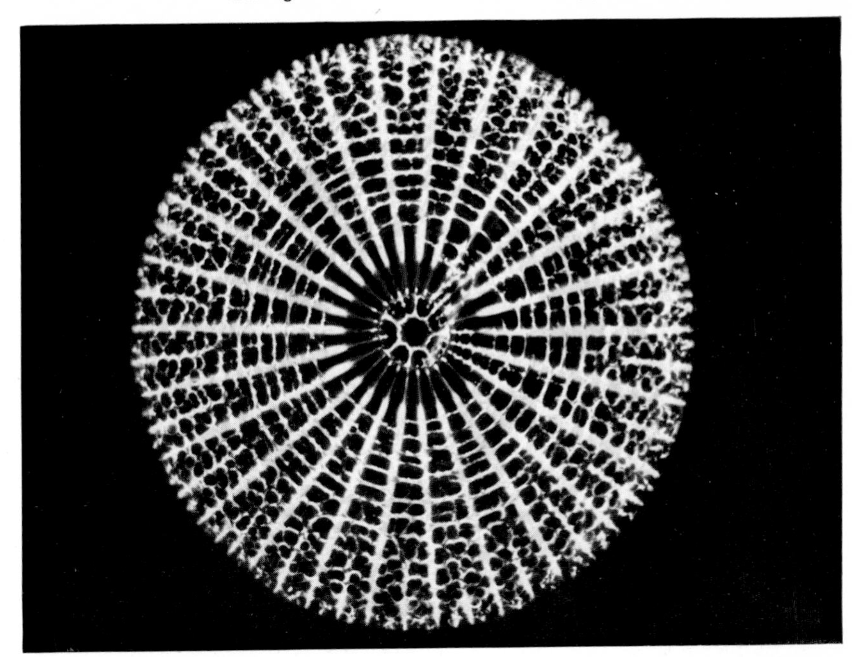

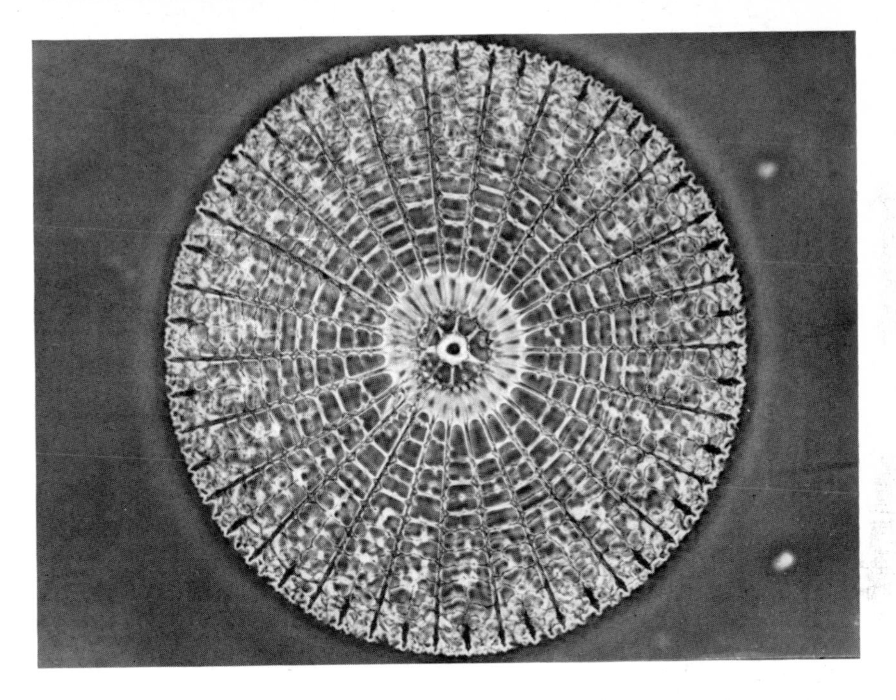

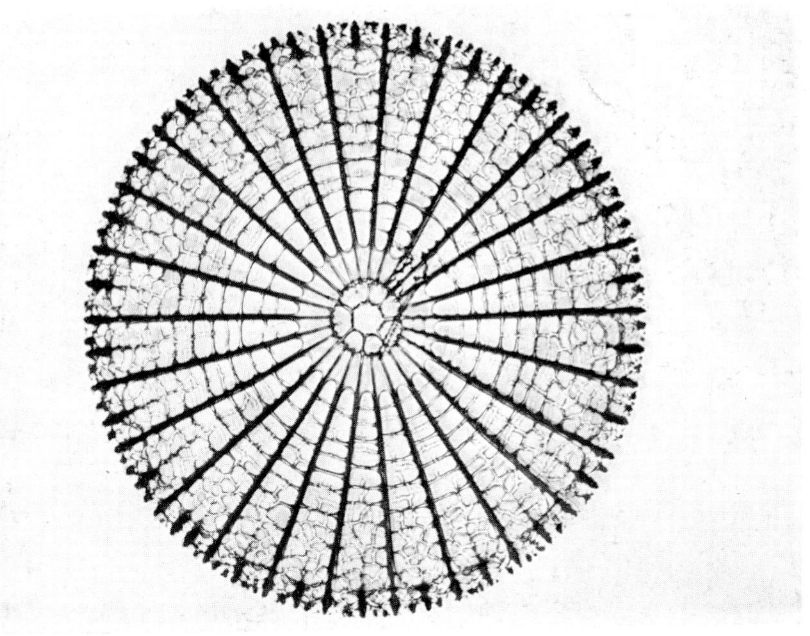

◀ This formation, resembling a bouquet of flowers, is a living colony of stemmed diatoms. Diatoms are microscopical protoplants with framework of silica. Two thousand years ago, the Roman naturalist Pliny the Elder recognized that "It is precisely in its smallest and simplest structures that nature shows itself most perfect and accomplished."

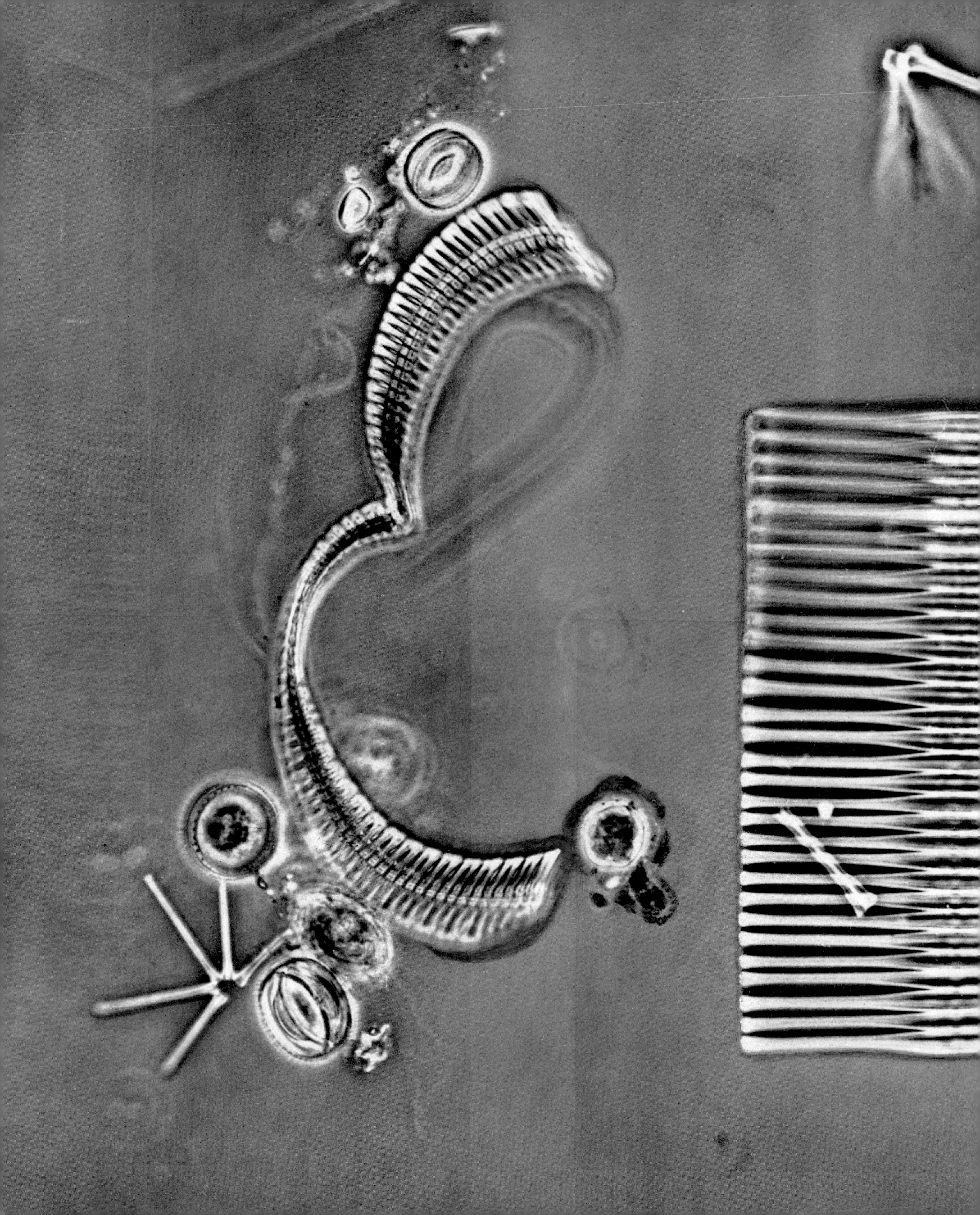

Dr. Ing. H. Reumuth, Institute for Applied Microscopy, Karlsruhe

◂ From the mud of an estuary: jewelboxes, chains, stars, bracelets—all of these are the siliceous cell walls of diatoms, the deposit of a floating world whose great variety became visible only under the microscope. Micronature styles its shapes out of the experience of eons.

Microscopic in size, these diatom skeletons in the shape of sieves once enclosed life. How nature creates these works of art remains a secret. The living protoplasm, a shapeless lump of slime, directs the silicic acid from which the cell walls are formed in such a way that in the end a structure of perfect symmetry emerges. ▾

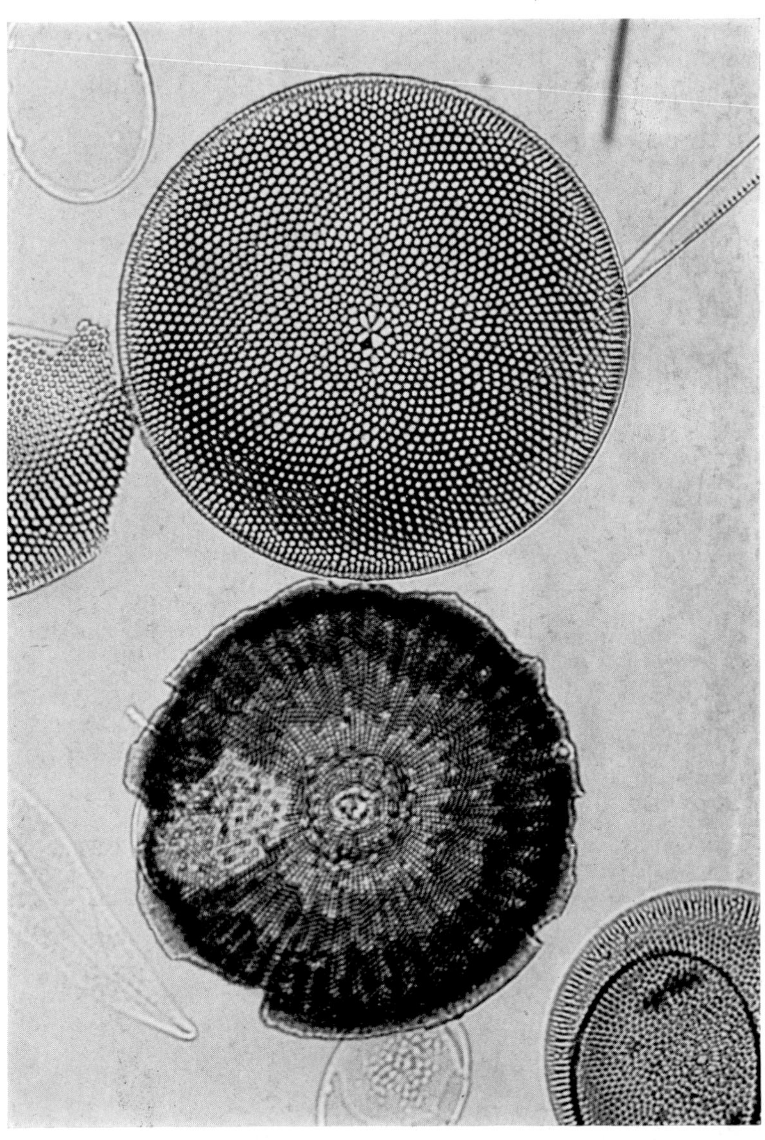

Dr. Ing. Horst Reumuth, Institute for Applied Microscopy, Karlsruhe

Dr. Ing. H. Reumuth, Institute for Applied Microscopy, Karlsruhe

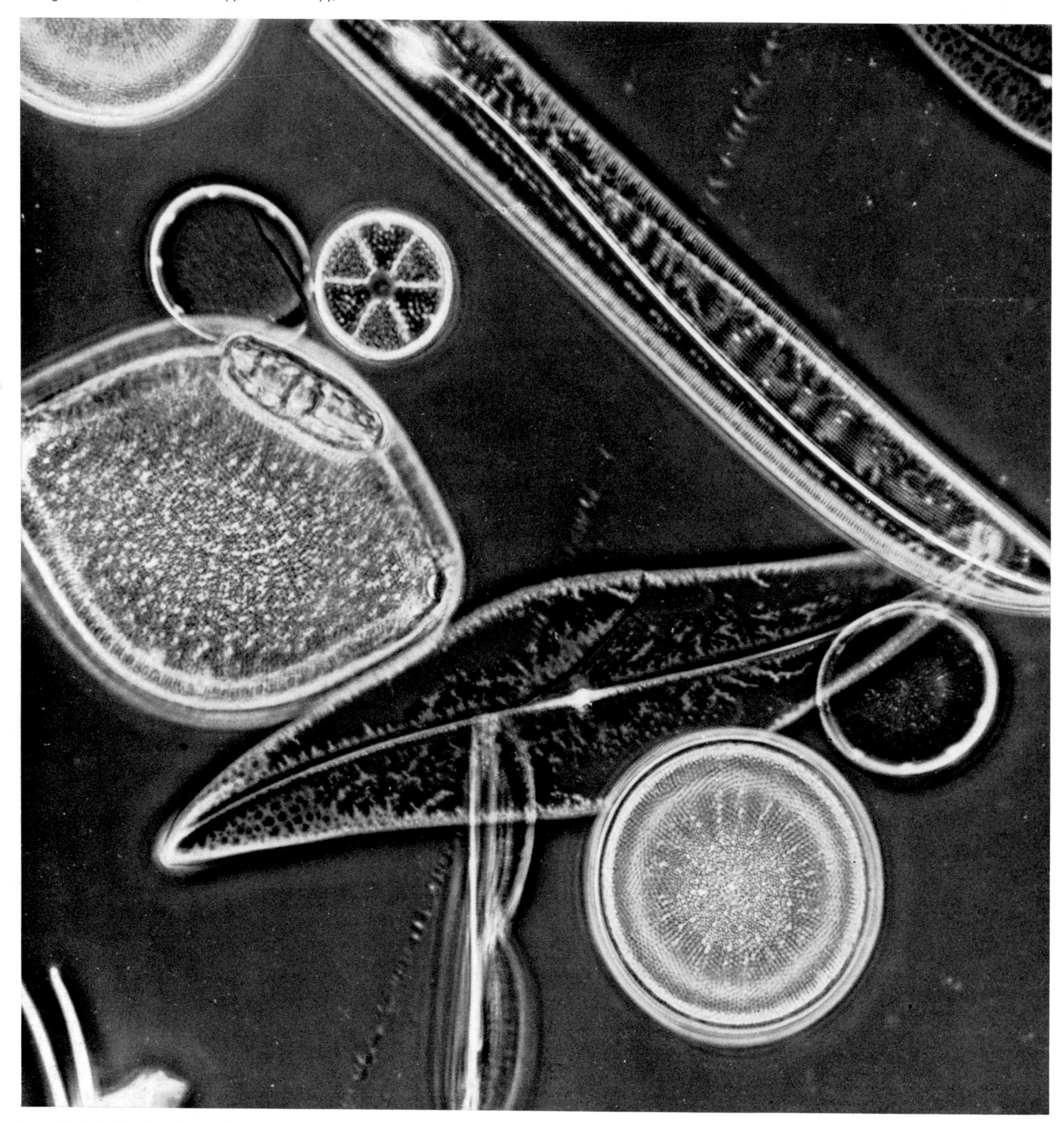

In dark-field illumination, objects are placed against a dark background and lighted from the side: they then stand out like dust specks in a slanting ray of sunlight.

Dr. Ing. Horst Reumuth, Institute for Applied Microscopy, Karlsruhe

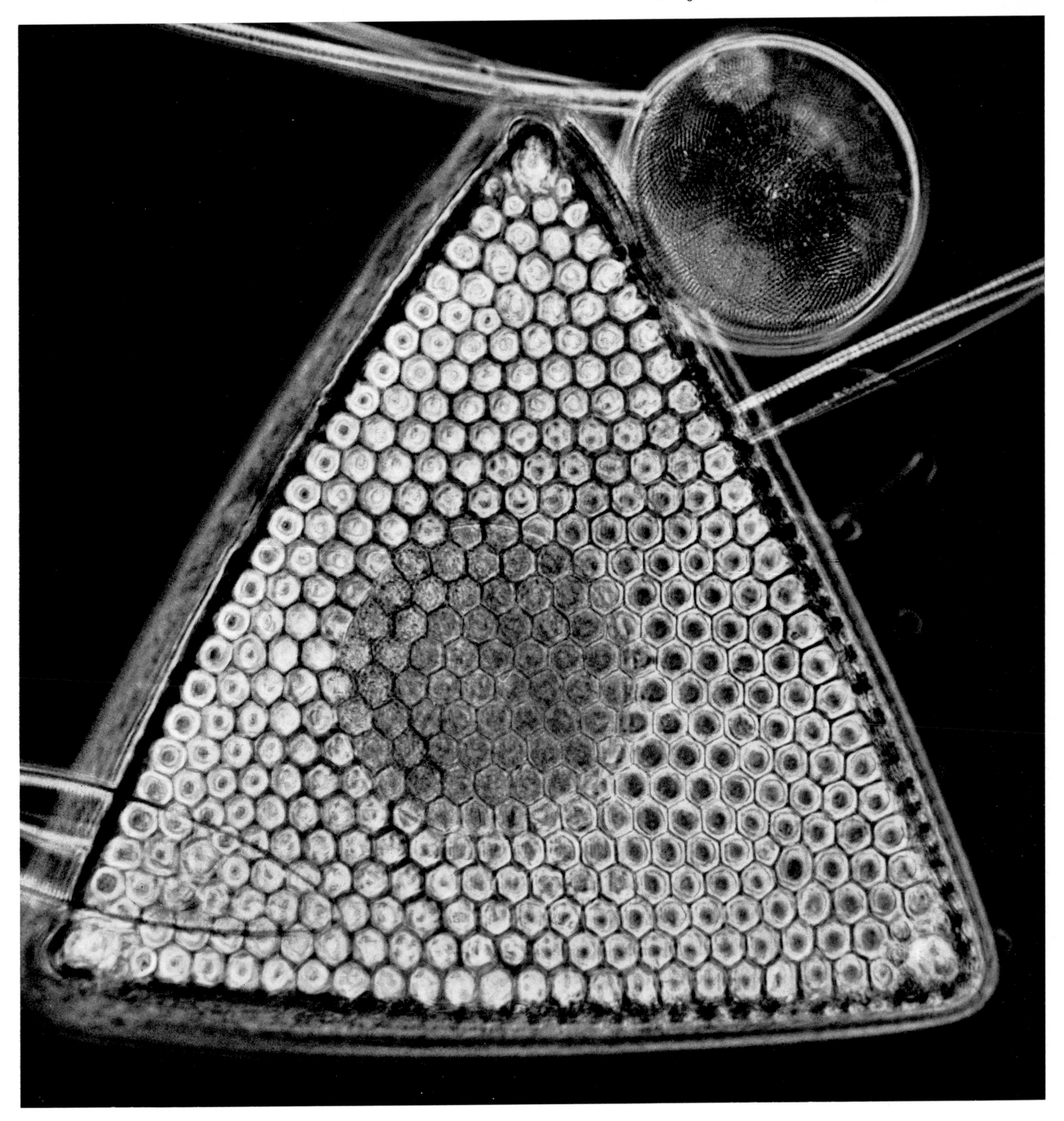

A *triceratium* diatom with a honeycomb pattern. The tenuous framework withstands the pressures at the bottom of the ocean.

Max Prugger, Munich

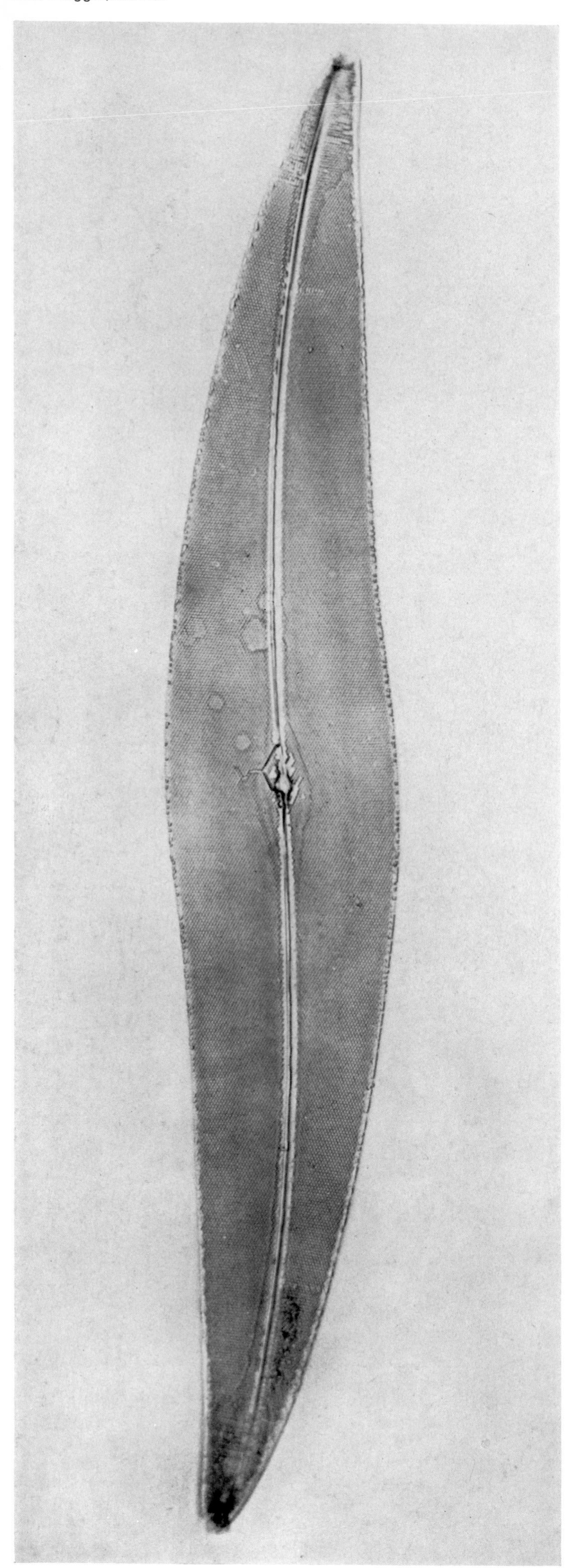

Nature is a master of filigree.

For the microscopist, diatoms are an inexhaustible source of enchanting finds. This bent airplane propeller is the best-known diatom: *pleurosigma angulatum*. It is the classical test object for the definition of optical lenses. A good light-microscope has to show the fine sieve of the shell sharply defined. The distance from point to point measures about five thousandths of a millimeter.

Max Prugger, Munich

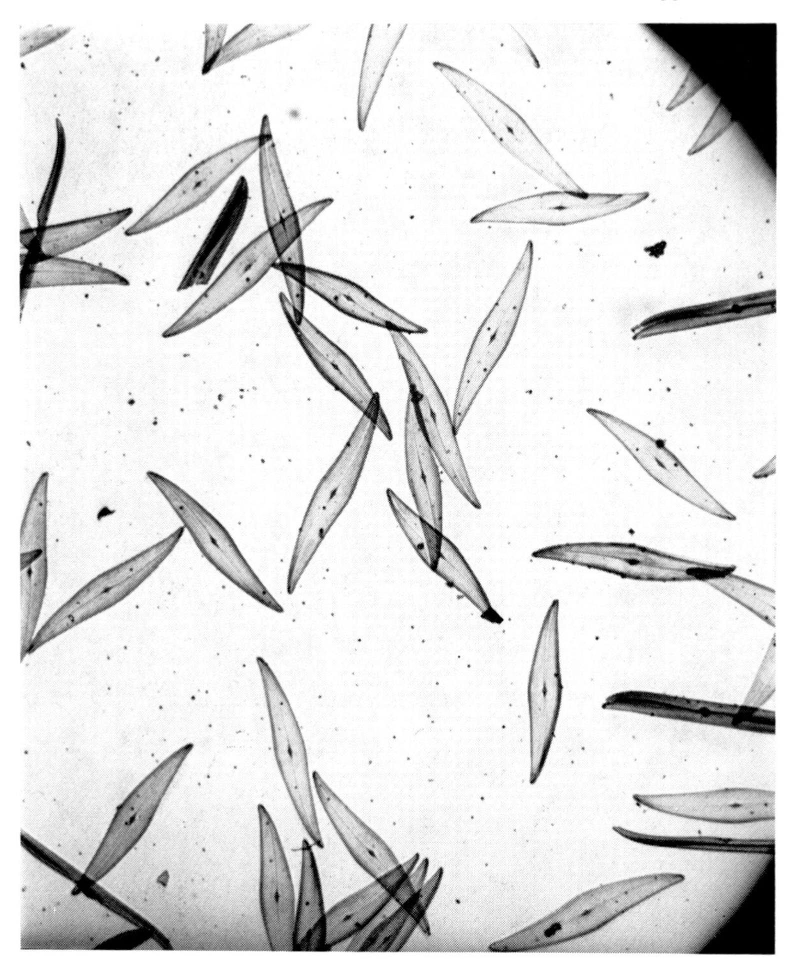

Karl Löfflath, Munich

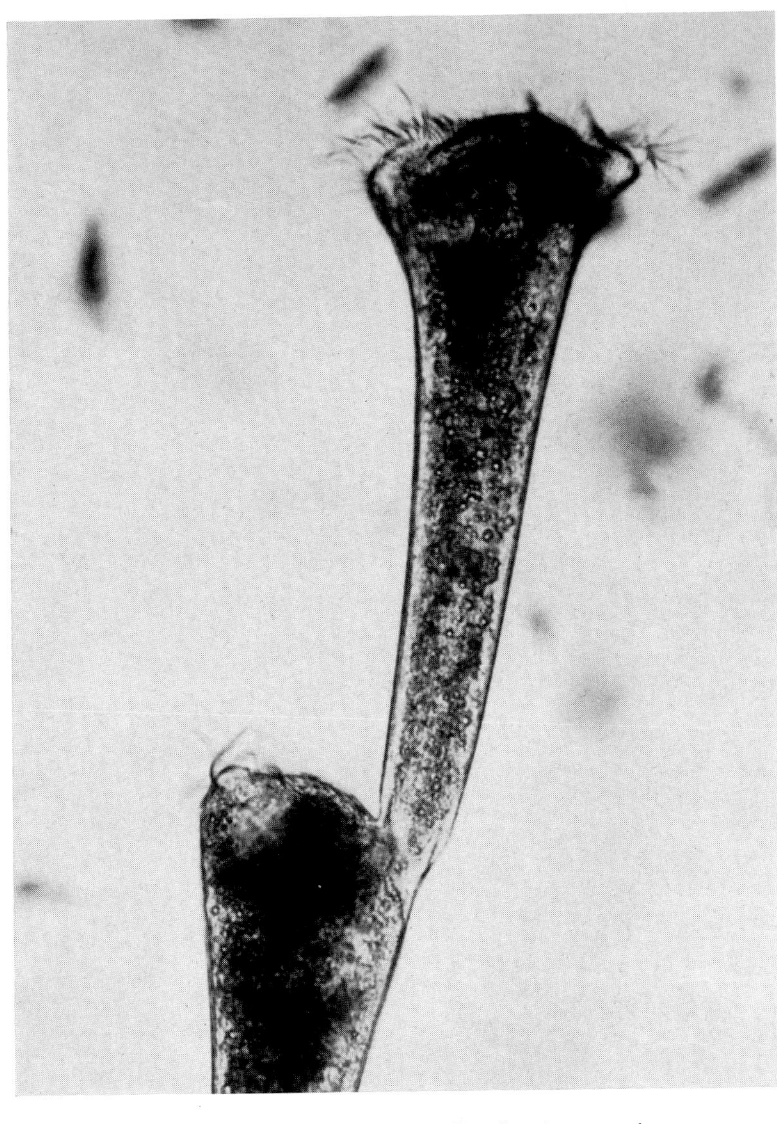

A trumpet animalcule splits. Single-cell animals reproduce by splitting. The "mother" lives on in two offsprings. There is no death by old age which would leave behind a corpse.

Karl Löfflath, Munich

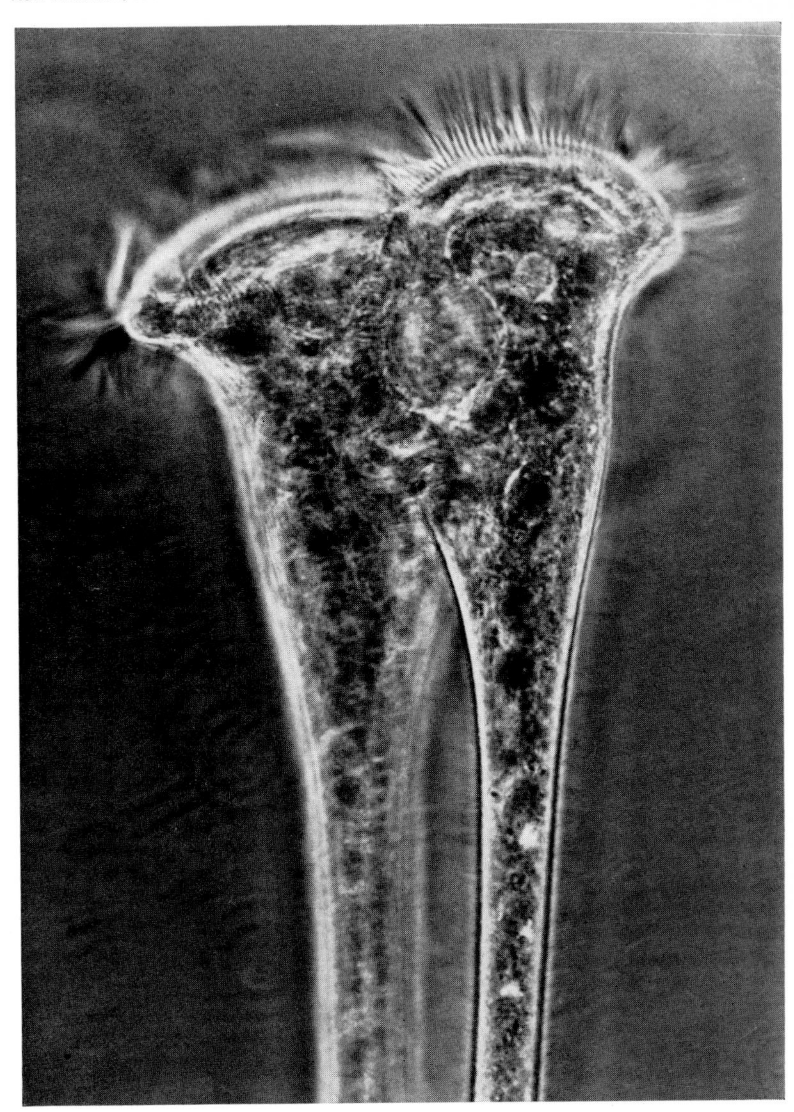

In conjugation, two single-celled animals exchange nuclei. Only with the advent of the microscope did it becomes possible to comprehend the long mysterious mechanism of fertilization, from that of single-celled animals to that of man.

Hans A. Traber, Zurich

Karl Löfflath, Munich

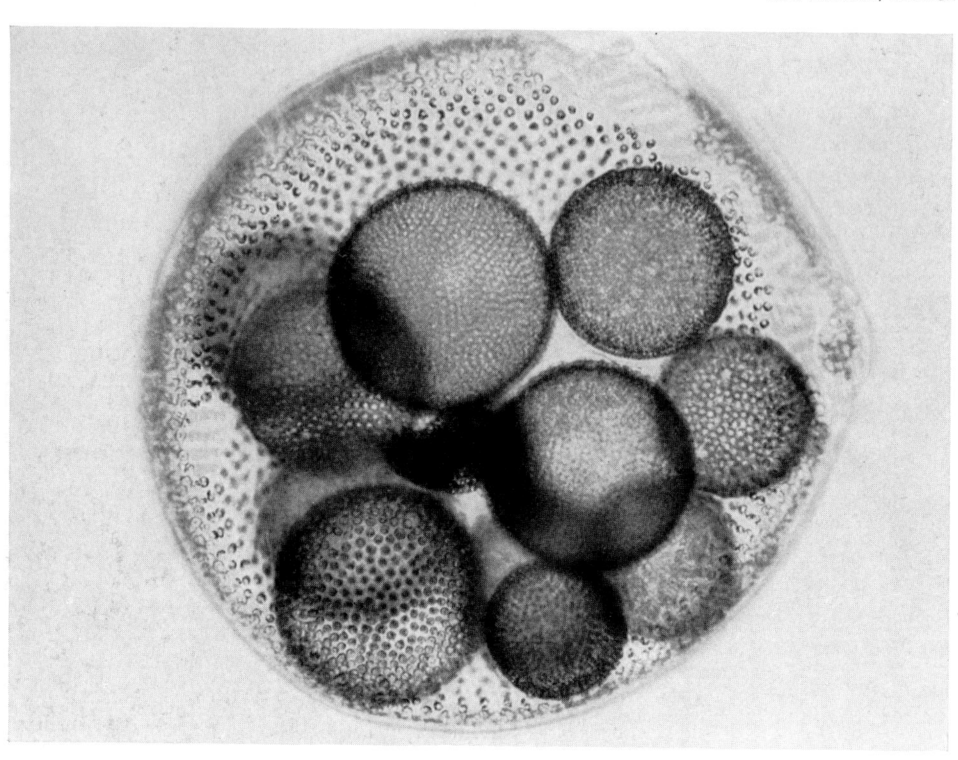

◄ A *volvox*, a colony of one-celled algae, slowly rotates, looking like a dotted glass ball in which other smaller balls are suspended. These daughter colonies enjoy protection inside the "mother," a beginning of parental care. The event pictured below, the breaking-out of the "daughters," may be compared to a birth.

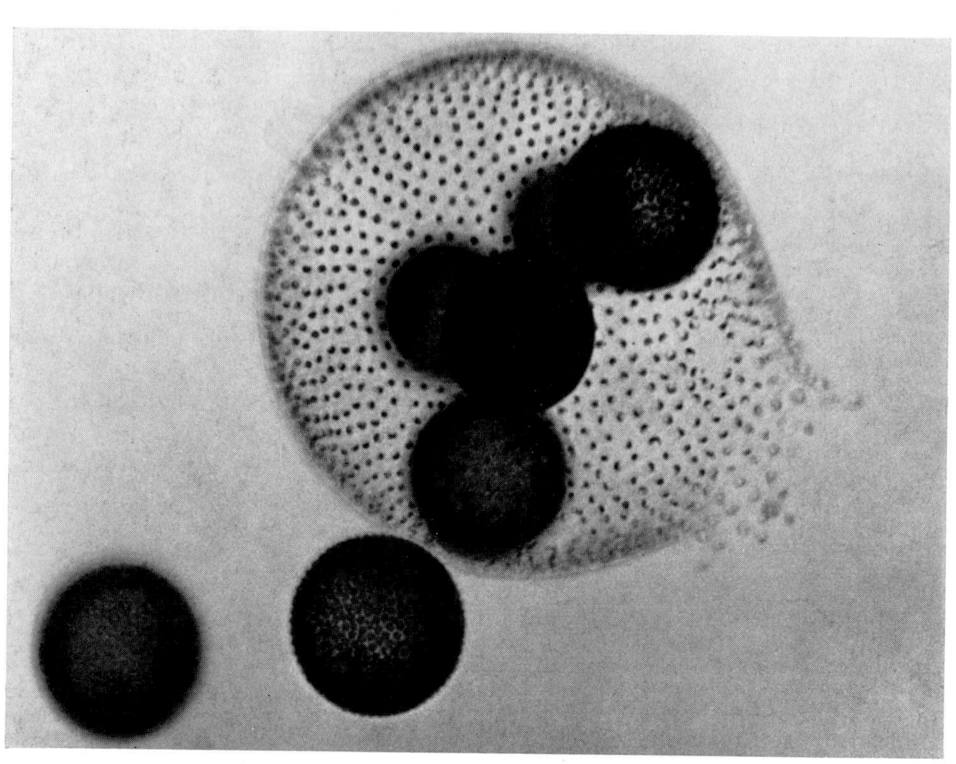

Each alga is secured in a ► cup-shaped receptacle of this branching colony of single-celled organisms. Magnification is about 1,500-fold. Such collectives are better able to survive in the plankton. Joining together results in a larger surface and greater buoyancy.

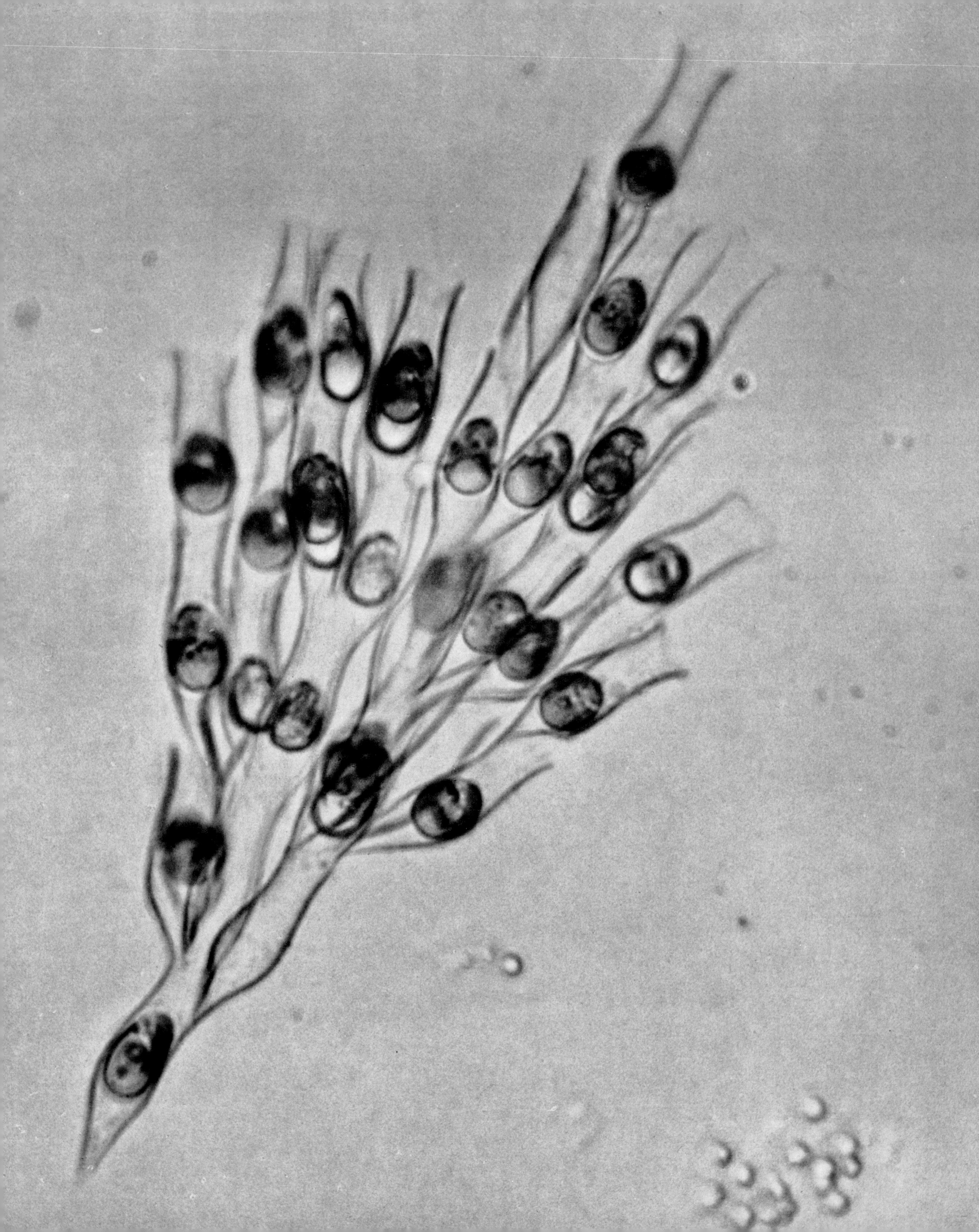

Hans A. Traber, Zurich

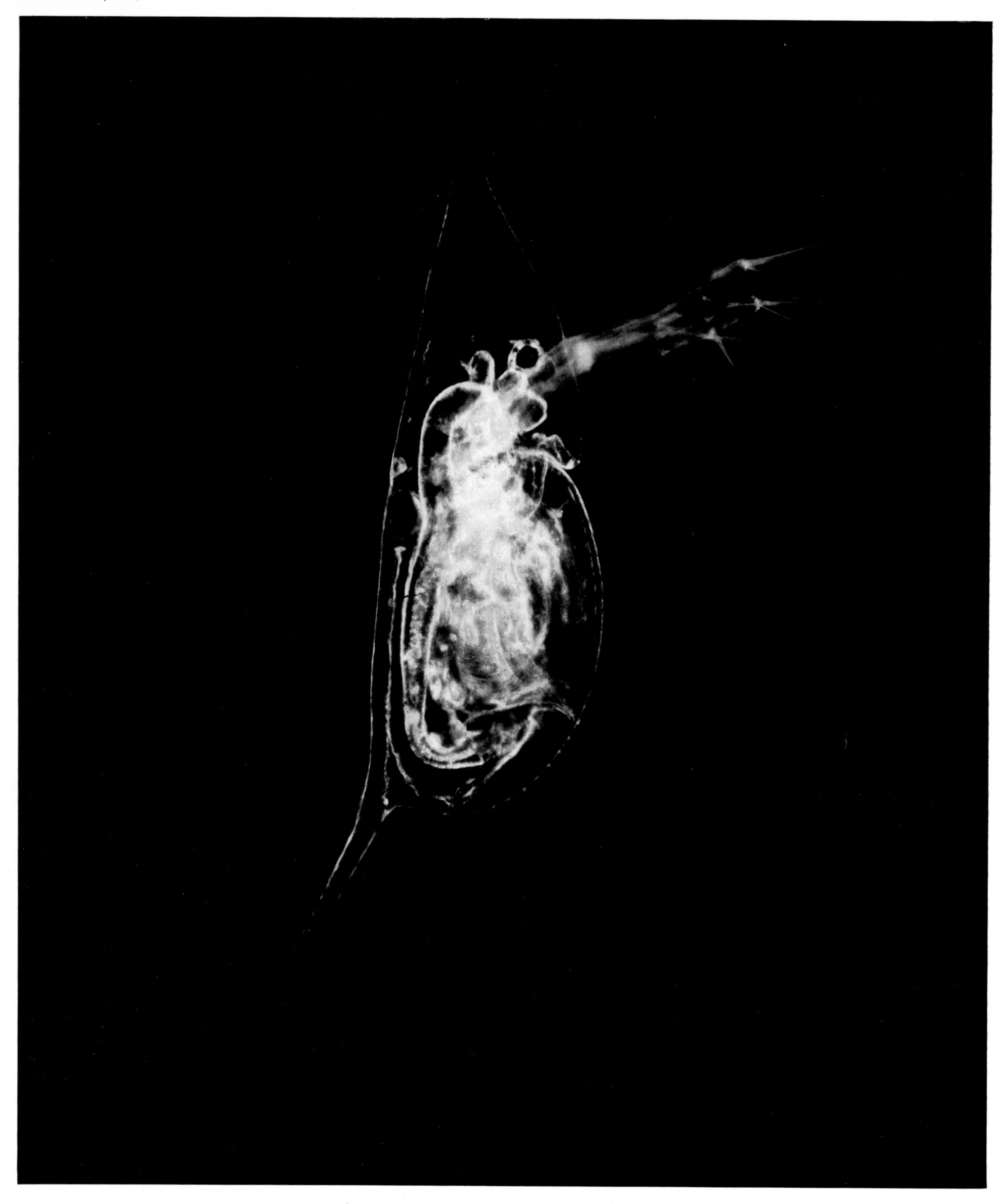

The tiny crustacean, shown here in 200-fold magnification, is one of the many kinds of water fleas. The pulsating bubble in its "back," behind the simple gut, is the heart. Carapace and spinous process are adjusted to the water viscosity.
These appendages facilitate floating.

Hans A. Traber, Zürich

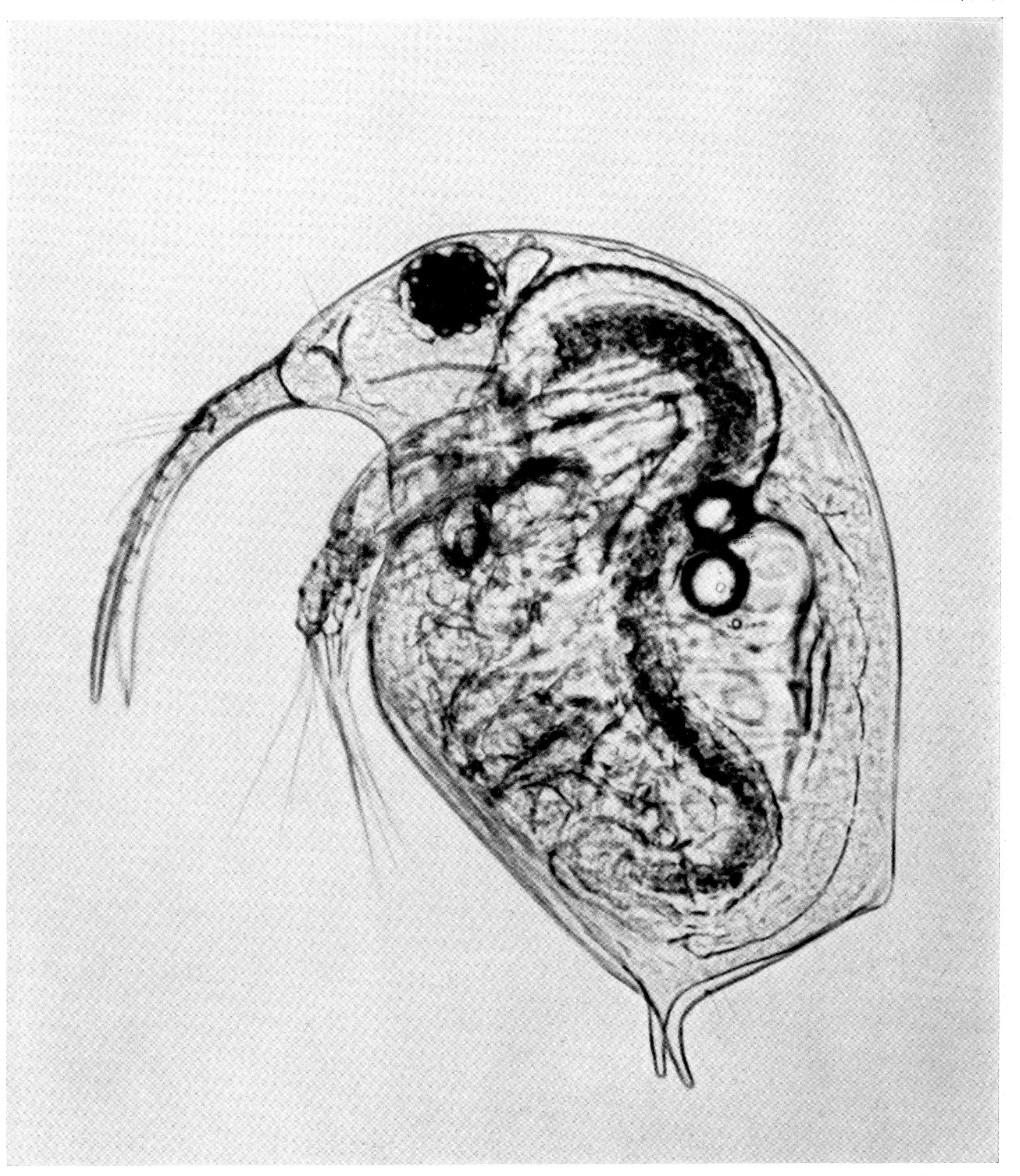

Bosmina, another crustacean (enlarged 700 times), gains additional buoyancy from two large drops of oil embedded behind the gut. The tiny crustaceans form part of the food chain in the plankton. They eat the protozoa and in turn are eaten by fish.

Needles of the stinging nettle magnified 70 times. They were among the first objects that the English naturalist Robert Hooke pictured in his *Micrographia* (published 1665 to 1667). Hooke was the first one to call the hollow spaces in cork "cells," a name that later came to designate the basic unit of organisms. ▼

Poetry in the realm of the minute: a moss. The pillow of the forest floor consists of innumerable structures of magical beauty. ▶

Rollei-Werke (Dr. Faasch), Braunschweig

Manfred Kage, Winnenden

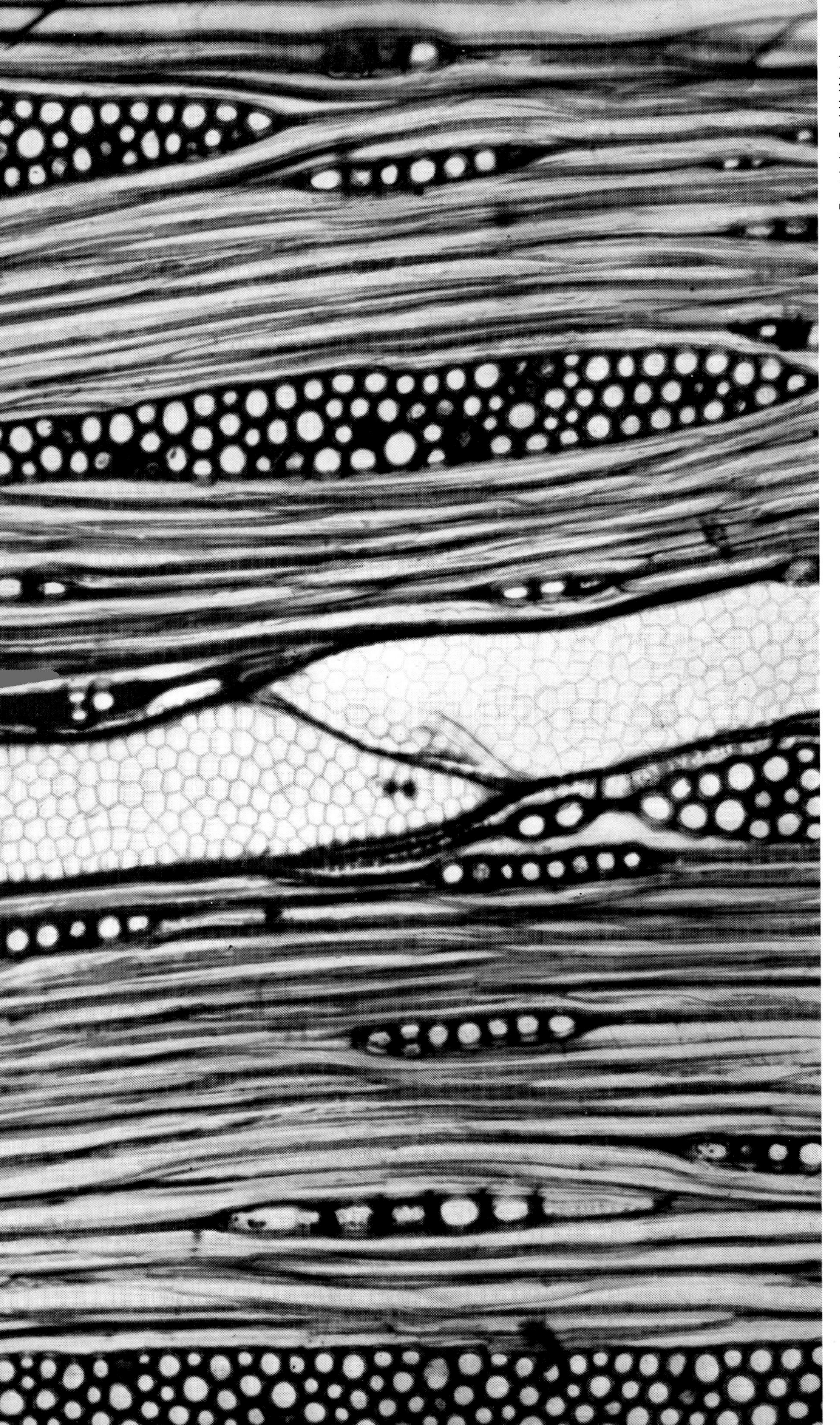

"I believe a leaf of grass is no less than the journeywork of the stars," Walt Whitman wrote. A graphically attractive picture: again and again it is borne out that nature has already designed and realized nearly everything that an artistic imagination could possibly invent and create. Shown is a cut across a maple shoot, enlarged several thousand times.

With proper lighting it is possible to make the invisible visible with a plain magnifying glass, as the dramatic photo on the right shows: a colony of ants defends itself. The air is filled with droplets of formic acid squirted out by the ants.

The compound eye of a honeybee consists of 5,000 crystal-clear facets. Jan Swammerdam, an Amsterdam anatomist, dissected one under the microscope as early as 1675. With his first book on insects, Swammerdam founded the science of entomology.

Harald Doering, Ebersberg

Harald Doering, Ebersberg

Institute for Applied Microscopy, Karlsruhe

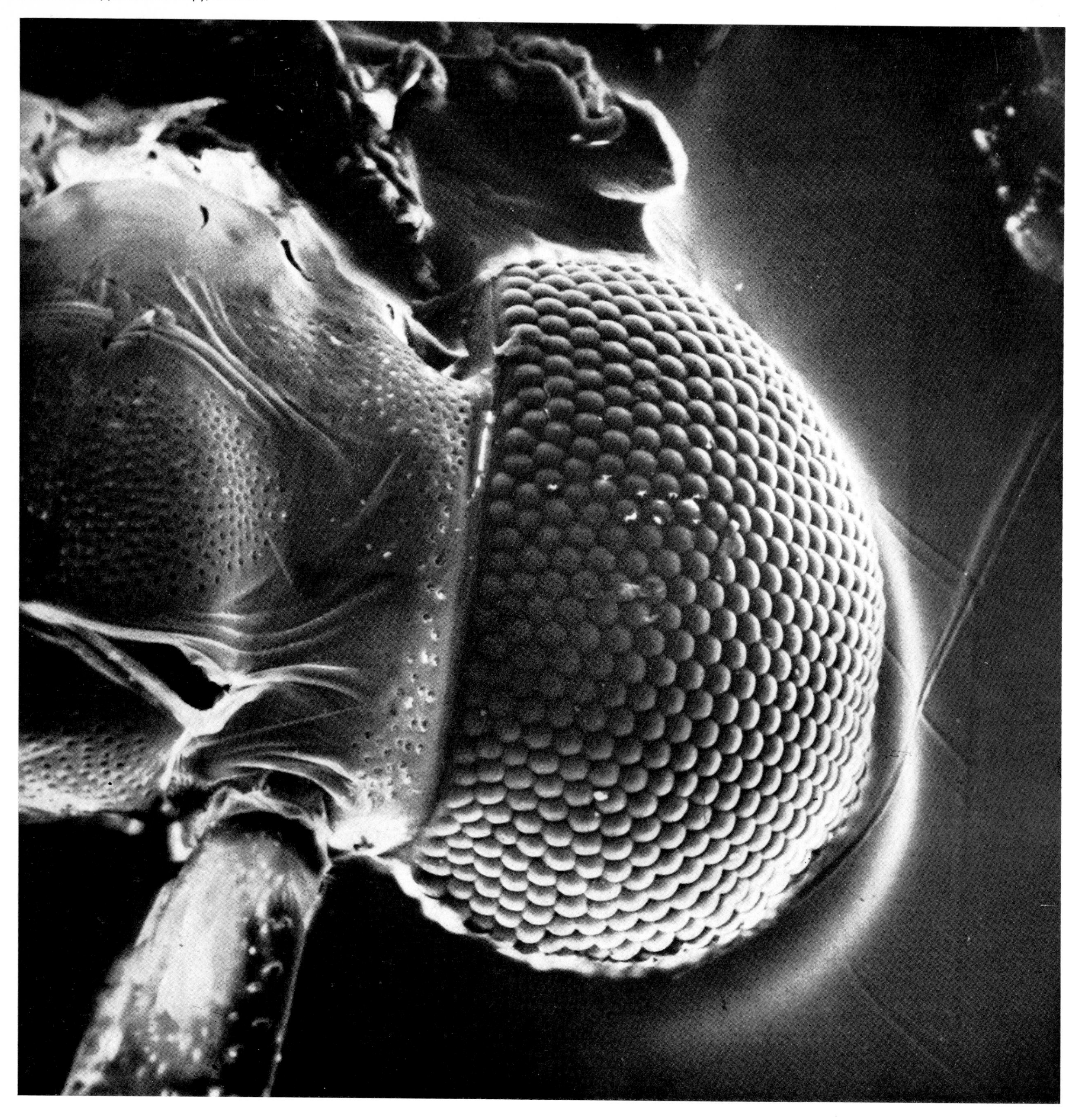

Eye of a clothes moth magnified about 500 times.
Part of each eye facet is a transparent tube in a dark casing that leads to the retina. Each facet picks up a tiny sectional image of the world in front of it. From the retina these bits of mosaic are transmitted to the brain. The picture fragmented at the surface of the eye is reassembled in the moth's brain.

Institute for Applied Microscopy, Karlsruhe

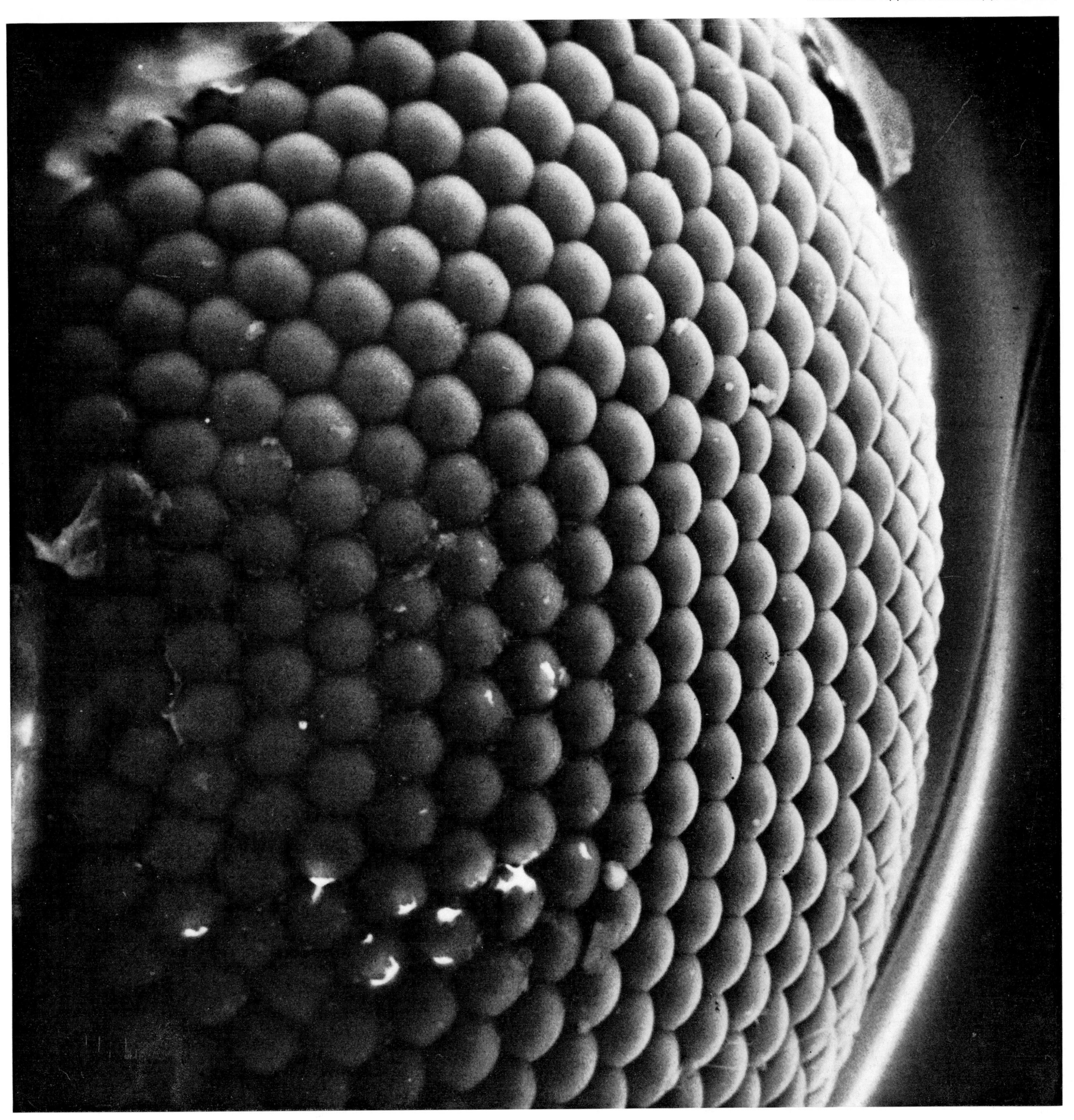

At a magnification of about 1,000 times, the facets of the moth's eye appear three-dimensional, like pinheads on a pincushion. The three-dimensional effect and the amazing depth of field have been achieved here, as well as in the photos of radiolaria on pages 92 and 93, through scanning by electron beams.

H. van Kooten, Zool. Labor. Rijksuniversiteit, Utrecht

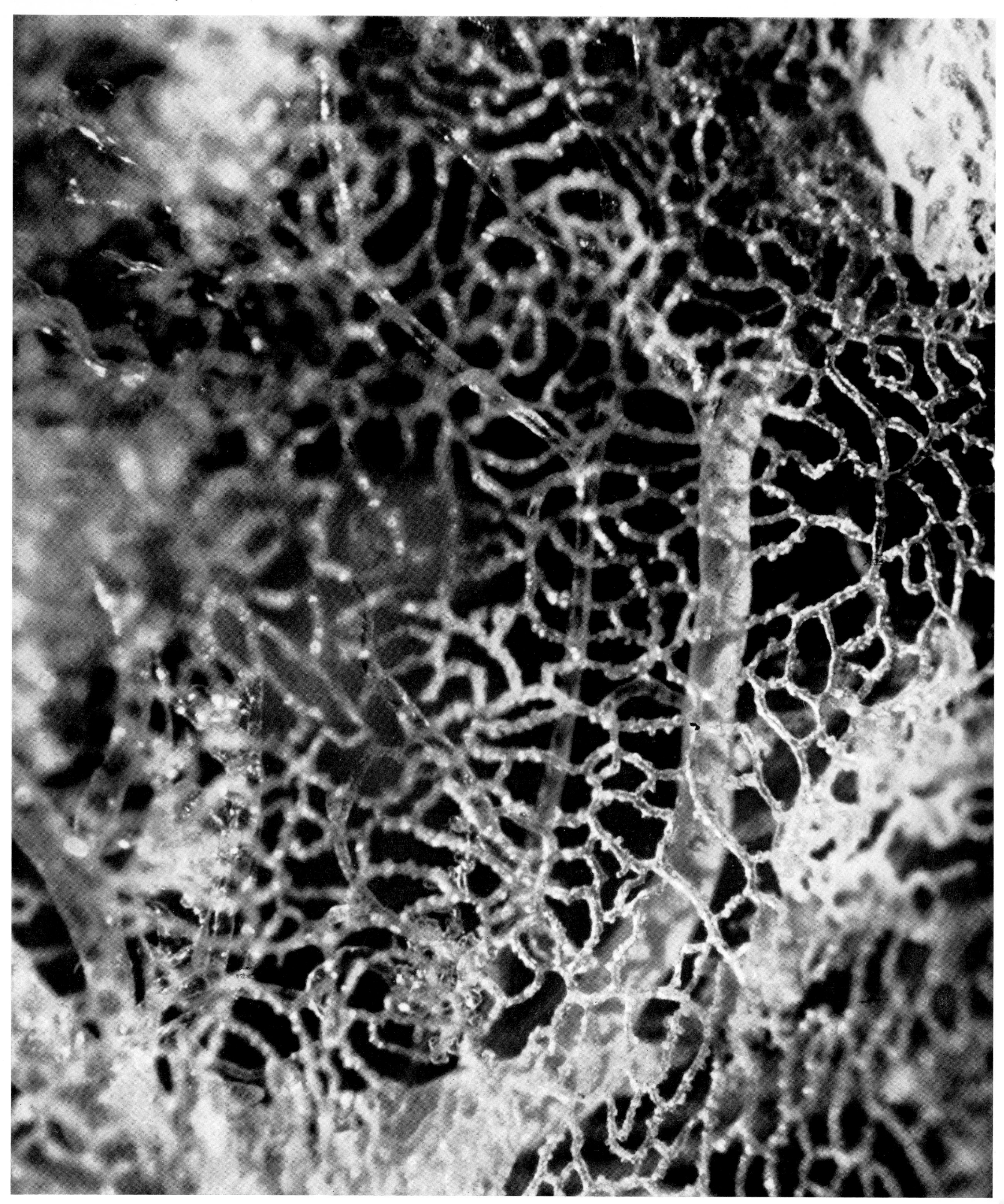

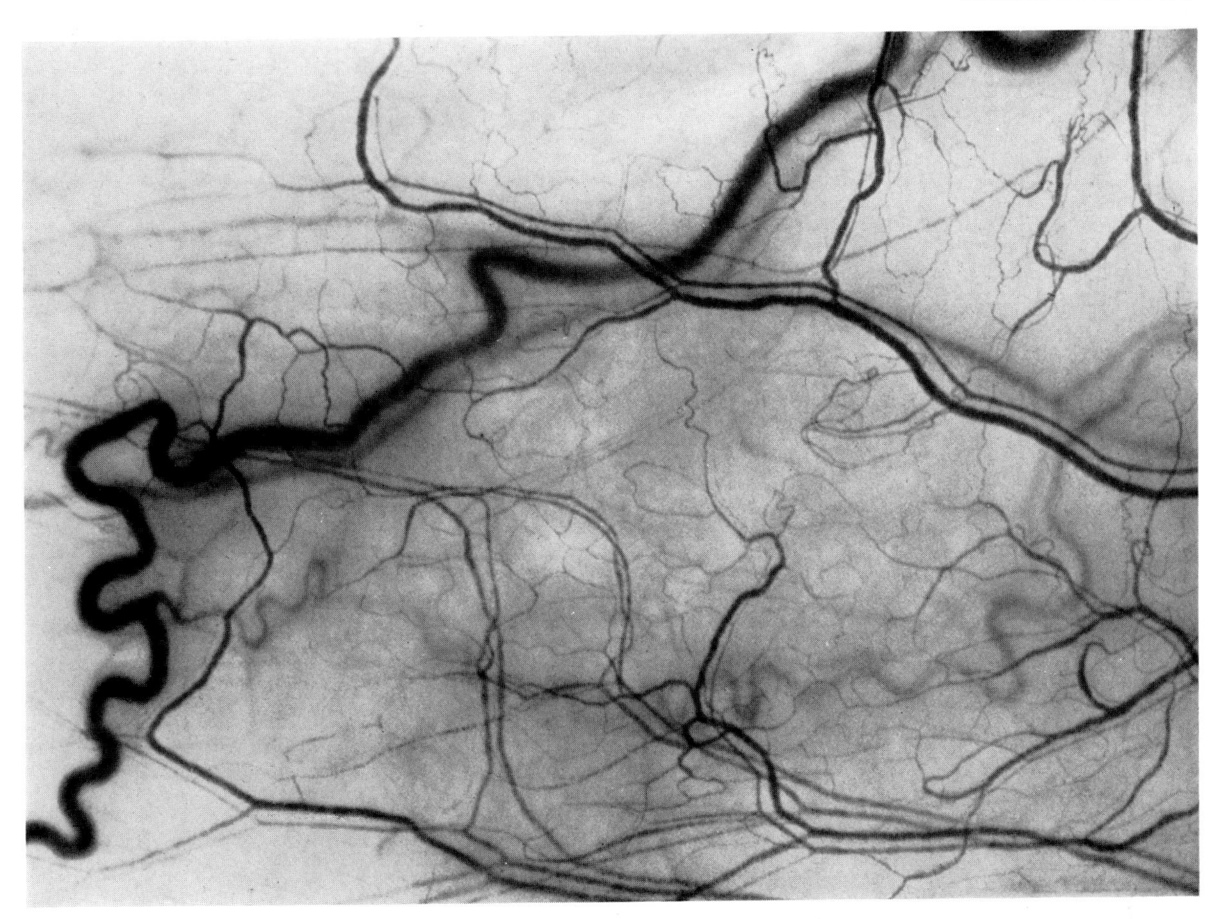

▲
In 1661, Marcello Malpighi, looking through the microscope, discovered capillaries, the finest of blood vessels that connect the arteries and veins. In 1628, the English surgeon William Harvey had published his discovery of blood circulation; but how the blood got from the arteries to the veins had remained unexplained. Our pictures show the capillaries of the eyeball.
▼

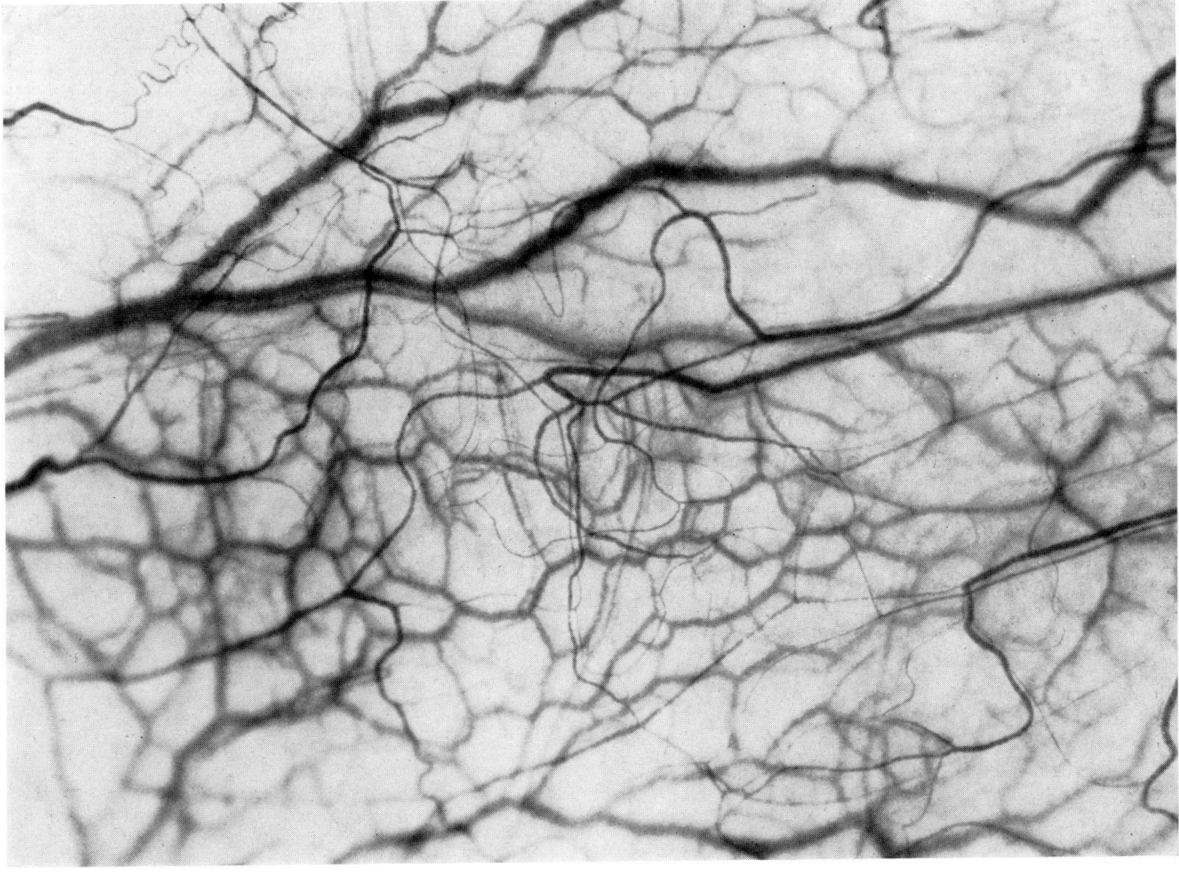

Part of the art of microscopy is the skill of the person preparing the specimens. Three hundred years ago, the Italian anatomist Malpighi (1628–1694) already had injected wax into tissue to render it more clearly visible.
The picture at left, in 100-fold magnification, shows a surprising sort of web: the blood vessels in the head of a frog, injected with a ◂ synthetic.

A view of life processes.

Ernst Leitz GmbH, Wetzlar

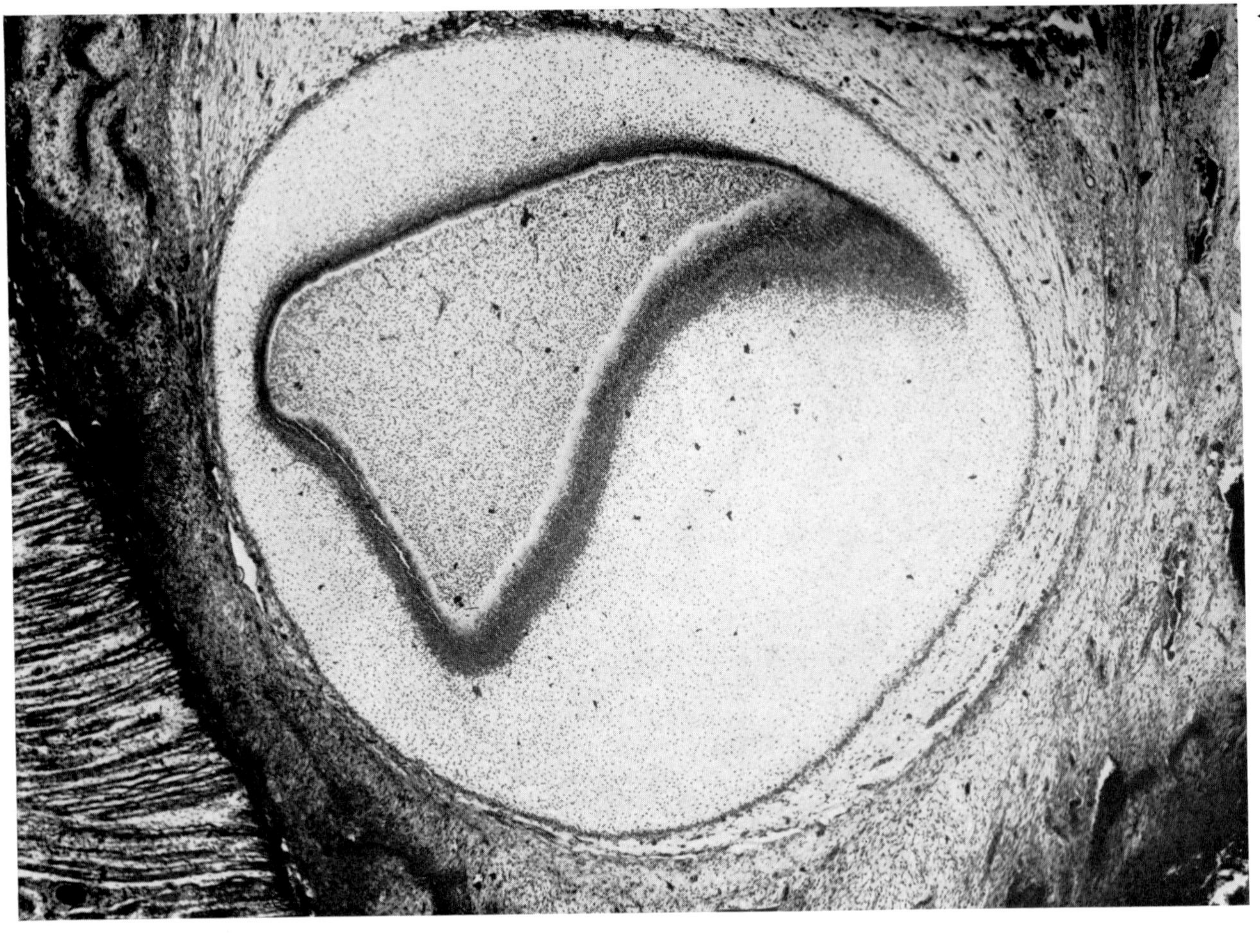

◄ Formation of a tooth in an embryo, magnified 50 times.
Marcello Malpighi was the first to follow the development of growing life under the microscope. He observed the growth of a chicken in the egg.

Capillaries in the nailbed ► enlarged 150 times. The capillaries have a diameter of about five to twenty thousandth of a millimeter. The finest capillaries permit the passage of just one blood corpuscle.

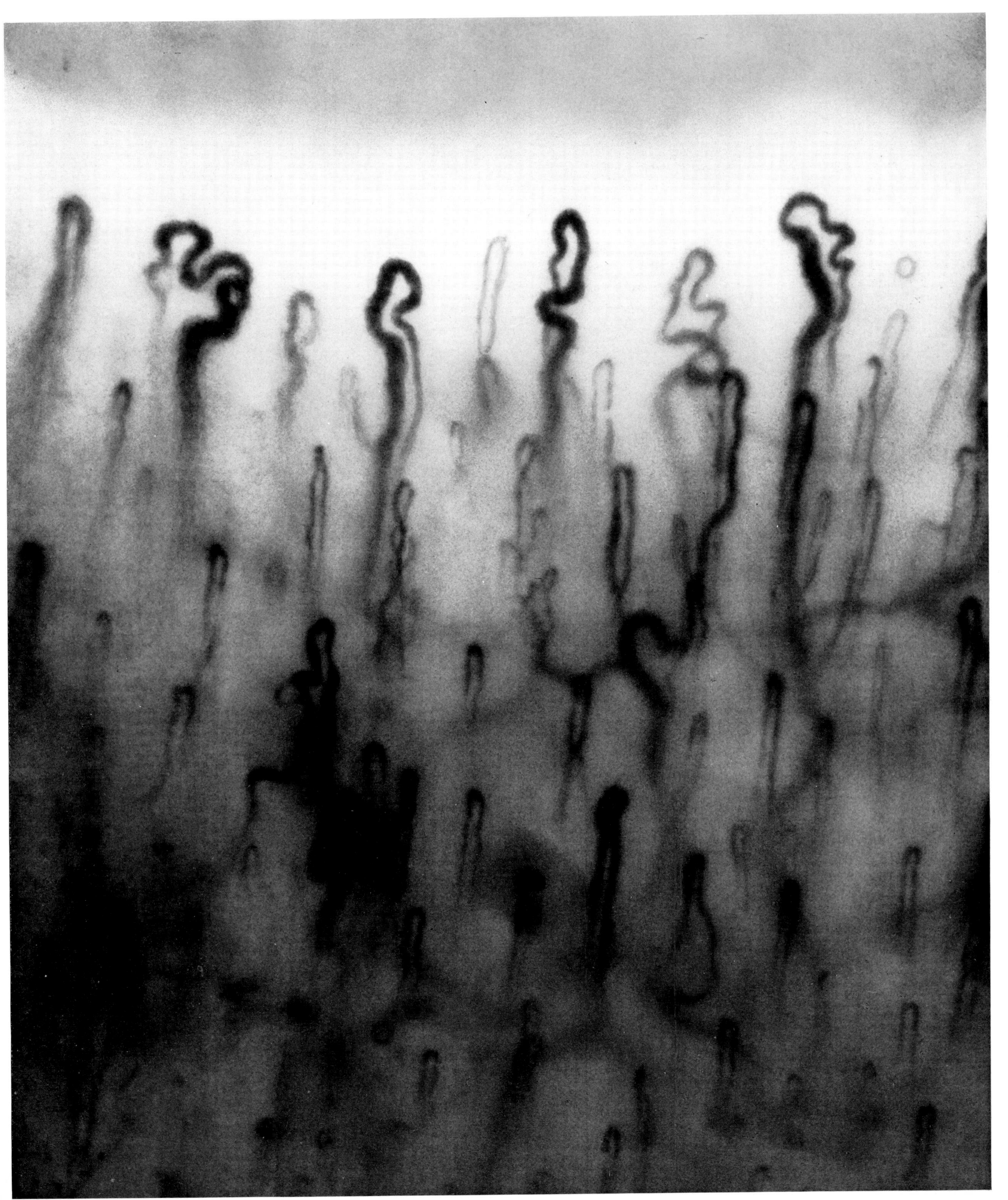

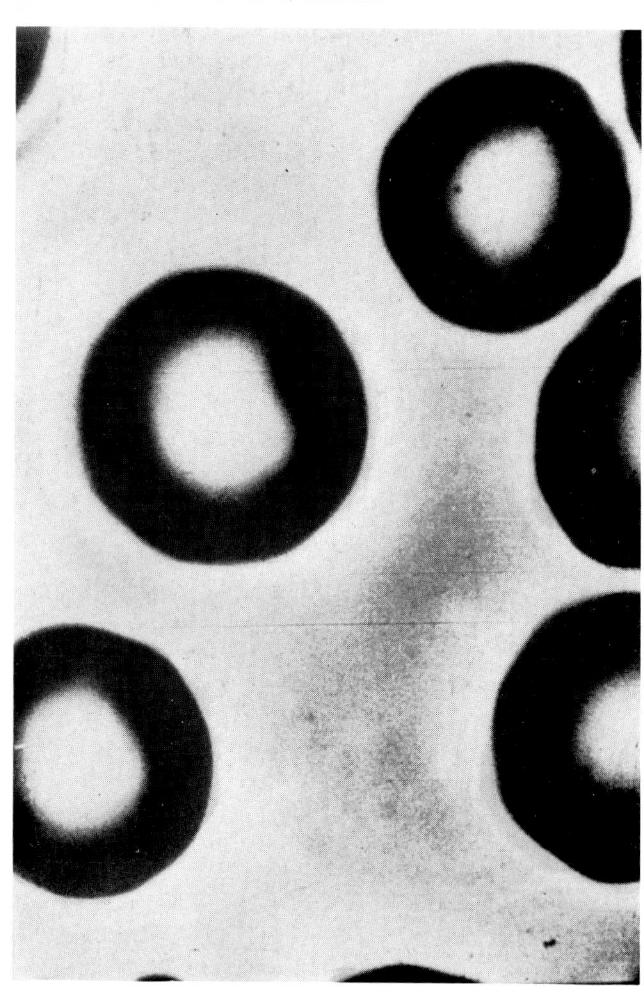

With the aid of the light-microscope it was recognized that blood is a "fluid tissue," an organ composed of cells. Earlier, blood cells had to be stained for closer scrutiny which caused them to die. Today, the phase-contrast method makes the study of living blood cells possible because a phase-shifting platelet in the path of the microscope's light emphasizes contrasts. The somewhat dented-looking disks in the upper photo, as well as the thorn-apple shapes below, are red corpuscles. Our blood carries five million of them in one cubic millimeter. They deliver the oxygen and carry away the waste carbon dioxide.

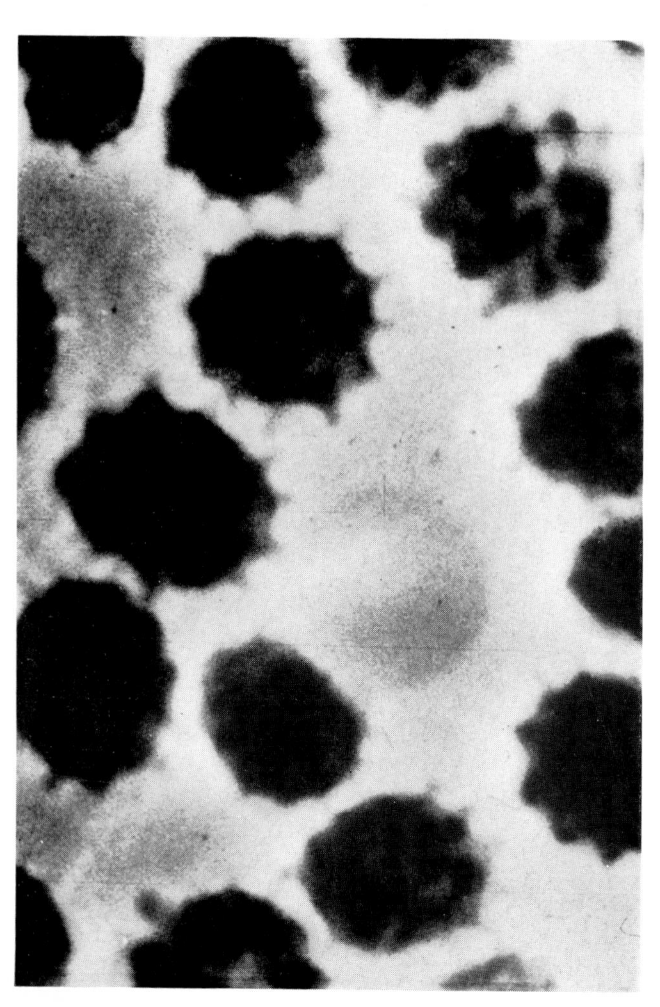

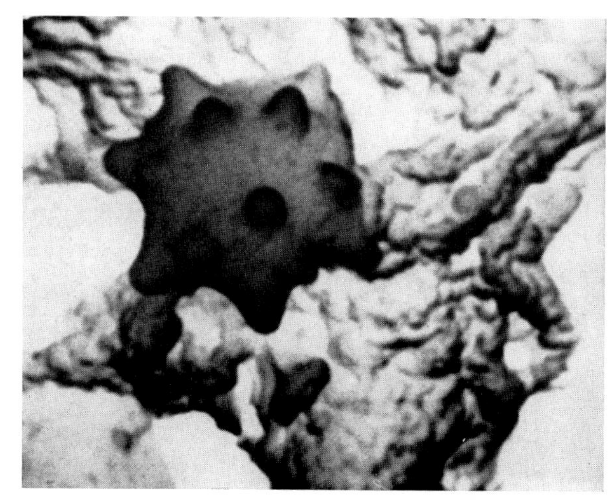

For comparison, a red blood cell magnified 5,000 times by a scanning electron microscope.

Deutsche Hoffmann-La Roche, Grenzach

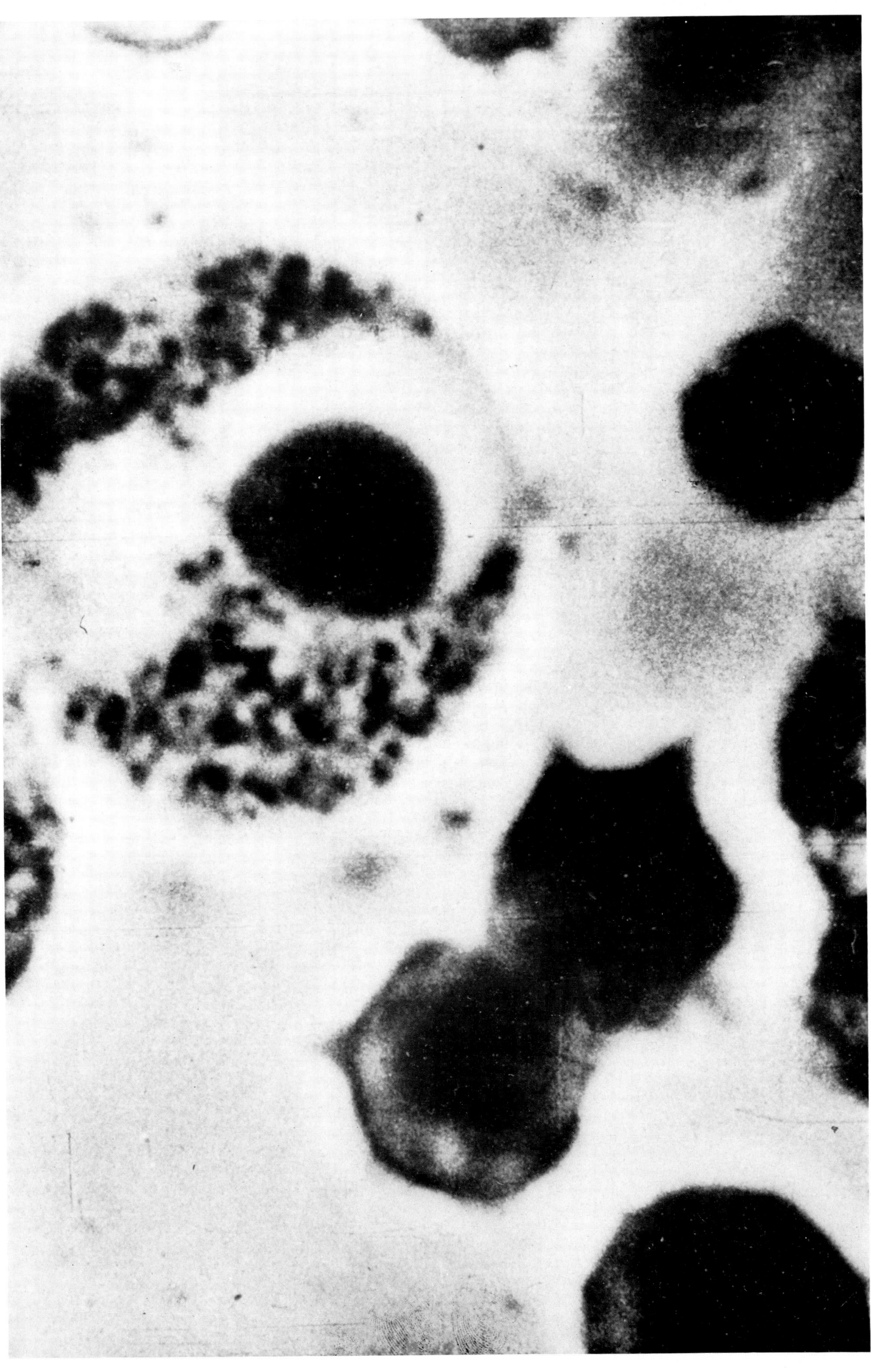

Occasionally light and dark bodies have been observed inside the blood cells. They can be seen floating back and forth in the transparent plasma. Their chemical composition is not known and their function is still a mystery.

A look into the realm of the blood is one of the strangest experiences. The realization that what we see here is simultaneously repeated a million times in our own blood stream causes feelings of awe similar to those experienced when looking into the depths of the Milky Way.

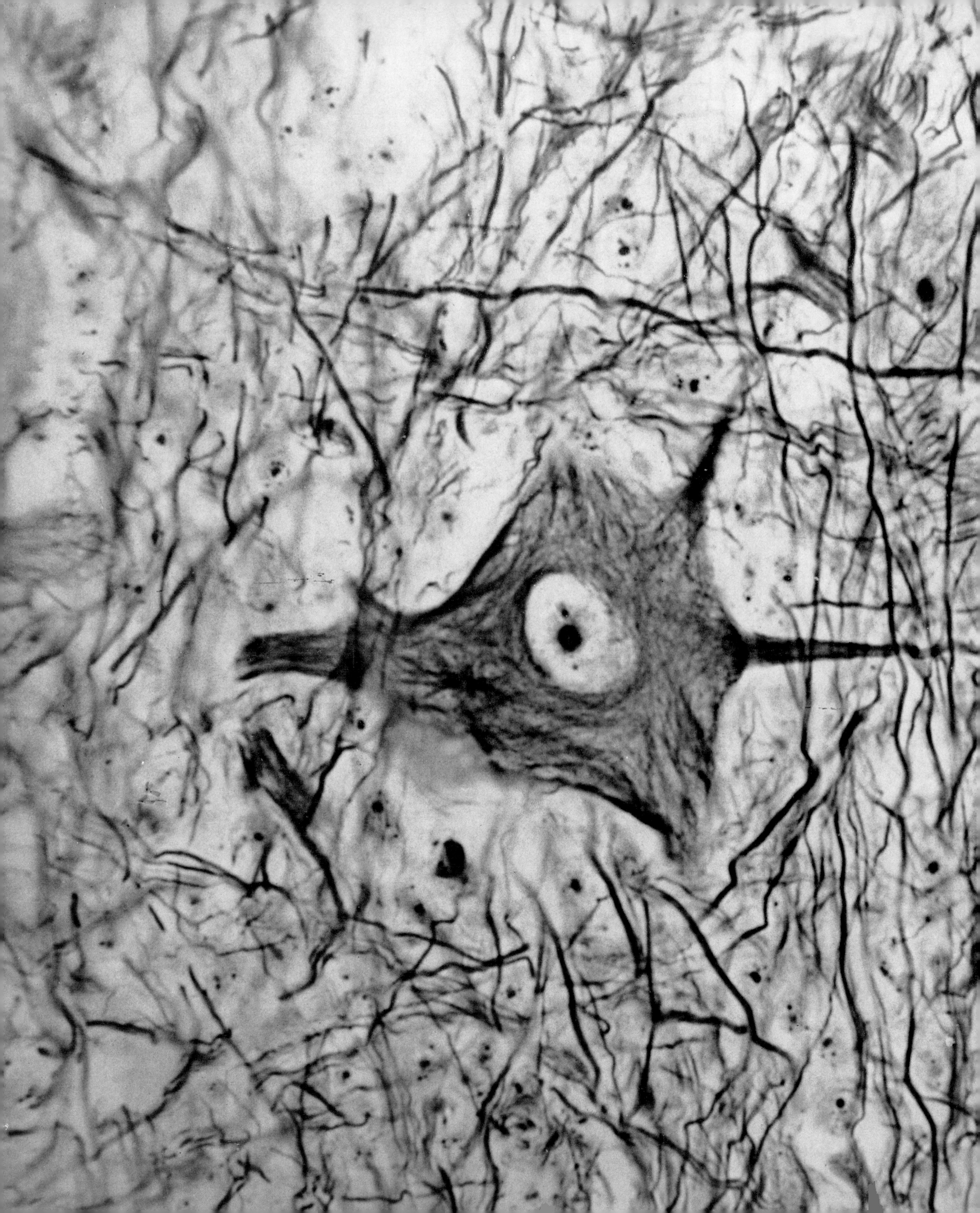

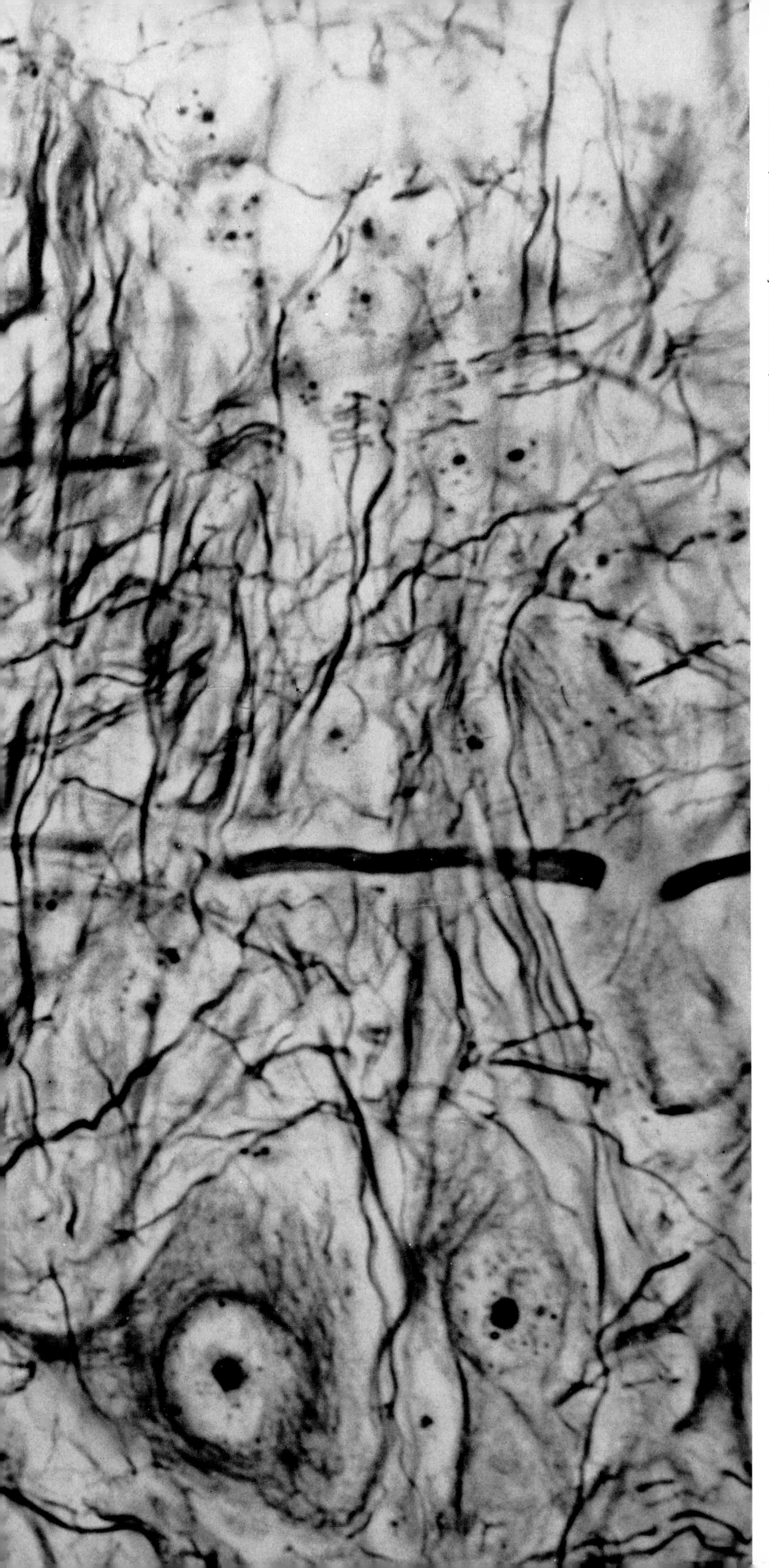

H. van Kooten, Zool. Labor. Rijksuniversiteit, Utrecht

◄ Experiments with nerve cells can be observed under the phase-contrast microscope. At left are shown nerve cells from the brain of a cat (2,000-fold magnification). In the bottom picture we see a human nerve cell. Its minute size is demonstrated by a statistic: the cerebral cortex of man, the "thinking part," consists of some ten billion nerve cells. In 1904, the U.S. zoologist Ross G. Harrison succeeded in removing nerve cells from living tissue and growing them in a culture medium. Today, nerve and brain tissue are grown in cultures.
▼

Carl Zeiss, Oberkochen

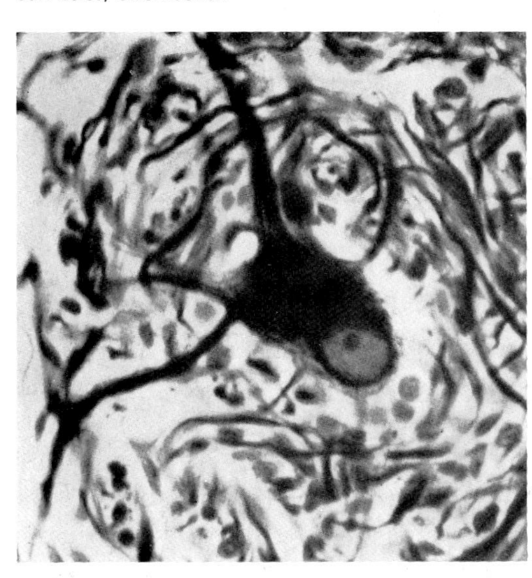

Prof. Dr. W. Beermann, Tübingen

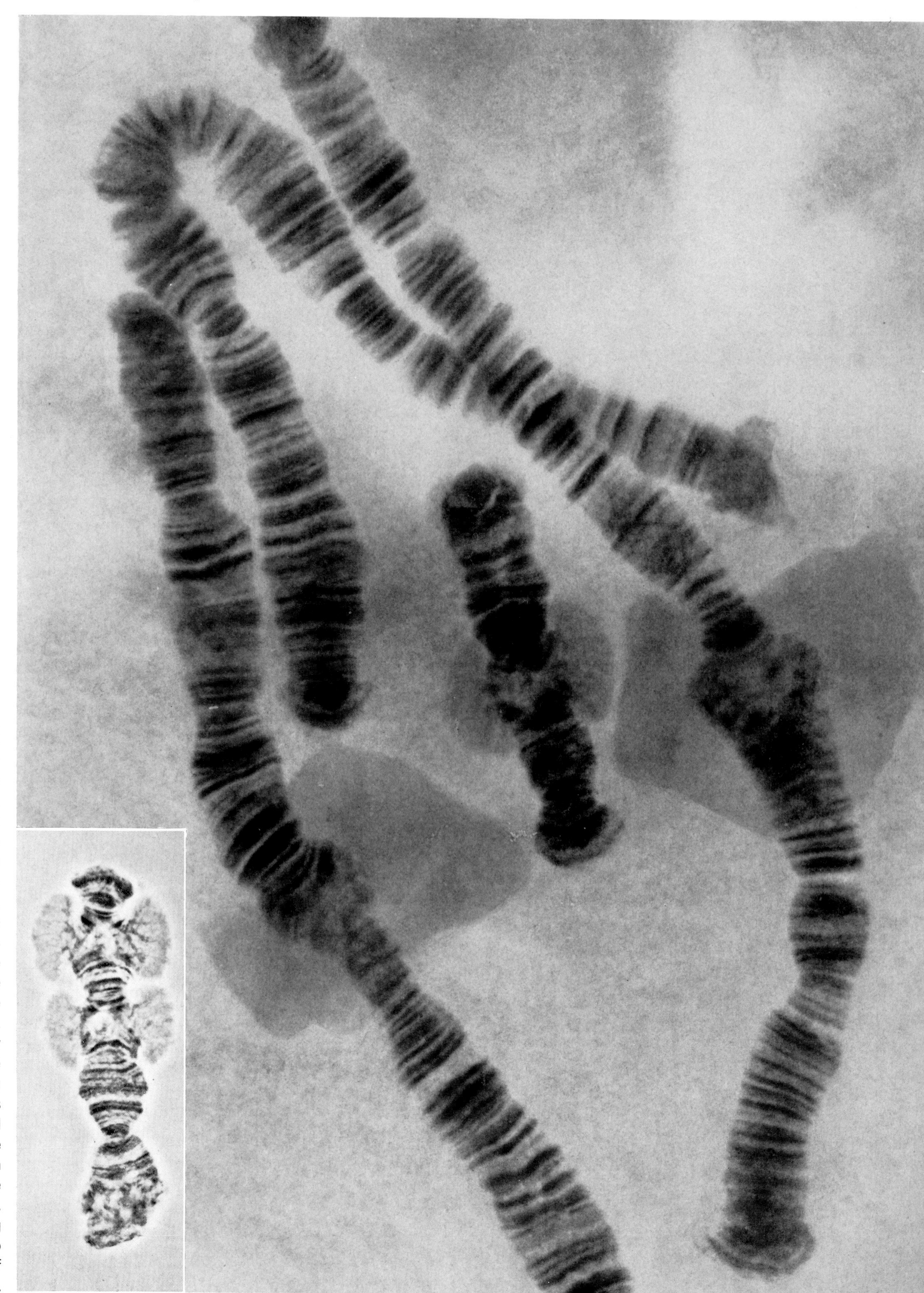

The laws of heredity, valid for all living creatures, were discovered to a large extent in the insects. The photo shows, 700 times magnified, the giant chromosomes from a salivary gland cell of *Chironomus tentans,* a midge. Within the chromosomes is contained the "information" for the composition of the saliva's protein. Geneticists are especially interested in a strange phenomenon, the forming of chromosome puffs (see insert, magnified 1,500 times). In the puffed-out areas certain genes seem to have been somehow activated. Researchers are facing here a possible clue to the understanding of heredity.

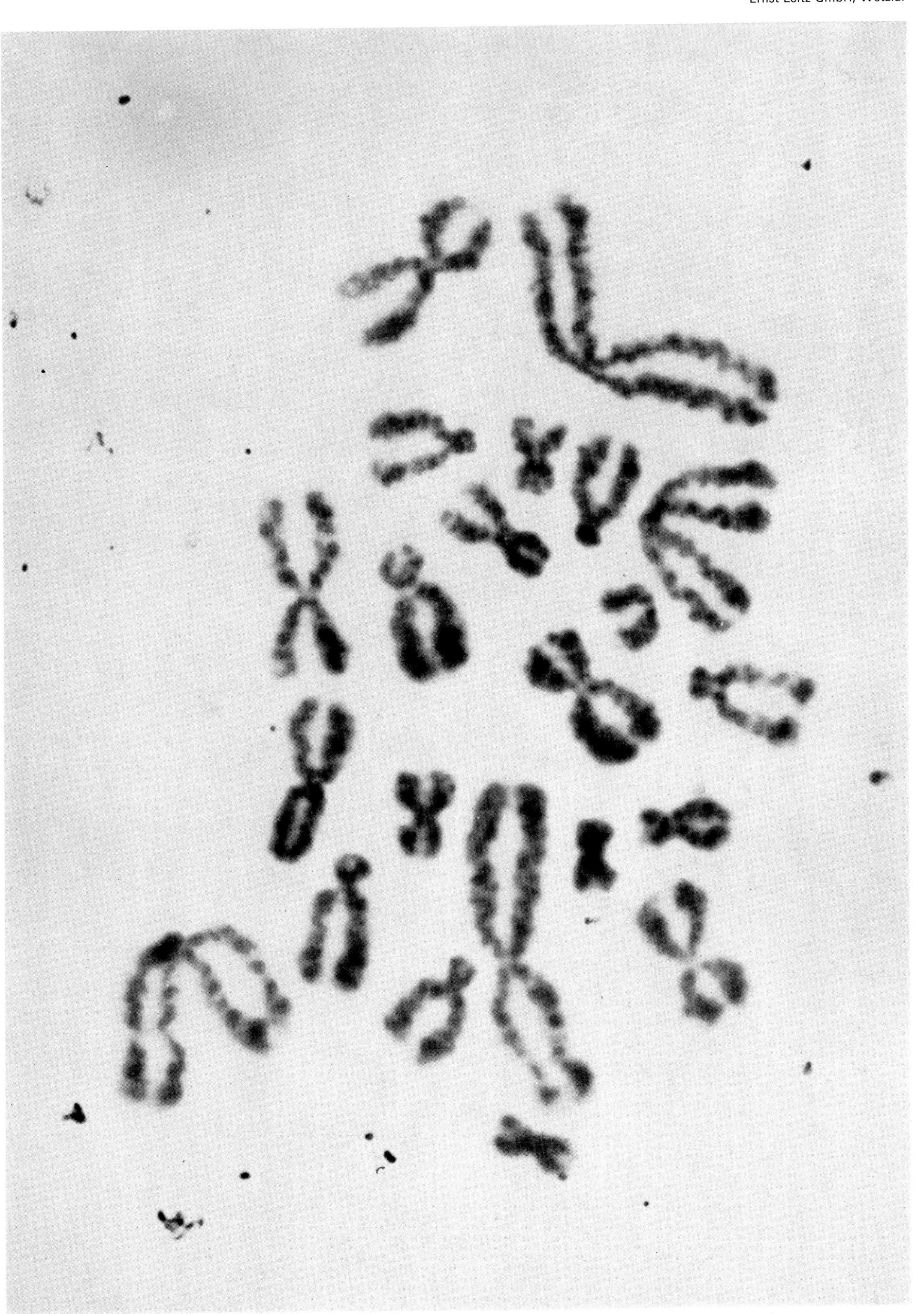

Ernst Leitz GmbH, Wetzlar

Man's twenty-three pairs of chromosomes contain the genes, the carriers of heredity. Genes determine the physical appearance and characteristics of the individual as well as his reaction to the stimulations and demands of his environment.

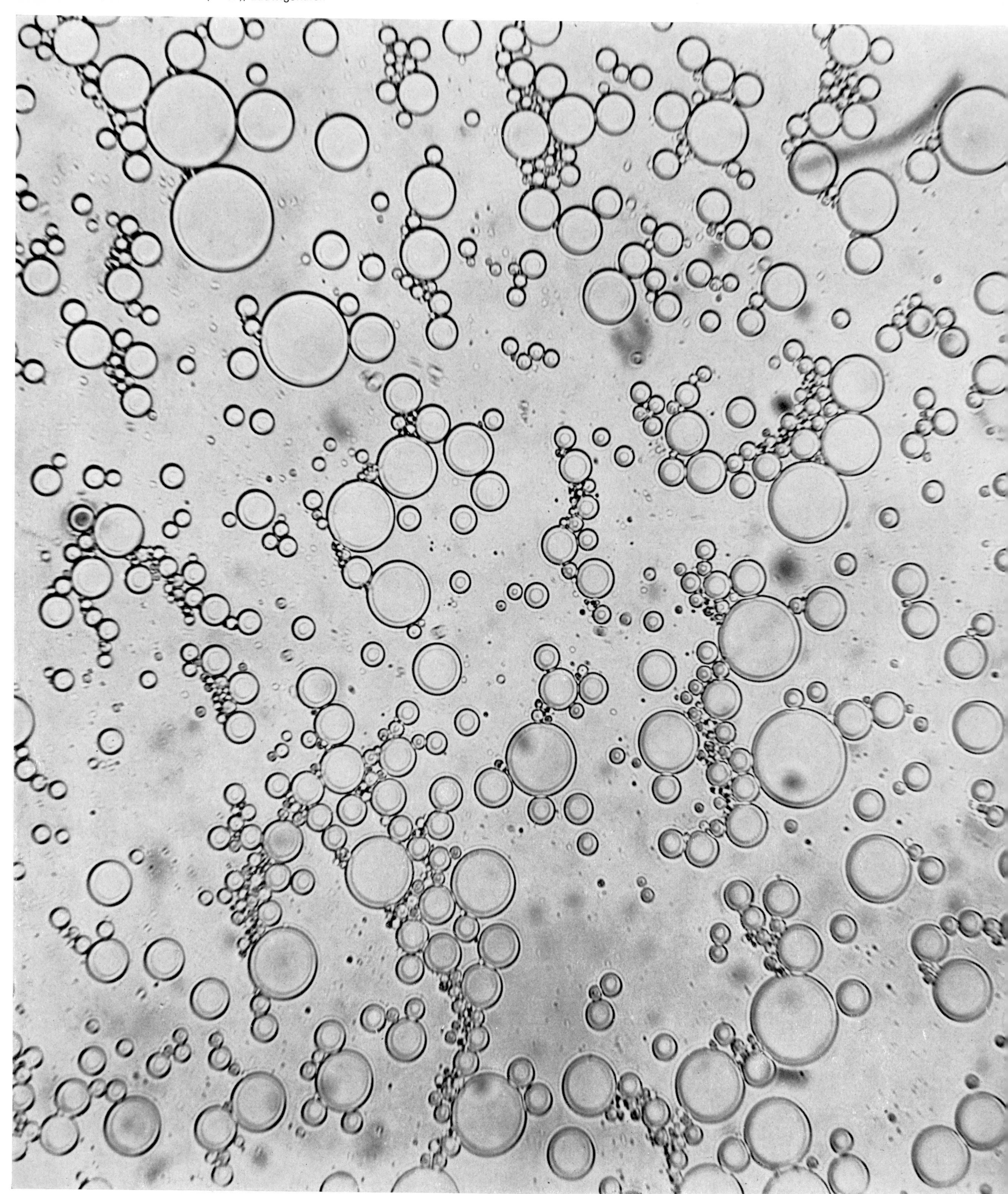

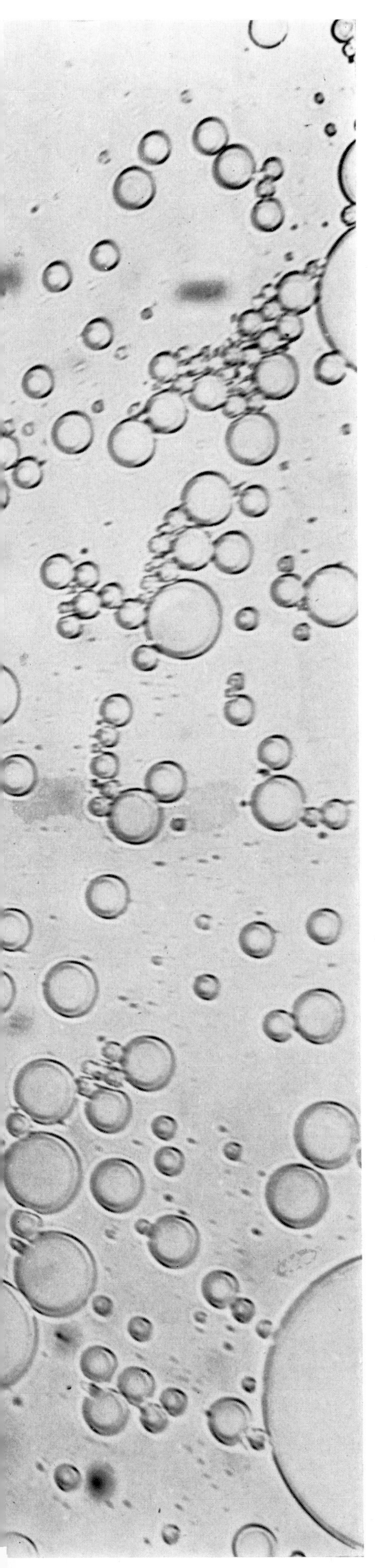

To investigate all matter on earth, to transform it, to create new substances out of nature's building blocks, is the task of chemistry. Goethe called its findings "a new, heretofore unknown, hardly considered possible, inconceivable world." The picture on the left shows an emulsion of paraffin oil in water magnified about 1,000 times. The picture on the upper right shows a watery solution of urea in the process of freezing (200 times magnified). Urea was the first organic compound, a product of living nature, to be synthesized from inorganic substances (Friedrich Wöhler, 1828). The picture on the lower right shows a dark-field photomicrograph of a thinned-down dispersion of acryl ester. Acryl ester molecules have a readiness to join into giant molecules, to polymerize. Synthetics like Plexiglas and artificial fibers originate in this way.

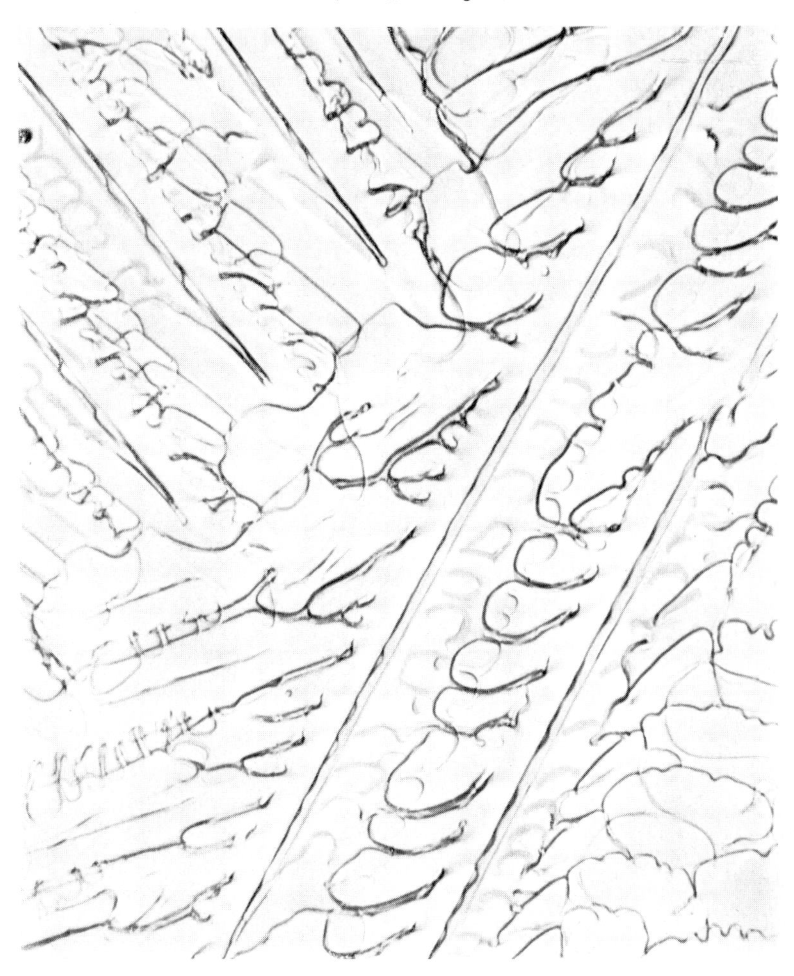

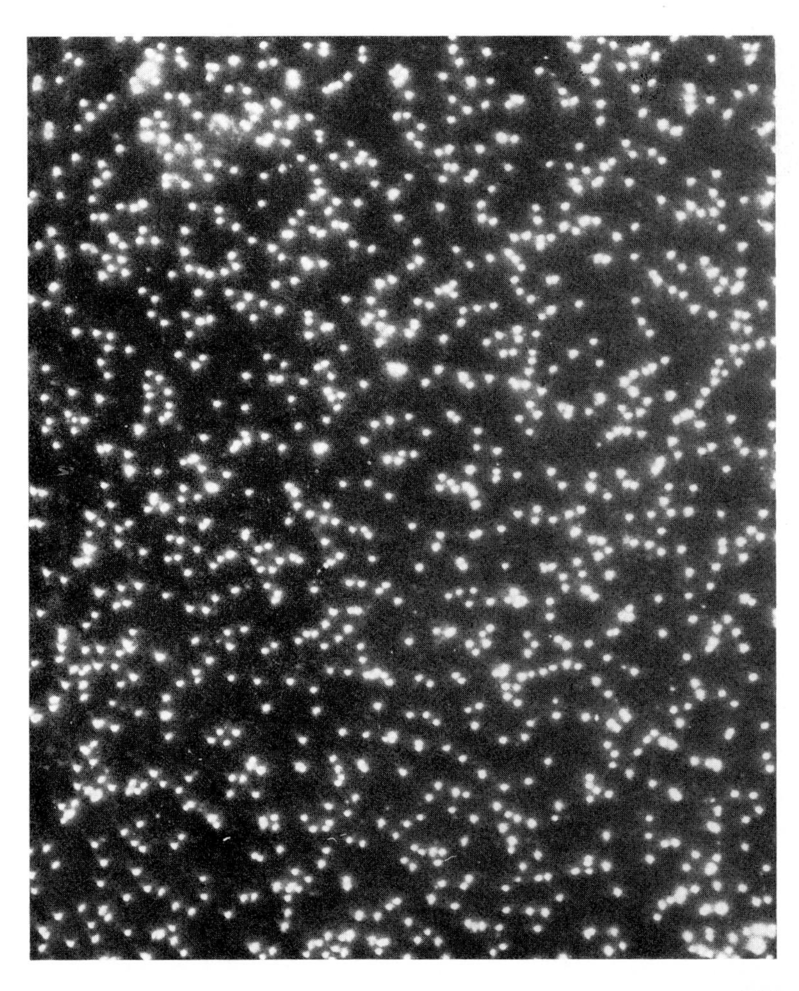

Institute for Applied Microscopy, Karlsruhe

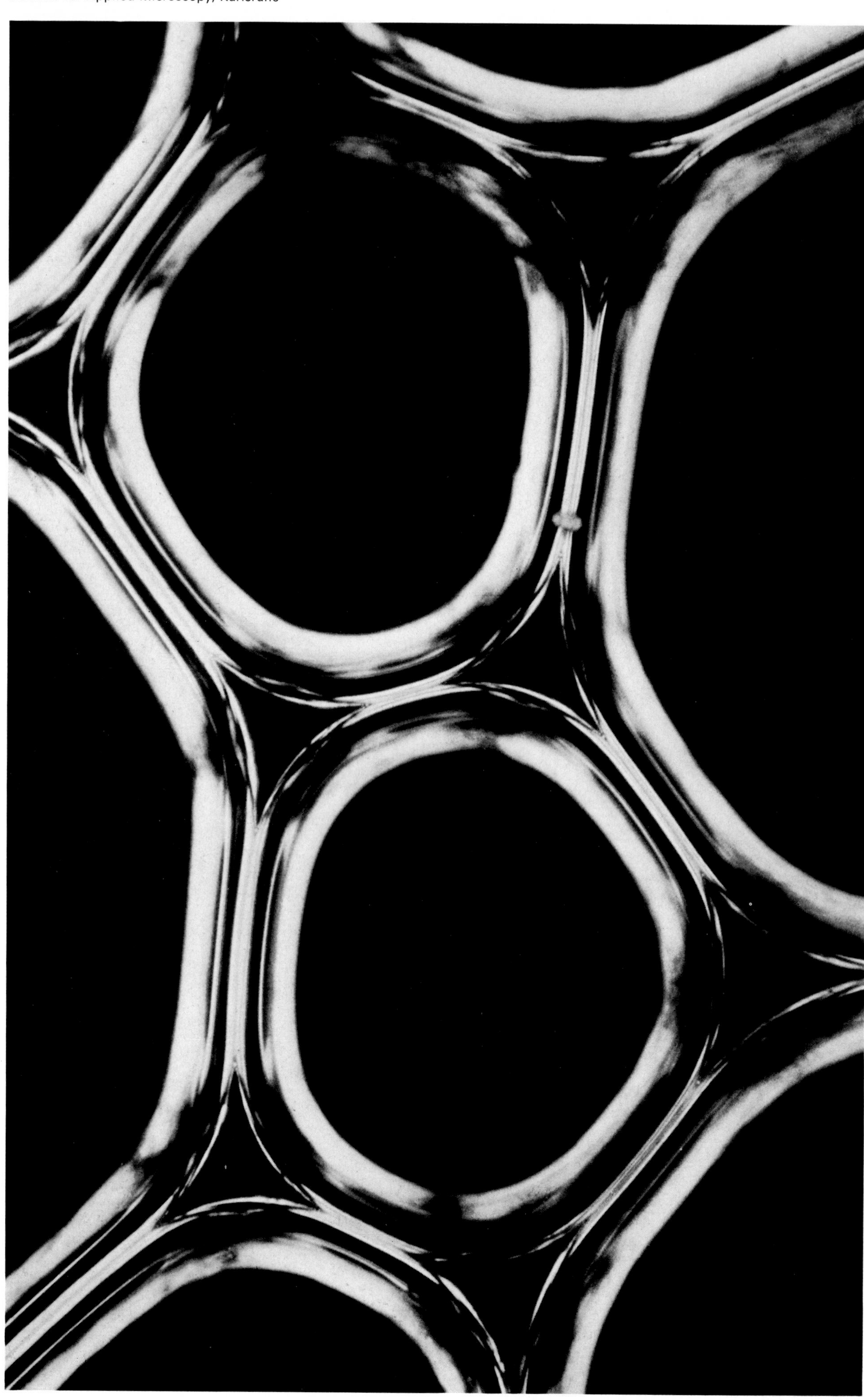

Soapsuds: a complex interplay of forces.

Soap bubbles are extremely tenuous structures; the thickness of their walls ranges from one tenth to one hundred thousandth of a millimeter. The dark-field photomicrograph shows the suds of a laundry soap.

Institute for Applied Microscopy, Karlsruhe

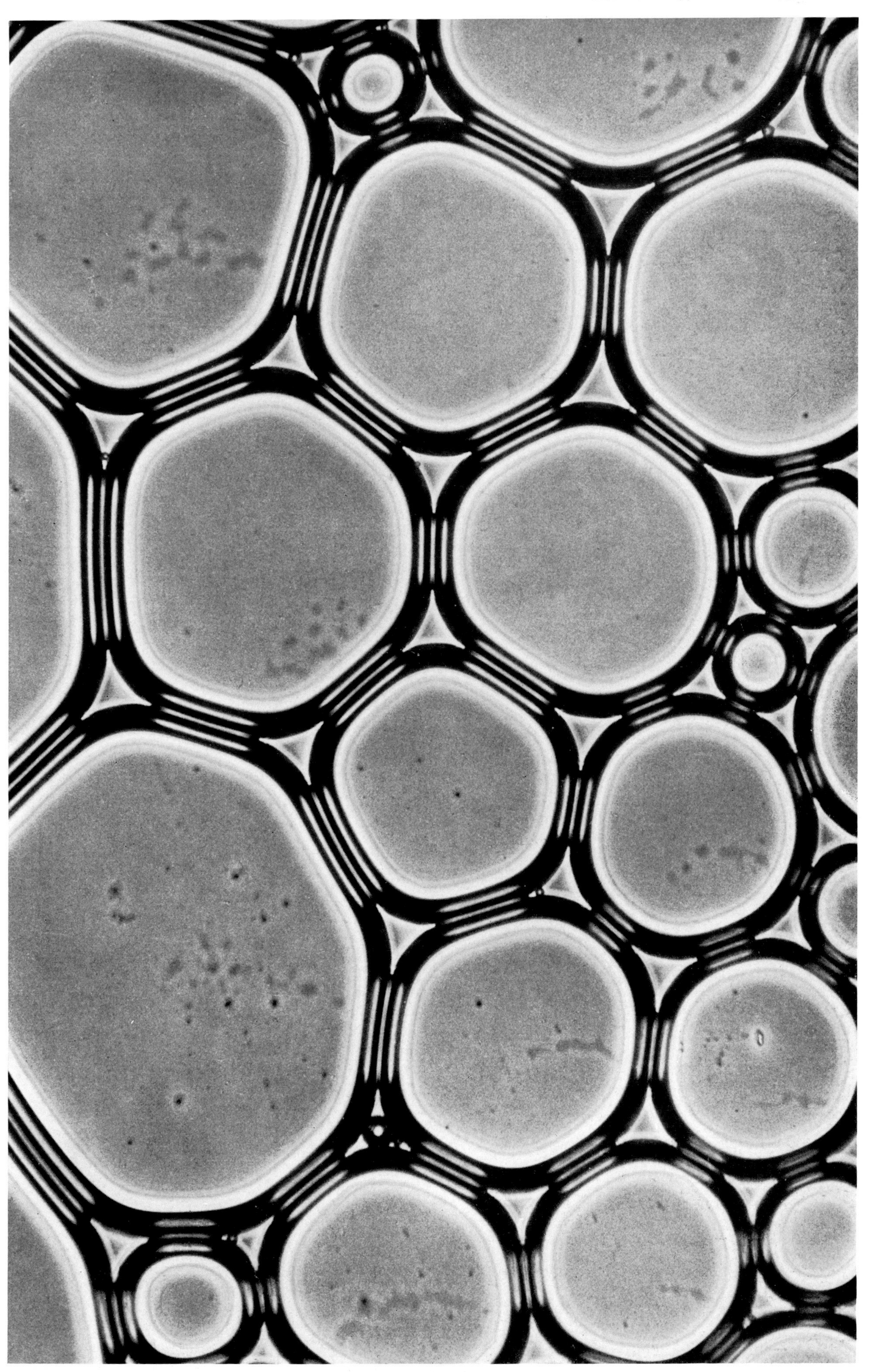

Tensions along the surface and adjoining walls interact. The multitude of soap bubbles within the suds forms a framework of laminae that channels the dirt and, through capillary attraction, even carries it upward against gravity.

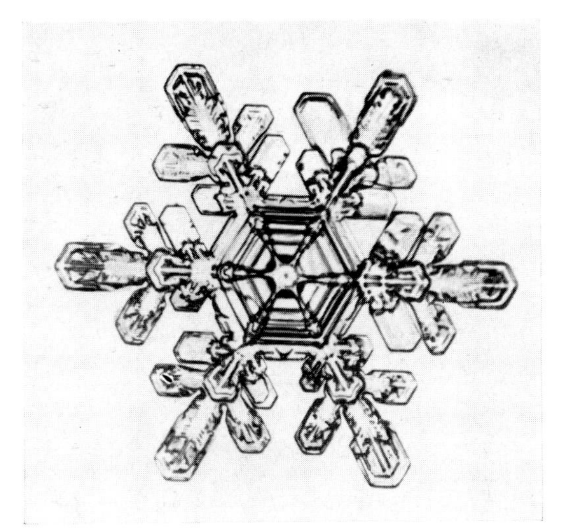

The great astronomer Johannes Kepler (1571–1630) praised the beauty of snow crystals in a poetic New Year's message. The first drawings of snow crystals under the microscope can be found in Robert Hooke's *Micrographia,* 1655. The first collection of photographs of 2,000 different snow crystals was published in 1931, a work by the Americans W. A. Bentley and W. J. Humphreys. Photographing snow crystals requires considerable skill; it is a challenging hobby that keeps some people under its spell for a lifetime.

Deutsche Hoffmann-La Roche, Grenzach

Vitamins, hormones, and enzymes are agents that have been compared by one biochemist to "a key whose small size bears no relation to the size of the portals it opens."

The photo shows vitamin B_{12} crystals.

Tiny keys open great portals.

Deutsche Hoffmann-La Roche, Grenzach

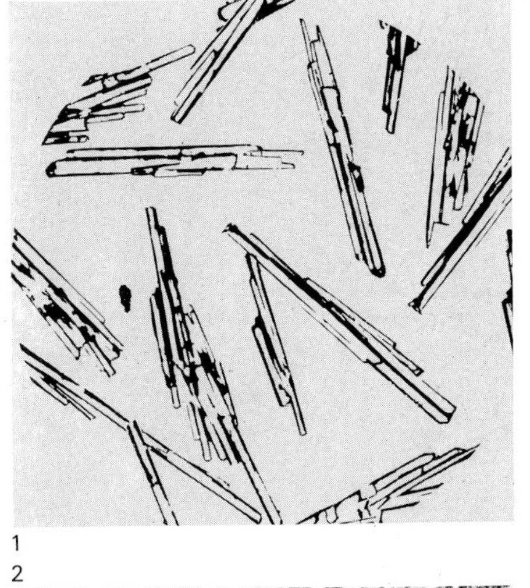
1

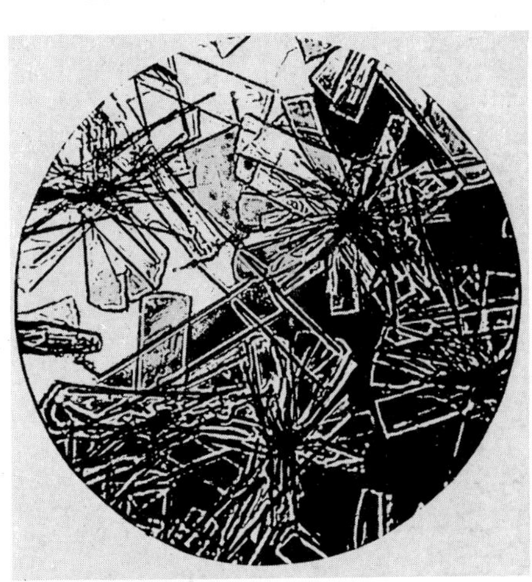
3

2
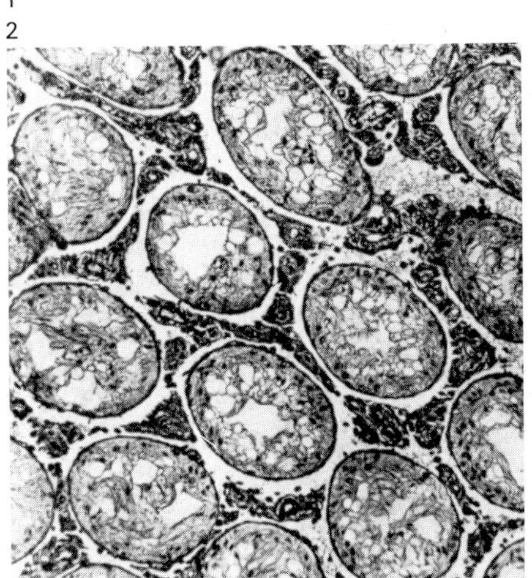

4
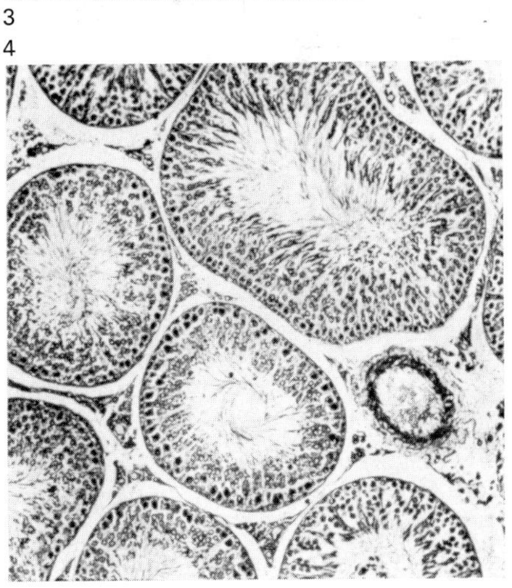

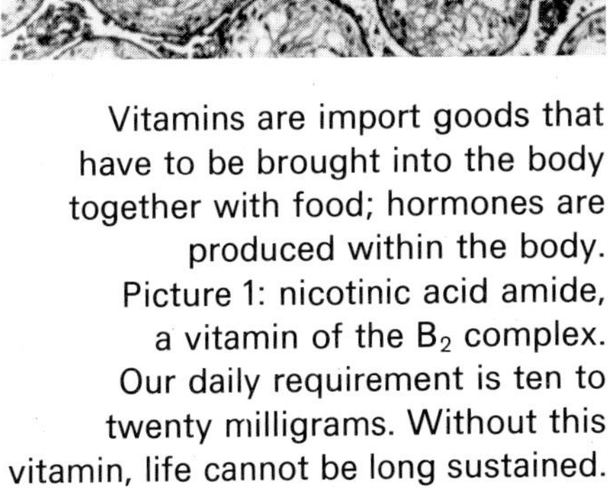
Vitamins are import goods that have to be brought into the body together with food; hormones are produced within the body. Picture 1: nicotinic acid amide, a vitamin of the B_2 complex. Our daily requirement is ten to twenty milligrams. Without this vitamin, life cannot be long sustained.

Picture 3: vitamin D, which prevents rickets. The daily requirement is around three hundredths of a milligram. Pictures 2 and 4: wheatgerm oil is especially rich in vitamin E. Lack of vitamin E is studied in rats; the sections of rat testicles show the decomposition of the tissue.

"Scientists change the world without knowing it." (Lord Balfour)

CIBA, Basle

The thicket of crystal needles is alizarine, the first colorfast artificial dyestuff. The rise of a large-scale chemical industry began 100 years ago with the synthesis of dyes. Plant and animal dyes were thereby dethroned. Over 10,000 dyes are now made from coal tars.

CIBA, Basle

Fritz Brill, Hofgeismar

◄ The saying about "treasures out of the test tube" is never more aptly applied than in reference to chemicals that kill pain, cure sickness, save lives. In 1932, the German biochemist Gerhard Domagk discovered among thousands of dyestuffs, the sulfanilamides, highly effective weapons against bacterial infections. After World War II, he discovered *neoteben* (shown in picture at left)—an anti-tuberculosis drug.

The British bacteriologist Alexander Fleming had noticed as early as 1928 that a mold could destroy bacteria. During World War II, Oxford scientists found a way to produce the effective agent in the mold in larger quantities. Penicillin became the first antibiotic with which to conquer some dangerous diseases. Later on it became possible to chemically synthesize the digestive products of these microorganisms. The picture ar right shows synthetic penicillin. ►

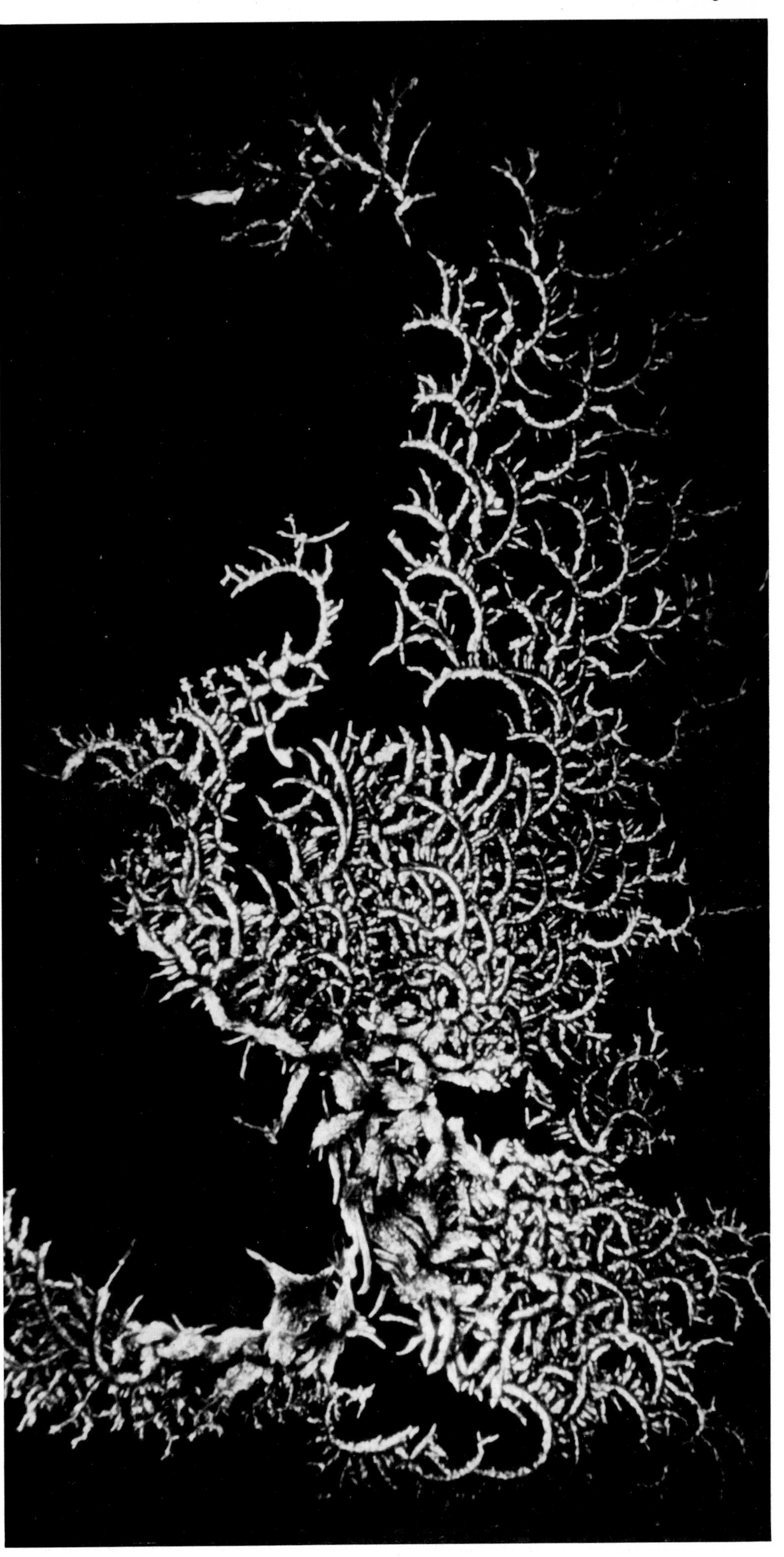

▲
Quinine, a white crystalline alkaloid (shown above in 100-fold magnification), has been known to the Incas for centuries. It is still in use to combat fever and to lower blood pressure. It is extracted from the bark of the cinchona tree, but it can be synthesized in the laboratory.

Caffeine, another vegetal alkaloid that crystallizes in bitter-tasting needles (shown 300 times magnified). First extracted from coffee beans in 1819 and synthesized toward the end of the last century, it is an ingredient in many drugs that stimulate, relax, or mitigate pain. ▸

Ernst Leitz GmbH, Wetzlar

◄ The desire for pain-killing and sleep-inducing drugs is as old as mankind. Drugs out of the test tube are available to modern medicine. Antipyrine, discovered in 1883 by Ludwig Knorr at Hoechst, was the first synthetic pain-killer; Pyramidon followed in 1896. Dimethyl phenol, an ingredient of such drugs, crystallizes in a pattern of peacock feathers.

▲
Rhodamine is one of the fluorescent dyestuffs. Rose-red fabrics dyed with rhodamine have an orange shimmer in their folds due to the fluorescence. Rhodamine B is prepared as green crystals or a purple powder. During crystallization amazingly plantlike shapes materialize.

Man searches for the utilitarian and longs for the beautiful.

Glanzstoff AG, Wuppertal-Elberfeld

Hippuric acid (from *hippos,* the Greek word for horse) derives its name from its origin. Justus von Liebig first isolated it from the urine of horses. Plant-eating animals excrete much of it; meat-eating, little. Humans produce not quite a gram per day. When heated (to 374 °F.), the crystalline structure softens; the photograph shows the melt.

Indanthren-blue, the first of the about 100 Indanthren dyes, crystallizes in this picture in the shape of a Japanese flower design. The name Indanthren composed of indigo and anthracene is a registered trademark. In 1901 René Bohn, from Alsace, discovered a dyestuff that produced more beautiful and luminous colors than indigo and that, like indigo, did not require a mordant: it became known as Indanthren-blue. It is remarkably colorfast; the dye outlasts the fabric.

Ernst Leitz GmbH, Wetzlar

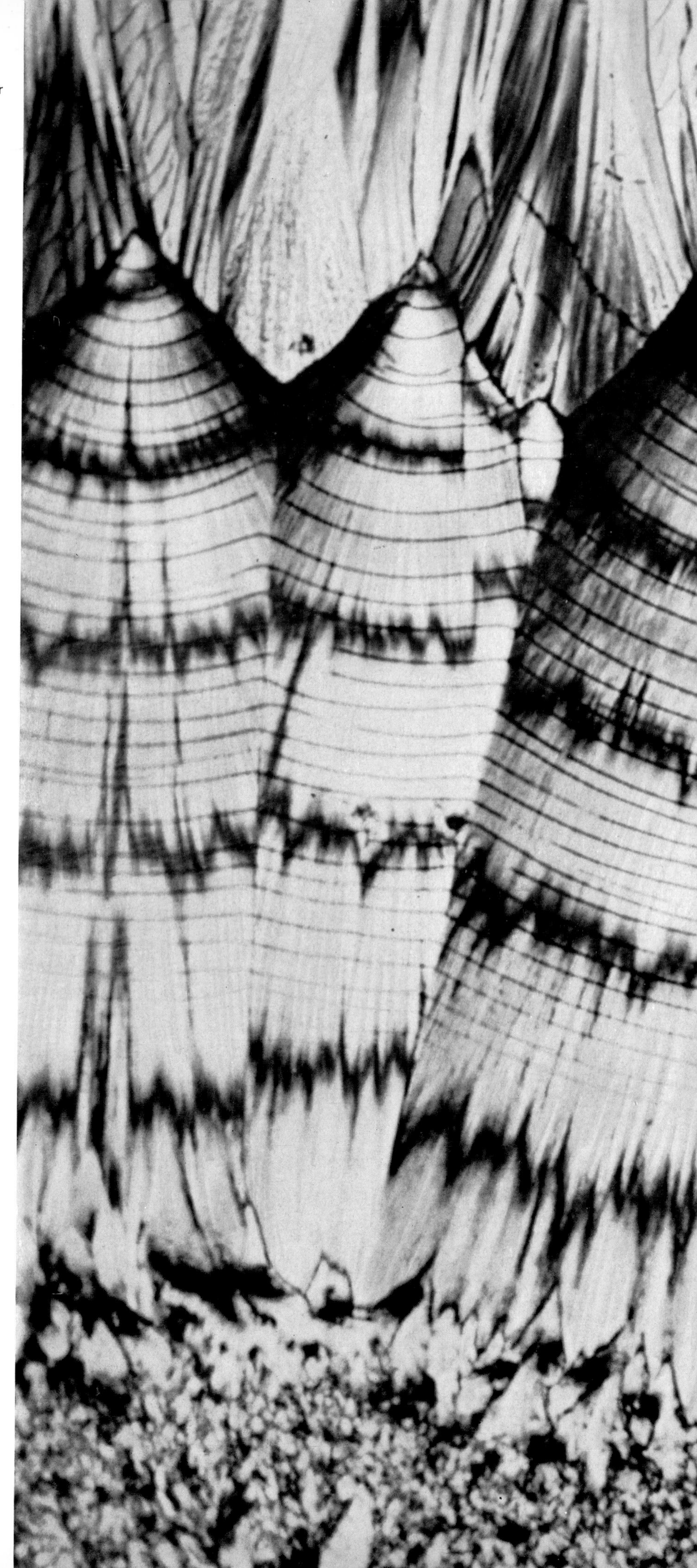

Ernst Leitz GmbH, Wetzlar

The growth of crystals, beyond all explanations, appears to be the imaginative playfulness of mysterious powers. Yet here, especially, nature's will to order becomes apparent; the mathematical plan, according to which our world is constructed, asserts itself. But within its laws, nature permits variations. It has quite obviously many choices and the freedom to opt for one. Both photos show the white, odorless crystals of hippuric acid under the microscope.

Ernst Leitz GmbH, Wetzlar

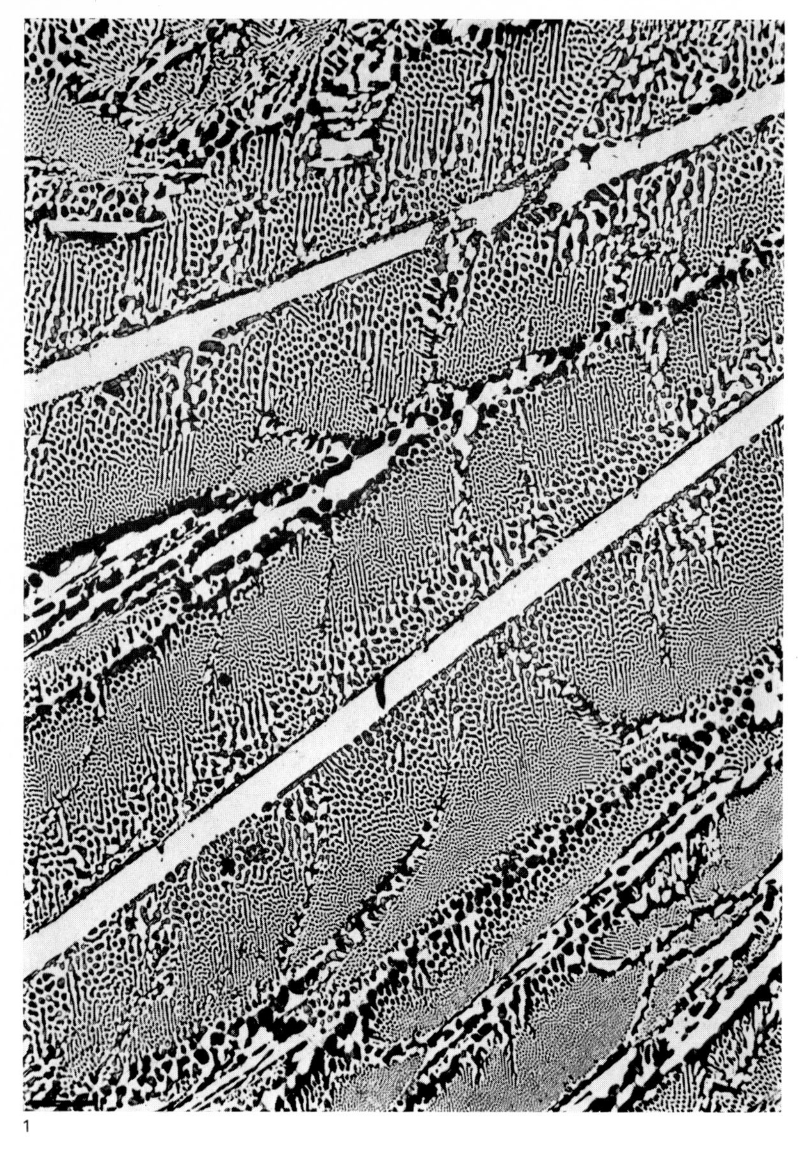

1

Ernst Leitz GmbH, Wetzlar

2

Under the microscope, the polished and etched surfaces of metals and alloys reveal characteristic patterns; the texture becomes prominent. Resistance to pressure and stress, elasticity, and other qualities are determined by the now visible structure.

Picture 1 shows speculum iron (from the Latin word *speculum,* meaning mirror) magnified 500 times. Some alloys of iron and manganese, carbon and silicon are pure white and of a mirrorlike sheen at points of fracture, thence the name. Picture 2: a cast-iron alloy equally magnified. The black pattern is formed by carbon, graphite platelets that loosen the structure. Cast iron is comparatively brittle.

Ernst Leitz GmbH, Wetzlar

3

Ernst Leitz GmbH, Wetzlar

4

Picture 3: thermit, a mixture of aluminum in fine grains, and iron oxide, is used for welding rails. Magnification also 500:1.
Picture 4: polished zinc, magnified 1,000 times. The regularity of the crystalline structure is revealed. All metals are crystalline, as is almost all of the earth's crust.

"Volumes can be read from a rushing brook and the stones are talking."

During antiquity it was believed that the world consisted of four basic elements, fire, water, air, and earth. Pliny, who perished during the destruction of Pompeii in 79 A.D., knew of seven metals. The alchemists believed that gold was linked to the sun, silver to the moon, quicksilver to Mercury, copper to Venus, iron to Mars, tin to Jupiter, lead to Saturn. Zinc has served for centuries to produce brass. Our picture, a 500-fold magnification of unrefined zinc, makes one think of abstract art. For the geologist it contains much information and for the imaginative layman it is a fascinating documentation of the fiery beginnings of the earth.

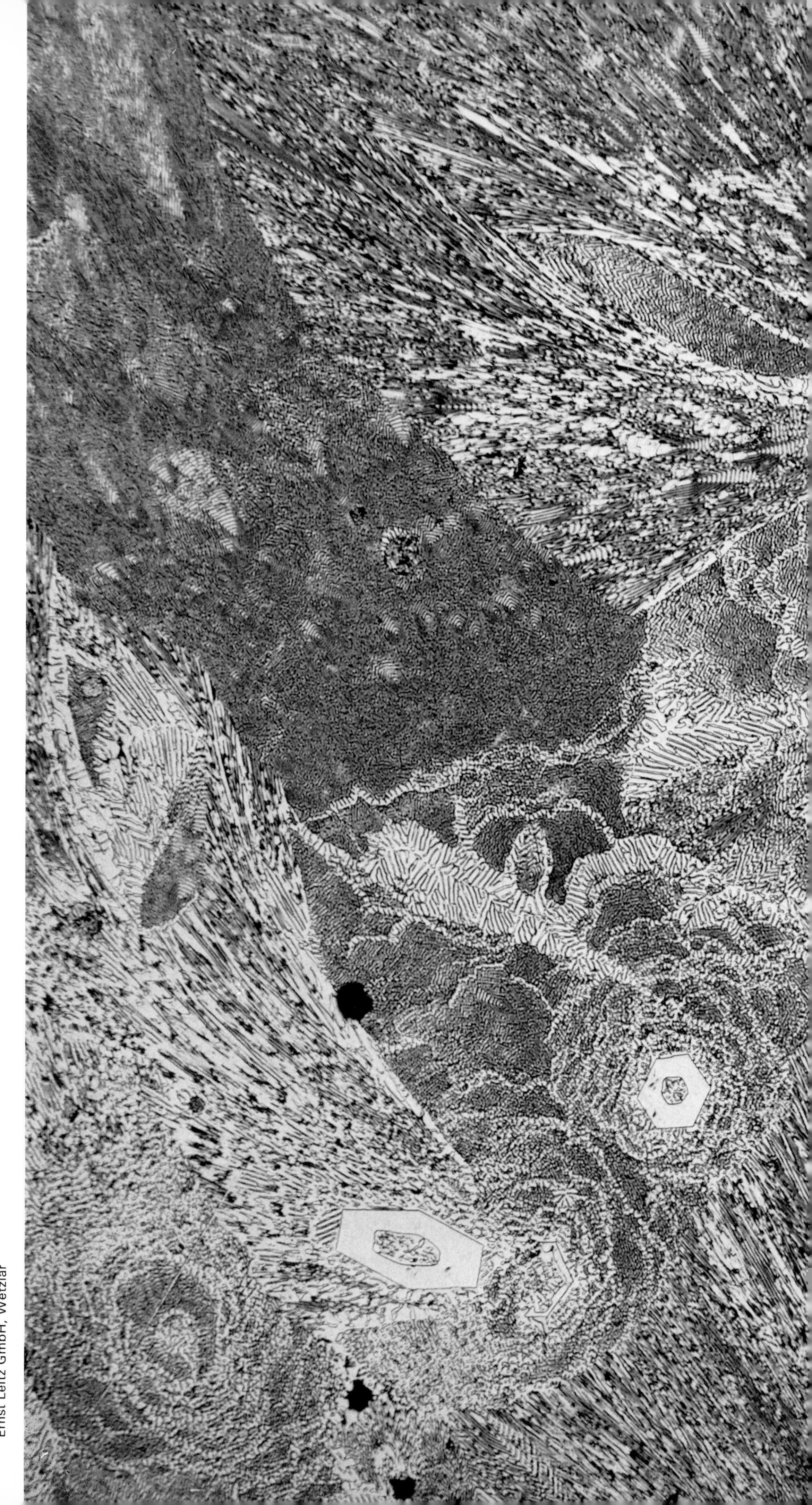

Ernst Leitz GmbH, Wetzlar

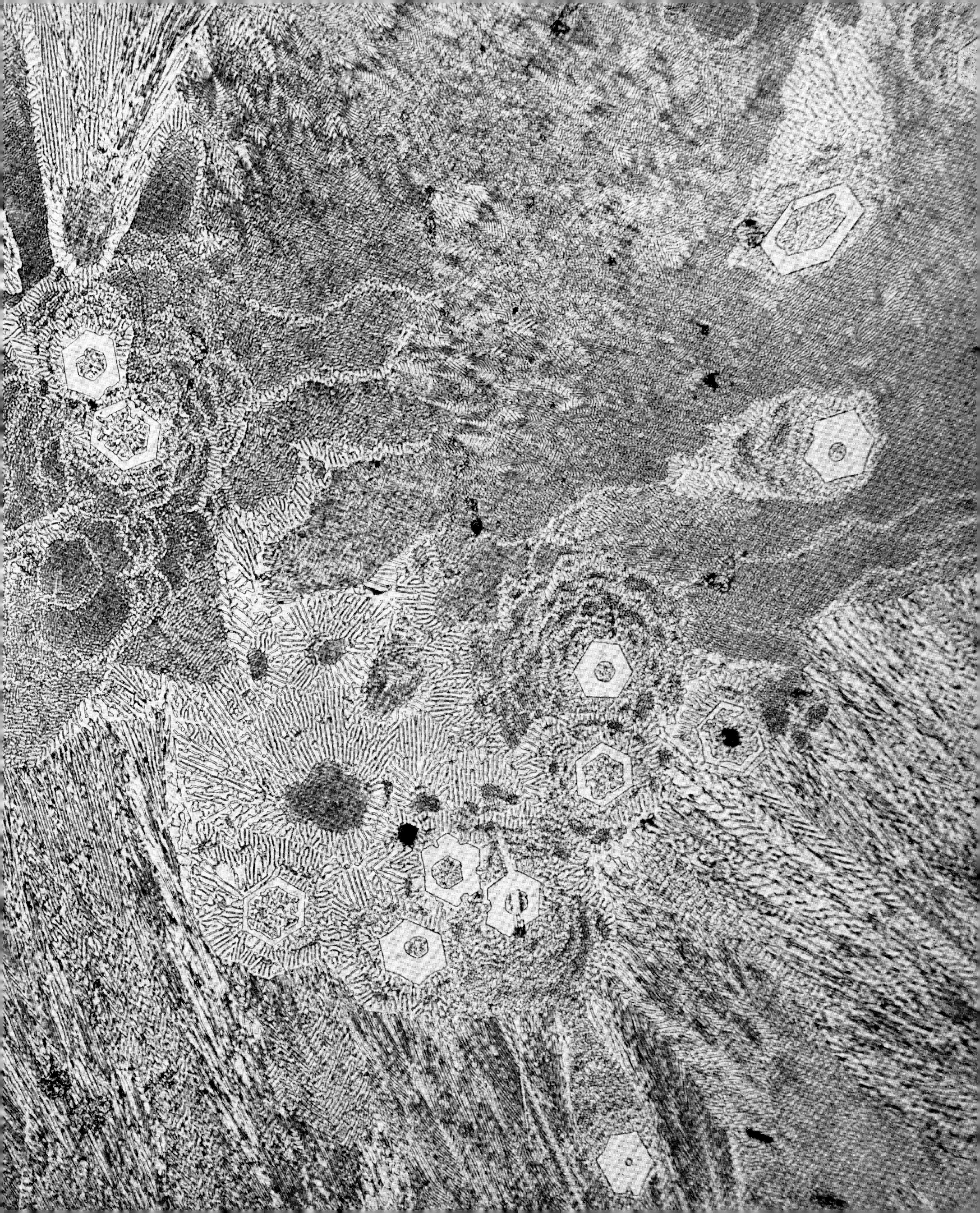

An interference microscope produced these odd relief pictures. They make it possible to recognize and measure surface roughness of as little as twenty-five thousandth of an inch. Two interacting light beams picture the unevenness as profile curves.

The left picture shows staggered rows of spirals on the etched surface of silicium. The crystal growth shows up not in closed steps as in a topographical relief, but rather as spirals, as a path that leads around and up to a mountain peak. In the case of the alum crystals at right, the crystals show that they are not spirally arranged but that they have grown in closed steps. Again the interference microscope has furnished a relief picture that permits measuring of surface unevennesses of less than one hundred thousandth of a millimeter.

Carl Zeiss, Oberkochen

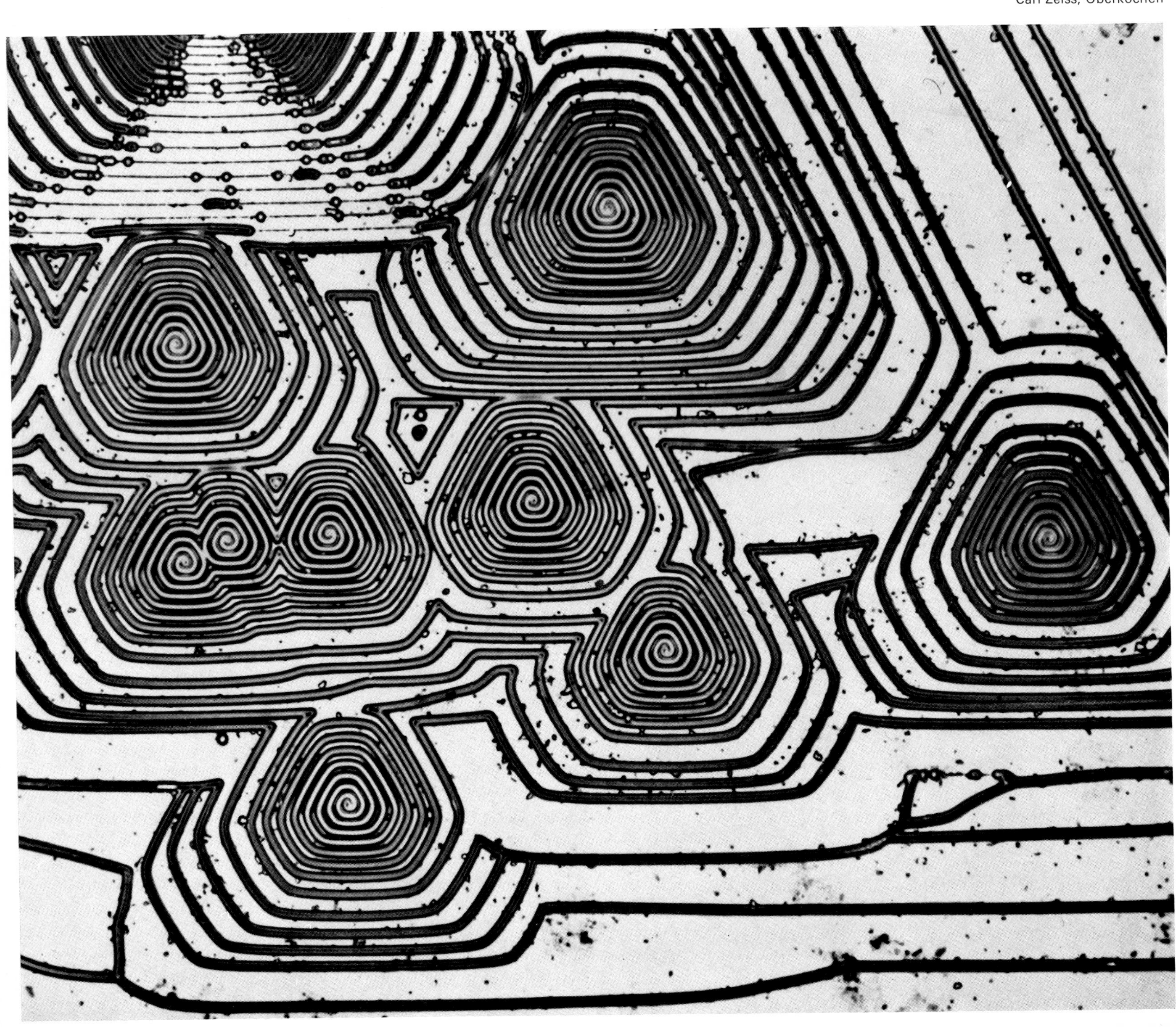

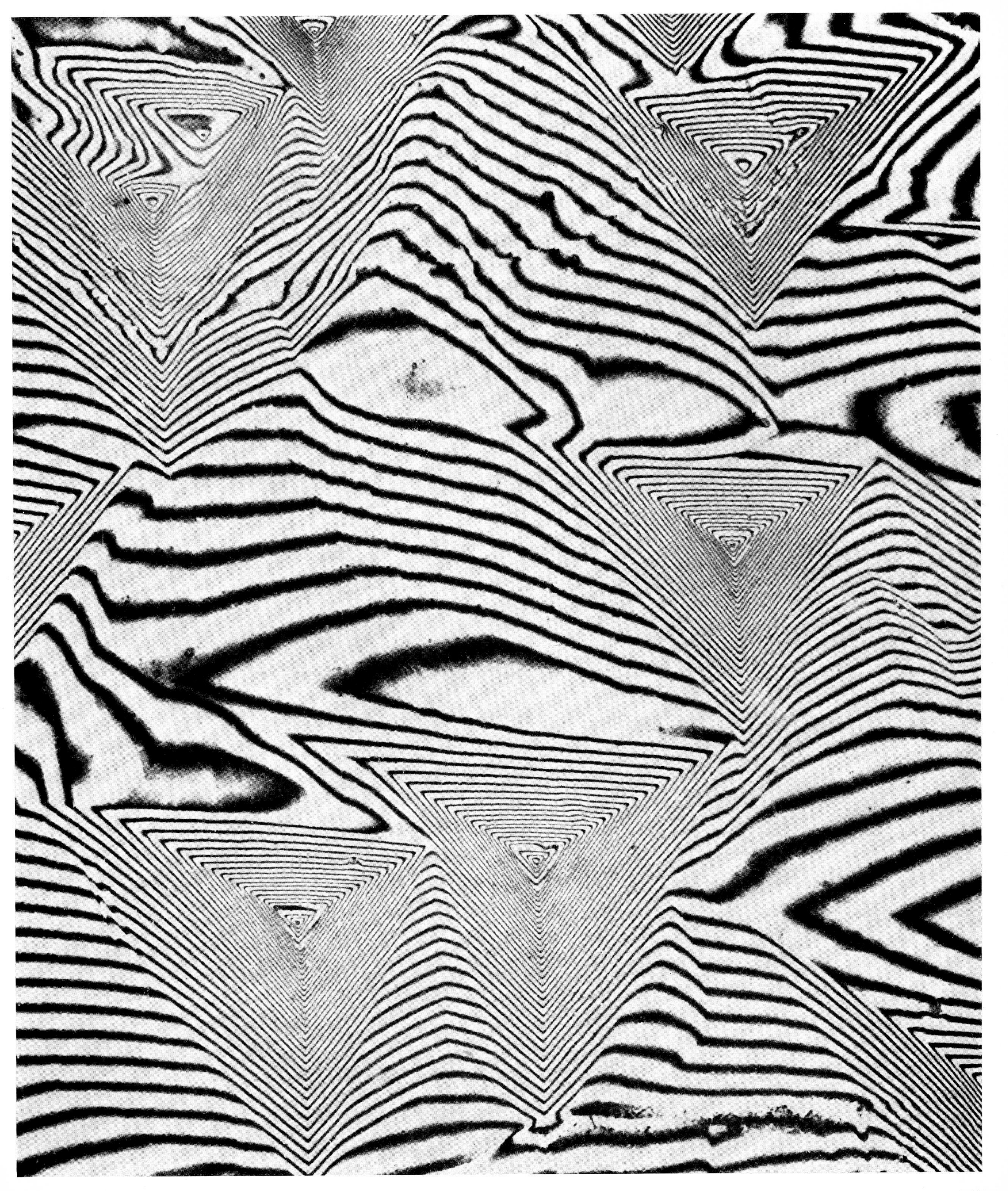

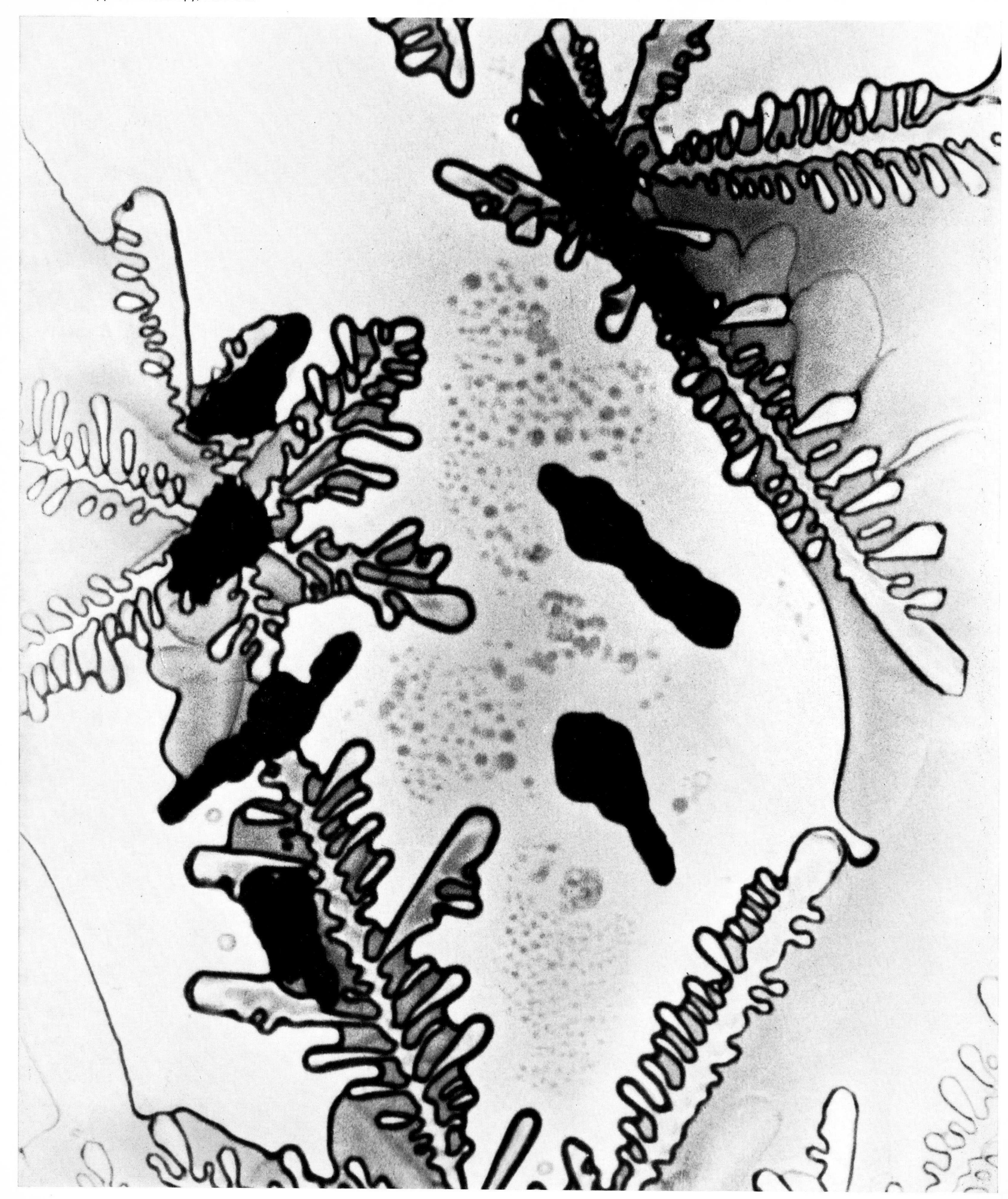

A small flaw in glass, due to a fault in production, shows up under the microscope (100-fold magnified) as a shape resembling an ice-fern. Such flaws are a characteristic of old glass, and also of early spectacles. The first spectacles were known only from illustrations until, in 1953, a few turned up at a monastery in the Lüneburger heath at Wienhausen near Celle, Germany.
The history of spectacles, leading to the development of telescope and the microscope, goes back a little more than 600 years. Spectacles and the printed book ushered in the modern age.

"If there were no lenses, how much would we ignore about nature."
(Lentilius, 1685)

The enigmatic medium of light forms a bridge between universe and man, between man and microcosm. With light begins the story of creation. As a physical magnitude it is of the first order.

According to legend, the young Narcissus fell in love with his reflection as he was bending over a pool to drink; it was a root experience from man's early days. The image in the water, science states, is just reflected light. A partly immersed stick that appears broken shows the effect of refracted light. Whenever light passes from one medium into another it is subject to changes. A coin placed on the bottom of a bowl in such a way that it is just hidden by the rim of the bowl, becomes visible, as if by magic, once the bowl is filled with water. Already in the second century, Ptolemy found the explanation: light rays are bent when passing from air into water.

The Properties of Light

Isaac Newton, three hundred years ago, showed that not only can light be refracted but that it also can be resolved into its component colors, that it contains all the colors of the rainbow. A glass prism causes the colorless ray of light to display its glowing spectrum. The primary colors cannot be divided any further. Green, for example, can be deflected by a prism but it remains green! In 1666, a generation after Galileo, Isaac Newton had this to say about the nature of light: "It is a shower of particles shot from a luminous object. Each color of the spectrum emits its own characteristic particles, and all particles travel at the same speed." A differing theory was propounded by Christiaan Huygens, a Dutch scientist of wide-ranging interests, the discoverer of the Orion nebula, the first to define the shapes near the planet Saturn as rings, the first to construct a pendulum-regulated clock, and the inventor of the balance-spring regulator for the watch. Light moves in waves, he wrote in his *Traité de la Lumière (Treatise on Light)*, 1690: "If it takes time for light to arrive at its destination, it follows that this motion, imparted to a medium, is a gradual one and that it progresses, just as the movement of sound, in spherical planes or waves; I call them waves because of their similarity to those that form when a stone is thrown into water."

What then is light? Waves like those caused by the impact of a stone on water, or an emission of fine particles? The eminent Swiss mathematician Leonhard Euler, who was called to teach at the courts of Catherine the Great in St. Petersburg and of Frederick II in Berlin, reopened the question about the nature of light and wrote: "The great Newton maintained that the rays of the sun actually flow out of the sun's body, and that extremely fine particles are shot out from the sun at such unimaginable speed that they reach us within about eight minutes." Euler himself advocated the wave theory of light, first in his *Nova theoria lucis et colorum* and later in his *Lettres à une princesse d'Allemagne sur quelques sujets de physique et de philosophie* (1768–1772), which, by the way, are a model for explaining science subjects to a lay person.

Today, we know that light is both, quickly moving particles and waves—it is both together, not one or the other! The phenomenon of light can be described either way, its properties can be calculated either way, and both methods lead to correct results. The true nature of light is beyond human understanding.

Beyond the Spectrum

In 1800, Frederick William Herschel (Sir William after 1816) published his notes on a discovery that, at the time, seemed most peculiar: There exist rays that cannot be seen by the human eye—heat-rays, immediately following the spectrum's red end. Herschel demonstrated that invisible, infrared light, just like visible light, obeys the laws of reflection and refraction, although at different angles.

Just a year after Herschel discovered infrared, the German researcher Johann Wilhelm Ritter found that beyond the opposite end of the spectrum other invisible rays made their presence felt. They blackened silver nitrate. About a hundred years later it was discovered that waves within that range have healing powers. From that discovery derived the sunlamp and the high-altitude sanitarium.

The findings of Herschel and Ritter were signpost and stimulus for further research. Did the range of wavelengths on both sides of visible light extend beyond infrared and ultraviolet? The ground had been broken for something wholly new: earlier, the telescope and the microscope had made use of the waves of visible light to bring the incredibly distant and the infinitesimally small within the eye's range. Now, invisible "siblings of light" had been discovered, related waves that perhaps could also be used to render visible that which so far remained invisible and unknown. Now could a searchlight be directed more sharply, more deeply into the world around us? Could, perhaps, new visions be wrested from the mysterious invisible? Here was a challenge to imagination and technical ingenuity. Scientists in their physics laboratories readied an arsenal of apparatus with which to query nature and perhaps receive an answer.

The Scottish physicist James Clerk Maxwell, in the years 1865-1871, undertook to express in mathematical formulas the relationship between magnetism, electricity, and light, a labor that brilliantly concluded what Michael Faraday had begun. Max Planck, himself a genius as a physicist, ranked Maxwell's equations among the "outstanding miracles worked by the human mind."

The Resonant Globe

According to Maxwell, light is only one special instance among a great number of electromagnetic waves, all traveling at the same speed—186,000 miles per second. It was Heinrich Hertz who applied Maxwell's theory and found that it worked. In 1887, the young professor from Karlsruhe, Germany, became the first to detect and study the electromagnetic waves that became known as Hertzian waves or radio waves. He caused an electric spark to jump between two terminals in his "transmitter," producing—wireless!—a spark in a "receiver." The two circuits stood a few yards apart in the same Karlsruhe laboratory. The spark was only visible under a magnifying glass, and yet, it was this tiny sprak that started off the age of wireless telegraphy, the era of radio and television.

Heinrich Hertz did not himself transmit any messages. But only ten years later a newspaper of May 19, 1897, reported that the Italian physicist Guglielmo Marconi had demonstrated to experts of the London postal administration an invention "that seemed likely to revolutionize the worldwide transmission of news. Marconi sent wireless messages. He carried out his demonstration in the Bristol Channel. His transmitter was located on the small island of Flatholm, his receiver not far from the coastal resort of Penarth. The Morse telegraph transmitted the letter "V": three dots one dash. The receiver's Morse apparatus clearly tapped out three dots and one dash. The impression made by the arrival of a message that had been carried by thin air was tremendous." Marconi had bridged a distance of three miles! Today the earth is ringed with sounds, words, music, and pictures.

Heinrich Hertz tried to render the wide distribution of electromagnetic waves comprehensible to his contemporaries with these words: "I make the outright claim that light is an electric wave, light per se, light of the sun, that of the candle, that of the glowworm. Remove electricity from this world and light will disappear.... We now notice electricity in a thousand places where, before, we had no certain knowledge of its presence. In each flame, in each luminous atom, we see an electrical process. Even when a body gives off no light, when it only radiates heat, it is the seat of electrical excitation."

After Hertz's death in 1894, at thirty-seven years of age, Max Planck, his junior by one year, said in a commemorative speech: "It was Hertz who in his experiments forced nature to reveal itself in manifestations not observed by anyone before him.... Who, among natural scientists, does not still recall how those first reports caused feelings of awe, and of admiration, for the infinite grandeur of nature whose laws apply equally to large and to small, and for the enormous capacity for abstraction of the human brain, made possible by a combination of exact logic and creative imagination...."

A New Kind of Rays

Barely two years later "a new kind of rays" was discovered in a scientific institute in Würzburg, Germany. As it turned out, they were rays of considerably shorter wavelength than those of visible light, in contrast to the Hertzian waves that were substantially longer. What was not known at the time was the wide range of electromagnetic wavelengths. Today, we know that visible light occupies only a narrow segment of the scale. After the infrared waves follow those waves that serve radar, television, radio and communications, and finally the wavelengths of electric household currents. After the ultraviolet rays, which can be produced artificially with a quartz lamp, follow X-rays and the rays produced by radioactive substances. In other words, the scale of electromagnetic waves begins at the many-miles-long waves of alternating current; it ends at the hundred-billionth of a millimeter wavelength of nuclear radiation.

During the night of November 8, 1895, Wilhelm Conrad Röntgen, then already fifty years old, made a crucial observation; it kept him for days thereafter in his institute checking over and over again whether what he had observed was really so. On December 28, he completed his report on "a new kind of rays." The famous meeting, during which Röntgen was to demonstrate his discovery, was called for January 23, 1896. Through an indiscretion the newspapers got hold of the story and published somewhat puzzling accounts. As a consequence, Röntgen's lecture about his experiment was awaited with more than usual anticipation. What Röntgen demonstrated had not been thought possible: a new kind of rays was produced in vacuum tubes by means of high-voltage electricity. They were invisible rays that penetrated just about any kind of matter and revealed hidden inner structures on photographic plates. Transilluminating rays! A new, deeply penetrating insight into nature had become possible. While the lecture was still in progress, the aged Würzburg anatomist, Privy Councilor Kölliker, placed his hand on a wrapped photographic plate. The picture was immediately developed and thereafter the skeletal hand with the seemingly free floating ring was to turn up in countless learned journals.

Kölliker evaluated the moment's importance correctly. He suggested naming the new rays, whose nature was not yet known, after their discoverer. Röntgen himself had labeled them X-rays and in many parts of the world, that is what they are called. Wherever German is spoken the spontaneous recommendation of the privy councilor was accepted; more than that, the name of the discoverer has become a universally used verb.

Strangely enough, *how* Röntgen made his discovery is a tantalizing secret. Walter Gerlach, the second after Röntgen to hold the physics chair at Munich, wrote in 1934: "How Röntgen investigated the Röntgen rays, we know; how he discovered them was and remains a secret." Röntgen, like Newton before him, destroyed many of his notes before he died. He stipulated in his will, for whatever reason, that additional documents, above all those from the period of the discovery, were to be burned. Friedrich Dessauer, one of the pioneers of X-ray technique, described Röntgen's feeling: "fame threatens a man in the possession of his own self, delivers him up to the public to play an alien role on the world's stage, and prevents him from pursuing his true calling." This same feeling that, by the way, also oppressed the Curies, may have prompted him to defend his privacy beyond the grave. However, since he had so thoroughly obliterated all records, rumors that started to circulate soon after his discovery and that he had left unchallenged, would not die down.

There is the story about the attendant at the institute, Marstaller, who supposedly was the real discoverer, since he "put Röntgen on the right track through a chance discovery that was to him unintelligible." Posterity, with its fondness for the "whodunit," has pursued this track. There is general agreement that Marstaller had a good head on his shoulders. As a matter of fact, among the students the new rays were referred to as Marstaller-rays rather than Röntgen-rays. All of which brings to mind the capable assistant of Faraday, a onetime soldier, who used to say: "I carry out the experiments and Faraday adds his two cents' worth of comments." It is on record that Marstaller, indeed, claimed that he had drawn Röntgen's attention to the fact that a ring in a small box had left its imprint on photographic paper in some inexplicable way. The later privy councilor, A. Dyrhoff, who himself heard Marstaller's comments on the affair, wrote: "Actually, Marstaller saw his contribution as only a chance observation of a chance effect." Can he, therefore, be considered the discoverer? Hardly. He may have triggered the discovery by noticing something puzzling (his attentiveness should not be underrated!). At the time many laboratories worked with vacuum tubes, and it was later found that in a number of cases, rays produced definite and observed effects. But no attention was paid to these effects! The thinking of the researchers was directed toward other goals, and what they saw could not deflect their train of thought. Whether Marstaller did, or did not, alert him, Röntgen, for one, stopped to think. He followed up, step by step. His imagination suggested possibilities. Some, he rejected; others, he accepted. He subjected his suppositions to scientific scrutiny. He devised experiments that eliminated some conjectures and made others appear more likely. He didn't give up until he was scientifically satisfied. Finally, he was sure of the facts that he presented in the seventeen points of his first report on the new rays. This, and only this, is rightly considered the discovery. By the way, Röntgen, the brilliant experimental physicist, had severely impaired vision in one eye and was, moreover, color blind. Nevertheless, he was able to see more "acutely," more "deeply."

In order to explain how X-rays come into being, it is necessary to digress a little: air is drawn almost completely from a glass tube that has electric wires sealed into both ends—a cathode or negative terminal, and an anode or positive terminal. When the wires are connected to a source of high-voltage electricity, an electric current flows through the remaining gas which begins to glow mysteriously. Gas ions, electrically charged atomic particles, travel toward the electrical terminals. As they hit the cathode, the positive ions free electrons that, greatly accelerated by the high voltage, shoot away from the cathode in a straight line. They bounce against the opposite wall and from there travel along the tube's glass wall toward the anode. Where cathode rays are suddenly slowed down, as by the glass wall, X-rays originate. If the voltage is raised, the volley of electrons becomes more violent, the energy level rises, and so does the X-rays' power of penetration. Later on, the hot-filament cathode was devised for the X-ray tube: it consists of a heated wire that emits electrons without the need for stimulation from the gas in the tube. This type of tube, commonly used today, is more easily regulated.

Before Röntgen increasingly withdrew from the onslaught of the public, he granted just one interview to a reporter. An excellent record of this meeting is the story, "The New Marvel in Photography," by H. J. W. Dam in *McClure's Magazine* (Vol. VI, No. 5, pp. 403–415). Dam was one of the first to have the experience of being X-rayed and he tried to pass it on to his puzzled readers.

"The rays flew through the metal and the book as if neither had been there. ... It was a clear, material illustration of the ease with which paper and book are penetrated. And then I laid book and paper down, and put my eyes against the rays. All was blackness, and I neither saw nor felt anything. The discharge was in full force, and the rays were flying through my head, and, for all I knew, through the side of the box behind me. But they were invisible and impalpable. They gave no sensation whatever. Whatever the mysterious rays may be, they are not to be seen, and are to be judged only by their works...."

The rest of the interview reads like a taped conversation.

"Now, Professor," said I, "will you tell me the history of the discovery?"

"There is no history," he said. "I have been for a long time interested in the problem of the cathode rays from a vacuum tube as studied by Hertz and Lenard. I had followed theirs and other researches with great interest, and determined, as soon as I had the time, to make some researches of my own. This time I found at the close of last October. I had been at work for some days when I discovered something new. ... I was working with a Crookes tube covered by a shield of black cardboard. A piece of barium platino-cyanide paper lay on the bench there. I had been passing a current through the tube, and I noticed a peculiar black line across the paper."

"What of that?"

"The effect was one which could only be produced, in ordinary parlance, by the passage of light. No light could come from the tube because the shield which covered it was impervious to any light known, even that of the electric arc."

"And what did you think?"

"I did not think; I investigated. I assumed that the effect must have come from the tube, since its character indicated that it could come from nowhere else. I tested it. In a few minutes there was no doubt about it. Rays were coming from the tube which gave a luminescent effect upon the paper. I tried it successfully at greater and greater distances, even at two metres. It seemed at first a new kind of invisible light. It was clearly something new, something unrecorded."

"Is it light?"

"No."

"Is it electricity?"

"Not in any known form."

"What is it?"

"I don't know."

And the discoverer of X-rays thus stated as calmly his ignorance of their essence as has everybody else who has written on the phenomena thus far.

"Having discovered the existence of a new kind of rays, I of course began to investigate what they would do. ... It soon appeared from tests that the rays had penetrative power to a degree hitherto unknown. They penetrated paper, wood, and cloth with ease; and the thickness of the substance makes no perceptible difference, within reasonable limits."

... In answer to a question, "What of the future?" he said:

"I am not a prophet, and I am opposed to prophesying. I am pursuing my investigations, and as fast as my results are verified I shall make them public."

If one follows the interview carefully, one notices that Röntgen mentions a certain observation, but answers only with inexact generalities to related questions. He speaks of a "strange black line" but doesn't want to go into the matter. But just this was the true focus of his interest, not the fluorescence of the screen, a phenomenon that had been observed before. What Röntgen had seen was most certainly the "bone shadow" of his own hand as he placed it by chance between tube and screen. The tremendous, revolutionary importance to medicine of the new rays revealed itself at the very outset as a fleeting shadow that presaged secrets yet to be uncovered. But no matter how fleeting the signal was, the man who made the experiment noticed it and—the discovery was made.

Röntgen discovered something that did not exist in "nature," at least not on earth. The X-rays from the sun, that were discovered later, are screened out by the earth's atmosphere. "Special apparatus had to be invented," Friedrich Dessauer once wrote, "to make it possible for such rays to appear." Their existence on earth is conditional upon an invention. They could be called "invention-discoveries" and they occupy, therefore, a special position. The same also holds true for the Hertzian waves. Dessauer thinks he can discern two types of scientist: "the classical and the romantic." Röntgen seems to him an example of the former; the silent, severe, and yet benevolent man. Paul Ehrlich, the founder of chemotherapy, whom he also knew personally, is regarded by him as representing the typical romantic scientist: the enthusiastic, scientific researcher, brimming over with ideas. On two occasions Röntgen proved to be the timeless professor: when he answered, "I did not reason, I investigated," and when he demurred, "I am not a prophet."

Transparent as a Jellyfish

The American physicist Arthur Compton, who won the Nobel prize for his X-ray research, estimated in 1957, two generations after Röntgen, that as many lives had been saved by X-rays since their discovery as had been lost in all the wars of that same period—a grateful and proud acknowledgement that should be included in every history book.

The importance of X-rays in present-day medicine, both for diagnosis and for therapy, can hardly be overrated. The possibility of making man "as transparent as a jellyfish" is in itself a magic gift full of blessings. Everyone knows that an X-ray picture gives the surgeon information of unmatched precision in cases of broken bones or imbedded bullets. Widely known, also, is the function of substances that will not permit the passage of X-rays, as for instance the chalky barium mixture swallowed to show the condition of stomach and intestines. Today it is possible to render visible the arteries of the kidneys, to find the blood clot that causes the dreaded lung embolism, to detect hidden damage to blood vessels, to follow the sluggish flow of the lymph fluid. Brain tumors can be located by injecting dyes or by introducing a few cubic centimeters of air into the spinal cord, letting the air rise into the ventricles of the brain. In the X-ray picture, the guided air bubble exactly traces the structure of the respective section of the brain. The air, because of its lower density, serves as contrast material. Special techniques make the vessels in the brain visible! The anatomy of this most mysterious of all organs, although encapsulated in bone, can now be grasped.

The Catheter in the Heart

That heart operations and even heart transplants are possible today, is due to an experiment made on his own body by a young German physician, Werner Forssmann! He dared to do what no one had dared before him: in an experiment in 1929, he passed a thin rubber tube, the sort of catheter used in the examination of the urinary tract, through the arm vein into the heart itself. "During the insertion of the catheter, as it slid along the wall of the vein, I experienced a mild sensation of heat, the same feeling that is caused by an intravenous injection of calcium chloride. This sensation of heat was especially noticeable back of the collarbone and was accompanied by a slight urge to cough. I introduced the catheter to a depth of sixty-five centimeters; I had calculated this to be the distance into the heart." With his arm hidden under a cloth, and the catheter in his heart, Werner Forssmann walked down the stairs to the X-ray laboratory in the basement of the Berlin-Eberswalde Hospital. "I stepped behind the fluoroscope and had someone hold up a mirror to the screen in such a way that I was able to check the experiment on the screen." Forssmann saw the tip of the catheter in the right ventricle of his heart. He had someone call in a doctor friend of his to take a few X-ray pictures, and for this purpose he withdrew the catheter part of the way and then pushed it back in again. It was apparently possible to examine pressure and flow in the heart, to withdraw some blood, and to investigate the transport of oxygen within the heart.

A few months later Forssmann went one step further. He now wanted to try introducing contrast material into the heart to render visible in the X-ray picture, not only the shadow of the heart, but also the ramification of blood vessels. He felt that it should be possible to inject drugs directly into the heart without opening up the chest at the threat of heart stoppage. But here Forssmann was no longer convinced that this was without danger. "A catheter can be withdrawn instantly. Contrast material, a chemical solution, would have to be expelled by the heart itself. How it would react was not known. There was a possibility of shock." Forssmann again introduced the catheter into his heart, and injected a sodium iodide solution. For a moment he experienced a slight dizziness. And for just that moment he saw the pulmonary artery clearly detailed on the screen. In less than a second the heart had eliminated the contrast solution. It did not take the trespass amiss. But, in order to get better pictures, a better contrast solution and better picture-taking techniques would have to be developed. A series of tests was necessary—animal tests from then on—and that would cost money.

Fortunately a cinematographic X-ray technique suitable for recording fast-moving body processes had been invented by the Berlin radiologist Dr. Gottheiner. By 1930, Werner Forssmann was able to show an X-ray film of the beating heart of a dog. The young intern presented his film to the fifty-fifth Congress of the German Surgical Society in 1931. His lecture, the last one of the day, was received without any comment. "The time was not yet ripe for this sort of thing. So much was still missing and had to be developed before my method could become what it is today: routine."

Werner Forssmann was unable to follow up on his initial experiments. After the war, news of progress in this field came from the United States. In Germany, hardly anyone remembered where and how it all had begun. Forssmann believed himself completely forgotten. Until the day a telegram from Stockholm arrived: together with André Cournand and Dickinson Richards, two widely known scientists, he had been awarded the 1956 Nobel prize for physiology and medicine. At that time, Forssmann was working as a public health insurance doctor.

X-rayed Crystals

The advances in medicine directly attributable to X-rays almost overshadow their importance for the natural sciences as a whole. What X-rays really were, their discoverer couldn't tell. He suspected that they were wave phenomena related to light.

But neither he nor the many other researchers who worked on the problem were able to prove this. One characteristic feature of waves is the so-called interferences that can be produced, for instance, by bending the rays; this could not be demonstrated with X-rays. Interference is based on the fact that waves which travel in the same direction either cancel or reinforce each other. Bending, diffraction, occurs at edges, fissures, and very small openings. The diffraction gratings used in optics are small glass plates on which parallel grooves have been incised with a very fine diamond point. The distance from groove to groove, the grating-constant, must not be much larger than the wavelength of the rays that are to be investigated; otherwise diffraction is difficult to achieve. Joseph von Fraunhofer, an optician and physicist who invented a diffraction grating in 1814, had already accommodated 300 lines within one millimeter; today, grids of 2,000 lines per millimeter have become possible.

There had been hints that X-rays might be some sort of light of very much shorter wavelength, but the then existing grids were too coarse to check this out. After sixteen years of the most intensive effort, the nature of X-rays was still uncertain. Then, in 1912, the German physicist Max von Laue, after a conversation with the then twenty-four-year-old P. P. Ewald, had a brilliant idea: nature herself was to supply a grid that was fine enough. As all substances ultimately are composed of atoms, it could be assumed that in regularly structured crystals these atoms would be arranged regularly—that is, in a grid pattern. What if one were to direct X-rays through crystals? Common salt, NaCl, crystallizes in cubes; it consists of an equal number of atoms of sodium and chlorine. The distance between atom and atom can be calculated and turns out to be 0.000,000,028 centimeters. Crystals, then, thought Laue, might provide suitable grating-constants for X-rays. He arranged for the appropriate experiments to be carried out; Walter Friedrich and Paul Knipping performed them. Through a crystal of copper sulphate that chanced to be lying around the lab, they directed a very fine X-ray beam onto a photographic plate. It was diffracted by the narrow passages between the atoms of the crystal. Where the rays reinforced each other through interference, black spots resulted. The experiment had been successful! Crystals through which X-rays are directed produce dark dots symmetrically arranged in patterns of esthetic harmony: the so-called Laue patterns.

"Whenever the nature of a radiation is in doubt, one tries to cause interferences; if this succeeds their wave character is proved," wrote Max von Laue. The wave theory of X-rays and the atom theory of crystals both fell into place in this one experiment of 1912, in "one of those surprising breakthroughs that makes physics so convincing."

Two things had been proved: first, that X-rays are shortwave oscillations, invisible light of very short wavelength; second, that the theories about the structure of crystals were correct. Crystals had revealed the nature of X-rays, and conversely, X-rays revealed the basic structure of matter. From then on it became possible to pry out of solids the secrets of their components by means of shortwave radiation that reached far beyond the microscope and into the realm of the atom. And now it was also possible to investigate shortwave radiation by aiming it through crystals of known structure. The wavelength of the softest X-ray radiation is about one ten-millionth of a centimeter, that of the hardest about one billionth.

Two Englishmen, William Henry Bragg and William Lawrence Bragg, father and son, working in part together, in part separately, laid the foundation for the X-ray analysis of crystal structures. Srangely enough, the stuff of nature, even metals, is mainly composed of crystals. X-rays were proving once more something that Lavoisier had shown to the king of France when he burned a diamond with the help of a concave mirror. The hardest of the precious stones, the diamond, consists of carbon. In the diamond, the carbon atoms are densely packed. That accounts for its hardness. In the graphite of the pencil, in the soot of printer's ink, the carbon atoms are present in a much looser arrangement. The varying hardness of a metal depends on whether it has been cast, extruded, rolled, or hammered. X-rays can reveal the reason for this. The crystalline structure is different in each case. For example, iron heated above 906 °C abruptly changes its qualities. It alters its crystal structure. Empirical knowledge in steel production could be confirmed, new and better methods could be explored.

Before the end of World War I, Peter Debye, a Dutchman, and Paul Scherrer, a German, found a new way to X-ray solids. They ground them into fine powders and thereby also achieved exact pictures, not symmetrical dot patterns anymore as in the Laue patterns, but concentric rings. The X-rays in this technique are aimed at a small tube filled with the crystal powder. The interference rings are registered on a cylindrical film strip. The position and intensity of the rings permit conclusions about the crystals' structure.

Crystal analysis made rapid progress: the younger Bragg determined the structure of simple crystals like zincblende and pyrite; together with his father he solved the question of the diamond's structure. A group of scientists in Manchester, England, studied the structure of the silicates. The group around the older Bragg at University College, London, researched organic crystals like anthracene. X-ray analysis was able to unravel even as intricate a filigree as the texture of wool. This led to the wide field of biology. After World War II, scientists began to analyze the structure of giant molecules with the help of X-rays. They succeeded in determining the structure of vitamin B-12, an extremely difficult task because an asymmetrical molecule is involved that is composed of six elements and 183 atoms. Max Perutz and John Kendrew successfully X-rayed proteins, the basic stuff of life composed of tens of thousands of atoms. Francis H. C. Crick and James D. Watson analyzed hemoglobin, the blood pigment, and deoxyribonucleic acid, DNA, arriving at the smallest, extremely complicated structure that can still be considered a "living unit."

The Heart of Matter

Today, the concept "atom" is used by everyone. One speaks, as a matter of course, of nuclear physics and nuclear chemistry. At the turn of the century one was not so sure yet that what some learned men called "atom" really existed. Great naturalists of the nineteenth century, such as Faraday and Liebig, considered the proposition a mere theory. The Austrian physicist Ernst Mach was known to interrupt any discussion about the atom with the remark: "Have you seen one lately?" It was hard to imagine that the secret of what binds nature together in its core would ever be revealed. And it seemed altogether hopeless to expect being able to prove some day the existence of these theoretical atoms by making them visible.

Otto Hahn, the man who opened the door to the atomic age with his discovery of nuclear fission, has likened modern atomic research to the placing of one brick on top of another to build a house. In order to be completely accurate, one must place its beginning back in the years 1895 and 1896, when X-rays and radioactivity were discovered.

A lecture at the Paris Academy about the penetrating rays discovered by the professor from Würzburg, prompted the forty-three-year-old Antoine Henri Becquerel to try further experiments. He soon made a curious discovery: minerals that contain uranium emit rays that fog a photographic plate through light-tight wrapping. The rays "photograph" themselves; they draw their own radiant picture.

Becquerel's observation became the theme for the doctoral thesis of the newlywed Marie Curie, born Marja Sklodowska, for whom this represented the fresh subject matter she had been looking for. As a young girl in Poland, she had saved money with iron willpower in order to study in Paris. "We must believe," she wrote to her brother in Poland, "that we are gifted for something, and that this thing, at whatever cost, must be attained." Marie Curie studied various uranium ores and then—in pitchblende—she found something out of the ordinary: the radiation did not correspond to the uranium content of the mineral. It was four times as strong. The young woman concluded that in pitchblende there must be a yet unknown substance of potent radiation and tiny volume, if smallest amounts of this element—more could not have escaped detection in chemical analysis—would produce such effects! Pierre Curie abandoned his own work and dedicated all his energy and experience to his wife's research. Their search was crowned with success. In 1898, the Curies discovered a new element. They called it polonium, after the former homeland of the young woman. And then they found a second one, the famous element that they named radium, "the radiant one." The Curies used up wagonloads of pitchblende, barren rocks from the Joachimsthal mines in Bohemia, pitchblende that was provided free of charge, except for shipping costs. "It was exhausting work to transport the containers, to pour the liquids back and forth, and to stir the matter boiling in an iron tub for hours," wrote Marie Curie. "We lived as in a dream, completely absorbed in this one and only task."

In the biography of her mother (*Madame Curie*, Doubleday & Co., 1937), the Curies' younger daughter Eve later described how her parents one day returned to their shed and found the room suffused with a magical glow.

"Do you remember the day when you said to me, 'I should like radium to have a beautiful color'?"

The reality was more entrancing than the simple wish of long ago. Radium had something better than "a beautiful color": it was spontaneously luminous. And in the somber shed where, in the absence of cupboards, the precious particles in their tiny glass receivers were placed on tables or on shelves nailed to the wall, their phosphorescent bluish outlines gleamed, suspended in the night.

What an element! Its radiation penetrates ordinary receptacles; it must be encased in lead. It mysteriously heats up. It causes the air to have the property of conducting electricity. It discolors glass. It radiates with an intensity two million times that of uranium. Medical doctors would soon discover how important this radiation is in assisting them in their fight against cancer. Had the Curies taken out a patent on their discovery, they could have become rich. They were aware of this and they decided against it—as did, by the way, Röntgen.

In December 1903 Henri Becquerel and the Curies received the Nobel prize for physics "for their discovery of radioactivity." Thirty-two years later, the Nobel prize would be given to the Curies' older daughter Irène and to her husband Frédéric Joliot for "the discovery of artificially produced radioactivity." If exposed to radiation from radioactive substances, certain elements become radioactive themselves.

A flood of scientific work has been released by the discovery of radiating substances. The investigation of these initially puzzling radiations finally led to the revolutionary event of our time: the reach for the atom. In August 1895, Ernest Rutherford, a young man of twenty-four came to Cambridge from New Zealand to work at Cavendish, England's most prestigious laboratory. Röntgen's and Becquerel's discoveries were made during Rutherford's first year at Cavendish. He immediately turned to the study of these baffling phenomena. He found that uranium emits at least two kinds of rays; he called them alpha and beta rays, after the first two letters of the Greek alphabet. He studied their properties. A third kind, the gamma rays, he was to discover later.

Before that hapened, an opportunity presented itself from across the ocean. A research professorship in Montreal, Canada, had become available at McGill University, which owes its generously endowed institutes to the Canadian tobacco king McGill. Here Rutherford succeeded in "discovering a certain natural phenomenon—namely, the fact that radioactivity is caused by a change in the nature of the atoms, nothing less than the realization of the alchemists' ancient dream of changing one element into another," to paraphrase Frederic Soddy, who was Rutherford's assistant and later on a Nobel prizewinner in his own right. "Nature can be a grim jester. To think that hundreds of alchemists sweated over their ovens trying to change one element into another, dying while trying in vain to reach their goal, while Rutherford and I were lucky enough to discover with our first experiment that in thorium and other radioactive elements the metamorphosis proceeds spontaneously and that it cannot be arrested by any means now or ever."

The Reality of the Atomic Nucleus

Before Rutherford's return to England in 1907, a young German chemist, Otto Hahn, paid him a visit. He did not stay long, but the two men later exchanged many letters. It was to be Otto Hahn who, one year after Rutherford's death, split the uranium atom and set atomic energy free. In Manchester, Rutherford had encountered his predecessor's assistant, twenty-five-year-old Hans Geiger. He later would invent the Geiger counter which became the atomic age's indispensable tracing and listening tool for the detection of radioactive radiation. Geiger's experiments led to heated discussions. Rutherford's creative mind grasped more and more clearly the likely structure of the atom. Rutherford received a very useful hint from the German physicist Philipp Lenard who had observed a curious fact while working with electron rays.

The nature of these electron rays was still an enigma. Their history goes back to about the time the young Röntgen was thrown out of high school (for refusing to disclose the name of a fellow student who had drawn the caricature of a teacher on the fire screen in front of the stove!). An ingenious glass blower, Heinrich Geissler, created his famous Geissler tubes at the time: glass tubes, pumped almost empty of air. If one directed a high-voltage current through the very rarified gas, it magically glowed violet or blue or green. The English scholar Sir William Crookes experimented with them. He observed how the luminous beam was pulled away from the negatively charged cathode toward the anode, or else, how it produced a greenish fluorescent spot on the opposite glass wall when one increased the voltage further and thinned out the gas residue within the tube.

The next task was to find the secret of these cathode rays. A magnet placed against the outside wall of the tube could deviate the rays. A small wheel began to turn when brought into their path. And, as another scientist, Johann Wilhelm Hittorf, had already determined: platinum foil, when exposed to the focal point of the radiation, begins to glow. In the opinion of Crookes, the negative current pulled along molecules out of the cathode through the (almost) empty space. He spoke of "radiant matter." In 1897, Joseph John Thompson, of Cambridge, recognized in this radiant substance the negatively charged particles that were later to be called electrons. "The true hero of our century, the electron," as Arthur Eddington, the English astronomer and physicist, once said, had thus been identified. Philipp Lenard used these rays as projectiles and found that apparently there were tiny yet immensely powerful force fields within the still invisible atom that these rays could not penetrate. Rutherford, for his part, experimented with alpha rays, a beam of helium particles. He determined that when they hit those "impenetrable zones" they literally bounced off the target.

Tentative ideas gradually firmed up: every atom apparently had a dense core, the seat of a positive electric charge, and it was this charge that repelled the positively charged helium particles. It was the year 1910 when the atomic nucleus was thus discovered.

Two years later the Dane Niels Bohr showed how the negative charge might be arranged around the positively charged nucleus. To Rutherford's nucleus he added a picture of madly circling electrons, their orbits constantly shifting while maintaining always the same distance from the nucleus so that they appeared in the end like spherical "shells." This classical portrait of the atom still serves today as a demonstration model. It shows as many whirling negatively charged particles as there are positively charged ones in the nucleus. Maybe "whirling" is not the right word—they actually "race" around the atomic nucleus with unimaginable speed, a few million billion circuits a second. Such an electron is a lightweight, about 2,000 times lighter than the protons in the nucleus. But it has to be kept in mind that the image of the whirling particles is no more than a crutch to help our imagination.

Electrons cannot be explained that easily. Sometimes they act as if they were particles, sometimes like an oscillating "wave." They are, as we have heard already, something unknown which can manifest themselves in various ways. In such cases, theoretical physicists cover their blackboards with strange symbols and we are better off leaving it to them. To determine exactly where an electron is at a given time is impossible. One can only deal with probabilities, as Werner Heisenberg, the great theoretical physicist, stated in 1927.

At almost any time, Rutherford could be found bending over a Wilson cloud chamber, "a most original and wonderful instrument," as he called it. The cloud chamber, developed in 1911–1912 by the Scotsman Charles Thomas Rees Wilson, is a glass vessel filled with air that is almost saturated with water vapor like that of a bathroom just before the mirror clouds up. In this glass container it is possible to show the trajectories of electrically charged elementary particles, the tracks of atomic nuclei and of their components. For a very short time they leave a track of droplets, similar to the vapor trails of airplanes, traces thin as a hair that can easily be photographed. They are, one might say, the "crackling in the texture of nature" made visible. These tenuous tracks vanish almost instantly. Then the bathroom atmosphere must be restored within the Wilson chamber for the next picture: once again there are the traces, fine as sablebrush hair, that can be caught only on film. Rutherford never grew tired of photographing these trajectories. They constituted documentation. What was it that Ernst Mach grimly used to ask whenever atoms were mentioned: "Have you seen one lately?" At that time the answer was still: Regrettably, not yet; but its traces can be seen, and from the traces certain conclusions can be drawn. To use the language of the criminologist: the suspect can be described but not yet apprehended. One could not say, when the camera had recorded such a track in the cloud chamber, "There *is* an atom!"—only, "Here *was* an atom!"

Reaching for Nuclear Energy

In 1919, Rutherford once again scrutinized such trajectory pictures. One such photo stopped him short: one of the hair-thin particle traces forked. A helium projectile obviously had collided with an atomic nucleus of the air's nitrogen, thereby shattering it. Rutherford had smashed an atomic nucleus. It turned out that the energy of the shattered parts was greater than the energy of the projectile. Part of the energy must therefore have been contained in the nucleus. True, such nuclear hits are a matter of chance and the energy released by the infinitely small is also tiny, barely measurable. Anyway: the minute power field within the atom, a fortress, can, after all, be penetrated—that much had become certain.

Ernest Rutherford (after 1931 Lord Rutherford, a member of the House of Lords) did not dare to predict that it would become possible to make use of the forces within the atomic nucleus. Journalists interviewing him always returned to this question. "Hold it, please," the sexagenarian would caution, "we will most likely never be able to do that." Indeed, Rutherford, whose ashes were entombed in Westminster Abbey in 1937, did not live to see it happen. The decisive experiment was carried out one year later in Berlin. By that time Otto Hahn, once Rutherford's assistant, had been director of the Kaiser Wilhelm Institute for Chemistry for five years. Otto Hahn is considered the master of "chemistry of the unweighable."

There are, of course, microscales that respond to even a fraction of a millionth of a gram. And there are spectroscopes capable of identifying chemical elements. The principle of spectrum analysis can be impressively demonstrated in the laboratory by spraying a solution into the stream of gas that feeds a burner. It colors the flame. Sodium turns it yellow; potassium, red. And the light emanating from the substance in the flame can be spectrally analyzed. Characteristic lines in the spectrum identify the substance even if there are only unimaginably tiny quantities involved. A finger, immersed in pure water, leaves a trace of common salt through the natural excretion of the skin. One part salt in ten billion parts of water can be detected today with the appropriate spectrometer. Here optics can reveal the hidden without there being any "magnification" involved.

Where even spectral analysis cannot find any traces, there is, in the case of radiant substances, the possibility of detecting the looked-for matter with the help of just this radiation. The discovery of "artificial radioactivity" by the Joliot-Curies makes it possible to render substances radioactive that do not originally have this property. Made radioactive, they become clearly identifiable even though they may be only a very few atoms, actually unweighable, and completely and totally invisible.

Toward the end of 1938, Otto Hahn and his young collaborator Fritz Strassmann set up on their worktable an experiment with a radiating substance. It continuously shot off uncharged nuclear building blocks, neutrons, and that was what it was meant to do—namely, serve as a neutron cannon.

Otto Hahn bombarded atomic nuclei of uranium with neutrons. They did not ricochet; they penetrated and caused the heavy uranium nuclei to split into two medium-weight atomic nuclei—as Hahn and Strassmann concluded after discussing and evaluating the results of their measurements—and thereby other neutron projectiles were expelled which in turn were capable of further splitting uranium nuclei. One explosion caused the next, as in the case of a fire in a munitions dump. Something Hahn and Strassmann initially could not fully appreciate, became increasingly probable: the fact of the "chain reaction." A giant, held captive by the infinitely small, can be freed; a new kind of fire has been gained. This new fire, in the form of a bomb, was to make the earth tremble—tamed, it would yield atomic energy. The worktable on which the splitting of the uranium atom was first accomplished is today one of the more valued possessions of the Deutsches Museum in Munich. In what modest circumstances a new era may make its entry! A few batteries, a few wires, clamps, tubes, a block of paraffin, a small tube holding the radiant substance which served as the neutron launcher, a small Geiger counter. That was all. The earth-shaking event took place in the silence of a laboratory, on a wooden table of less than two square yards.

Electron Microscopes

In 1924, the French Nobel laureate Prince Louis de Broglie postulated the following: the electron that spins around the atomic nucleus is not just a particle as in Niels Bohr's "planetary model." It also has characteristics of a stationary electromagnetic wave—it vibrates instead of whirls around the nucleus. A short time later, the validity of this theory could be demonstrated, and it became evident that electron rays have a much shorter wavelength than visible light. This suggested the construction of an instrument that would make use of electron rays to show the most minute structures.

The brilliant basic concept for the electron microscope is the substitution of the much finer shortwave electron rays in place of the far too "coarse" light rays of the light microscope. Glass lenses are replaced by electric or magnetic fields.

In 1926 the German physicist Busch invented technical equipment for concentrating electron rays on one point, the way a burning glass focuses light rays. Inventive technicians worked out the best way to direct the electron rays for achieving the mathematically calculated optimum result. In fact, the new instrument's resolution is about a hundred times that of the light microscope.

Essential equipment for an electron microscope is a vacuum pump and a source of high-voltage electricity. Briefly, the microscope consists of a vertical metal tube, pumped practically free of air. The specimen is introduced through an air lock. The beam of electron rays, invisible to the eye, is directed downward from the top. A fluorescent screen excited to glow—just as during an X-ray checkup—pictures the specimen. The telescopic magnifying glass for viewing is directed toward the screen: a new world emerges. It may be more alien than anything we will ever encounter on the planets and moons of our solar system. If we want to find shapes comparable to those appearing under the electron microscope, we must look for images in our everyday world, although it has an entirely different scale of magnitude. The lightly etched surface of aluminum looks like a bird's-eye view of a slate quarry; zinc oxide, that we know as an ointment, looks like a thorny thicket.

Regrettably, electronic rays are incapable of deep penetration; they can, therefore, transilluminate only extremely thin and tenuous objects. The roller-shaped body of a rodlike bacterium appears only as a dark sack. But other features do reveal themselves: some bacteria look "hairy"; whiplike appendages appear as sharply defined as if they were incised by a knife.

Meanwhile, technology has achieved even more finely differentiated pictures. The protein inside the bacterium may be drained out, or one waits for it to be eliminated by "digestion." Only the shell is left and that then can be transilluminated. It looks like a delicate net. Or the bacterium can be embedded in plastic and slices as thin as a fraction of one thousandth of a millimeter can be cut off. In 1952, a microtome was constructed that is able to cut ultra-thin slices of no more than one fifty-thousandth of a millimeter in thickness. The specimen is fixed on a steel rod; the latter is heated and its expansion brings the specimen up to the blade, a "specially ground razor blade." Today, one cuts with plate-glass splinters or diamond knives. For the unaided eye the slices are invisible; only when they float on top of the liquid in the receptacle do they become visible. A bacterium of a one thousandth of a millimeter thickness can be cut into fifty slices. Such fine cuts are penetrated "effortlessly" by electrons. One can perceive that bacteria are not just enzyme-filled "sacks" but that they have an extremely interesting structure.

The Enigmatic Viruses

The electron microscope has also revealed those strange pathogenic organisms which can penetrate bacteria-proof filters and which, in contrast to bacteria, cannot be grown in an artificial culture-medium, but multiply only in living tissue. These agents, which remained undetected under the light microscope, were given the name of virus (from the Latin "poison"). In 1939 it became possible for the first time to show under the electron microscope the virus that causes a plant disease, the tobacco mosaic virus (TMV), a rod-shaped crystal, similar to a tipless matchstick. Since then, the electron microscope has revealed the most diverse virus shapes: the smallpox virus which looks like a rounded-off cube; the polio virus, small balls of one fifty-thousandth of a millimeter in diameter that form latticed clusters. Unusual three-dimensional pictures have been achieved: the rods and balls cast deep shadows, as if they had been surprised by the brilliant light of a sun and become transfixed in that instant. Here a special trick is used: the specimen inside the vacuum tube is coated with an unimaginably thin layer of metal, generally platinum or gold. The thus encrusted specimen stands out sharply against the background, but it is dead. Heat and vacuum exclude the observation of living organisms under the ordinary electron microscope. However, the first successful observations of living microorganisms have been made. For the first time bacteriophages ("bacteria eaters"), the presence of which was long suspected by biologists, have become visible under the electron microscope. They belong to the viruses. The little stem with which they attach themselves to bacteria is not a "suction hose" as was initially assumed, but a hypodermic needle with which the virus injects its own substance into the body of the bacterium.

The electron microscope makes possible a magnification of 500,000 times. A resolution of about 3 angstrom (1 angstrom, named after the Swedish astronomer Ångström,=two-hundred-fifty millionth part of an inch) has been achieved. That goes far, but not far enough, if one keeps in mind the ultimate goal—the atom.

At Last: Atoms Made Visible

A new instrument, called the field electron microscope, appeared at the Paris International Congress for Electron Microscopy in 1950, and caused a sensation there. Slides of the giant molecules of polystyrene, a synthetic, had just been shown: everyone was greatly impressed by the advances in electron microscopy. Immediately after that, Dr. Erwin W. Müller from Berlin-Dahlem stepped up to a small makeshift apparatus, his field electron microscope, the plans for which he had worked out as early as 1936. He attached a clamp for the anode wire; he switched on the high-voltage current. Only those seated close by could see what appeared on the picture tube of the instrument. Another manipulation. Then Dr. Müller projected onto a large screen what simultaneously took place on the picture tube: the swarming molecules of a dyestuff became visible, each a thousand times smaller than those of polystyrene. The audience was electrified. Shortly after, in an atmosphere of hushed tension, Erwin W. Müller showed during another experiment—atoms! Barium atoms! For the first time—and this in a lecture hall—the building blocks of matter had become visible.

Erwin W. Müller, a professor at Pennsylvania State University since 1952, calls the principle of his instrument "surprisingly simple." The apparatus is essentially only a vacuum glass tube, the base of which shows the picture much as does a TV picture tube. Inside the container, located opposite the fluorescent screen, there is a very fine tungsten wire point. In front of this point, field strengths of up to fifty million volt per centimeter may be created that cause electrons to shoot off the wire point; they spread out and produce, where they hit the fluorescent screen, a correspondingly enlarged picture of the events at the wire tip. This consists of one single crystal; it is a thousand times finer than the point of a needle and appears enlarged to the size of the fluorescent screen. Onto this wire tip one can vaporize traces of those substances whose building blocks one wants to see. Dr. Müller had vaporized barium, the ghost of a trace of this silvery-white metal; its atoms can be seen as points of light that become greatly agitated when the wire tip is electrically heated. It is a glimpse into the inner structure of the universe.

But even this was not yet the final advance. Meanwhile, Erwin Müller has created the field ion microscope. It is the crowning achievement of a twenty-year effort. Again the metal point performs its function, but now it is positive ions, split-off atomic particles, that picture the events at the wire tip on the fluorescent screen. With this instrument it soon became possible to distinguish points at a distance of 4 angstrom from one another. Professor Müller then found further means for increasing the capability of his microscope; he cools it with liquid nitrogen. He placed it into a transparent thermos bottle. At a temperature of minus 192 °C, pictures of fantastic magnificence reveal themselves. There appears the atoms' latticelike arrangement within a crystal: chains of light, strings of pearls, radiant reflection of the order that sustains the world's innermost structure. Let us observe Erwin Müller as he makes his preparations for such photographs: he selects a wire with an extremely fine point; he introduces it into a Braun tube. The necessary implements seem relatively simple. The high-voltage wire is connected; the vacuum pump is switched on to evacuate the picture tube. Liquid nitrogen for refrigeration is applied; the processes are slowed down and a sharper picture at the base of the tube results. Helium gas is introduced; the camera is set up; the gauge for the pressure inside the tube is being watched. Lights out! A last fine-tuning and—on the picture tube appear the building blocks of matter, revealed by the pattern of the crystal lattice. In the field ion microscope, it is the ionized helium atoms pushed by the atoms shooting out from the wire tip that hit the fluorescent screen and make it light up, revealing the atomic structure of the metal point magnified ten million times! Nature sketches itself. It reproduces the blueprint for the structure of a metal on the picture screen.

The learned men of antiquity "saw" the atom in their mind's eye. In the twentieth century, scientists have been able to reassure themselves that atoms are a reality, even if they are not, as their name "atomos" would indicate, indivisible. Only now can they be *seen*; their existence is proved to the eye. The most minute structure in the world, the arrangement of atoms, is a thing of mathematical beauty.

Once again we have come to a—temporary—halt. Microscopes, beginning with the magnifying glasses around 1600, have brought the microcosm layer by layer to the level of the human eye. Or, to use another metaphor, we have descended the rungs of a ladder: the light microscope reaches a magnification of about 1,000 times, the electron microscope up to 500,000 times, field electron and field ion microscopes up to five million and even ten million times! It began with the hair of a flea and within three and a half centuries arrived at the depth of the atom.

Man-Made Storms of Energy

Today, physicists who study the atoms use mighty machines. Gigantic installations are employed to force nature to yield up her secrets. Modern nuclear physics calls for heavy "artillery" and electronics provide it. The speed of the radiating particles can be increased by sending them through electric fields. A proton passing through a field of ten million volts attains a speed of 44,000 kilometers a second; it could chase around the globe in one second. The greater the speed of such projectiles, the greater their power to penetrate their targets of heavy atomic nuclei. Physicists speak of electron volts. An electron volt is the amount of energy that a single electron gains when it is accelerated by an electrical force of one volt. A tiny amount of energy concentrated within a tiny particle. And yet: at the firing of an infantry bullet each single lead atom receives about that amount of energy. Atomic physicists require very much more, they call for a million, even a billion, times that power of penetration. A billion electron volts could be achieved with a billion volts. But also in this way: on a circular course, atomic particles are accelerated. When they have traveled a thousand times through a field of a million volts, the result—one billion electron volts!

The European atomic center, a joint effort of twelve nations, located in Meyrin near Geneva, Switzerland, is called CERN (Centre Européen pour la Recherche Nucléaire). The institute has some of the biggest accelerators in the world: a synchrocyclotron (600 million electron volts) and a proton-synchrotron, which accelerates up to 28 billion electron volts. Particles, building blocks of atomic nuclei, are accelerated in a subterranean "race track" of 200-meter diameter. They are bounced into the circular evacuated tube. A hundred electromagnets accelerate them more and more, until they race around the track 500,000 times a second, at almost the speed of light. Finally a window is opened to the racing particles into a space where they hit their small targets; the effect on matter, of an attack by such enormous energy, can then be studied. The power of the magnets is fantastic; it fills the whole test-room. One gets a timely reminder to take off one's watch. A nailfile placed vertically on the tip of a finger remains upright. A bunch of keys in one's pocket springs to life. The coins in a change purse cluster together like grapes. The layman considers it fun, but this display of technological might serves ambitious scientific aims. The energy storms that roar around our sun are being copied here.

Not long before World War I, the physicist

Albert Gokkel, of Freiburg University in Switzerland, returned from a balloon ascension with data that broke ground for one of the greatest discoveries of modern physics. He had found that out of the depths of space comes radiation more energy-laden than the hardest X-rays and so penetrating that its presence can be established at hundreds of feet under water and in the galleries of mines. Today it is thought that the question of the mysterious rays' origin can be answered. They originate, theorizes the English Nobel laureate Cecil Powell, in the stellar explosions called supernovae; they are the atomic "hailstorms" of cosmic catastrophies. Fortunately, this radiation is being attenuated by the gas cushion around our earth to such an extent that it not only does not harm life but may actually be beneficial. Cosmic rays trigger a series of nuclear processes in the atmosphere that result in the so-called mu-mesons that reach far down into the earth and water. Similar mesons are found in the atomic nuclei where they oscillate between protons and neutrons, activating the nuclear force and functioning as a sort of "nuclear glue," holding the building blocks of the atomic nuclei together. Free mesons live but a short time, some of them not even a millionth of a second. That is why they were hard to find; but two years before their discovery, the Japanese physicist Hideki Yukawa had already predicted their existence on the basis of mathematical calculations.

Balloons, carrying photographic plates shielded from light, have been launched to record processes on the threshold of space. The plates were returned by parachute. Radiation particles had registered their transits in the emulsion. Sometimes the particles smashed an atomic nucleus in the emulsion; the scattering fragments left a "star" imprint and from the shape of these "stars" it was possible to reconstruct what had taken place. We are familiar with the fact that matter is transformed into energy; our sun burns up millions of tons each second (and does not lose even one thousandth of its mass in ten billion years!). Now, the opposite process has been registered in "black on white"; energy "condensed" into matter. From energy-rich radiation, particles of matter originate. The process could be considered an act of creation. Time and again, photographic plates and cloud chamber pictures show that there exists quite a number of "elementary" particles. More than thirty kinds have been traced "in which energy manifests itself," as Werner Heisenberg puts it. And this catalog seems still incomplete. One may well conjecture that in the end there will remain one single "proto-substance" from which the diverse elementary particles in various ways derive. The bewildering multiplicity in physical nature would then be restored to a higher simplicity.

Blueprint of the Universe

Werner Heisenberg and the late Professor Wolfgang Pauli of Zurich jointly tried to formulate "a unified field theory" of elementary particles, a "world formula" as it was called. The mathematical symbols were immediately reproduced with due respect, but even university professors of physics were not able to interpret the formula at the first try. Carl Friedrich von Weizsäcker, for his part, elucidated his theories in 1967. These are the first, as yet tentative, steps of the great theoreticians groping their way toward an answer. The discovery of the mathematical blueprint of the world could become the greatest triumph in our time of human capacity for abstract reasoning and ingenious experimenting. The solution of this riddle may just possibly succeed with the help of those gigantic energy boosters. In 1958 in Geneva, Werner Heisenberg said, it may happen "within a few years or at the most in a few decades." Should it come about, then surely, photographic documentation is bound to follow. Pascual Jordan, a German physicist and one of the pioneers of quantum mechanics, once said: "Seeing and understanding are inseparably linked. The intellectual conquest of nature is largely accomplished by making visible things that previously were invisible, and that can be made visible only through the perspicacity and the skill of experimenters and technicians."

"Light is an electric wave, light per se, light of the sun, that of the candle, that of the glowworm. ... Remove electricity from the world and light will disappear. ... We now notice electricity in a thousand places where, before, we had no certain knowledge of its presence: in each flame, in each luminous atom, we see an electrical process...."

(Heinrich Hertz)

ESSO AG, Hamburg

The Hamburg swimming–pool reactor. Around the block of uranium rods, the heart of the reactor, the water shimmers in an unearthly blue: the so-called Cerenkov radiation. The discovery of this effect by Pavel Cerenkov and two other Russian physicists earned them a 1958 Nobel prize. Electrically charged elementary particles, building blocks of the atom, are being scattered in all directions; the particles move through the water at a higher speed than that of light in water. As a boat moving through water at great speed creates a bow wave, so the speeding particles cause a wave of Cerenkov radiation. A remarkable sight: atomic energy, whose eruption we fear, gives the impression here of a tamed animal performing a trick in its cage.

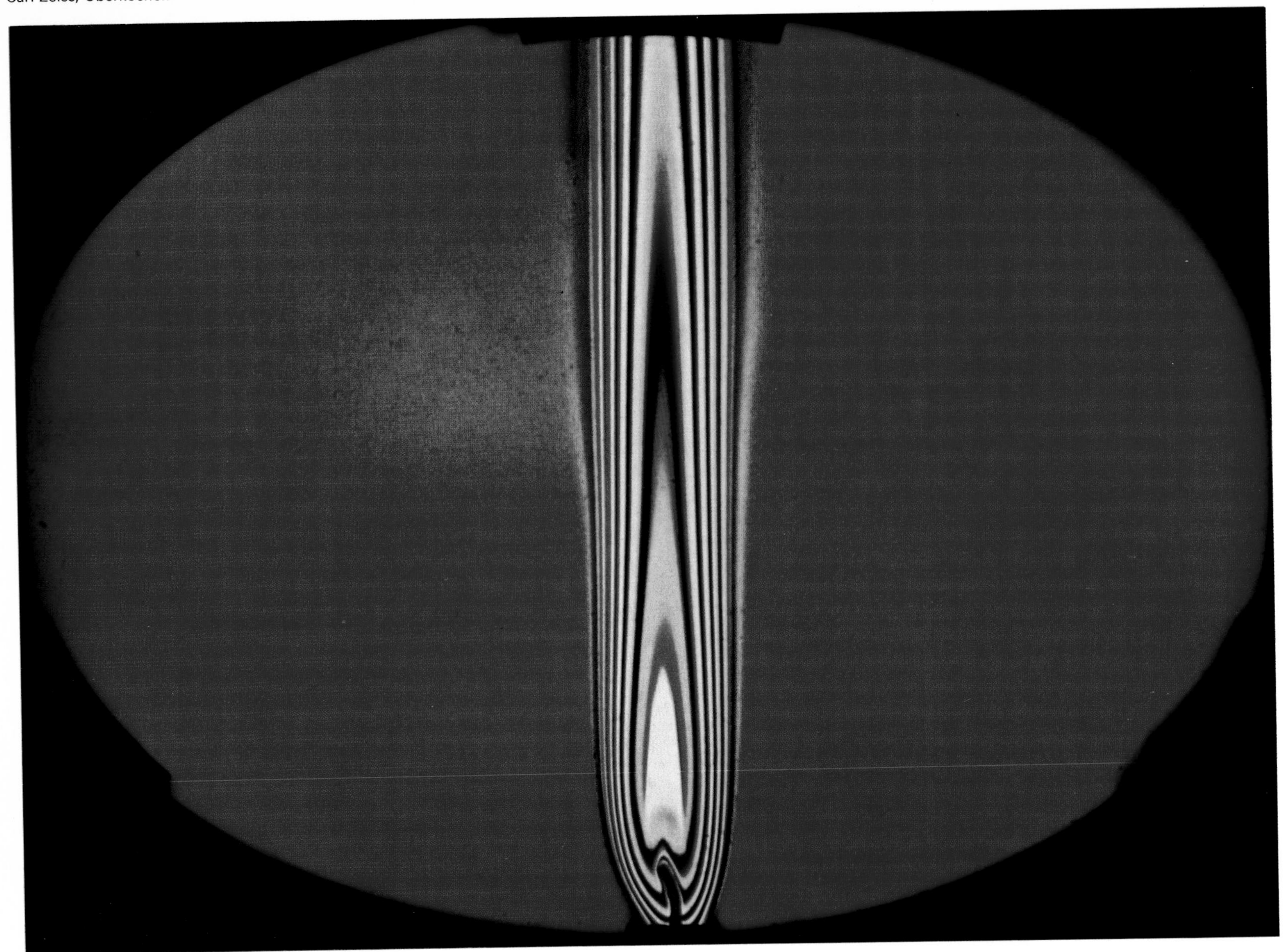

Flame of a candle in the interferometer. In an interferometer the beam of light is being divided; the separated beams are directed along separate paths and then reunited. On a screen the effects of overlapping light waves, the interference, appear as stripes. With an interferometer, gas density, pressure, and temperature can be made visible and measured.
The formerly invisible emerges as if by magic: the hot gas stream of the flame is being analyzed by the interferometer through captivating designs that are layered in zones of radiant color.

S. Looser, Munich

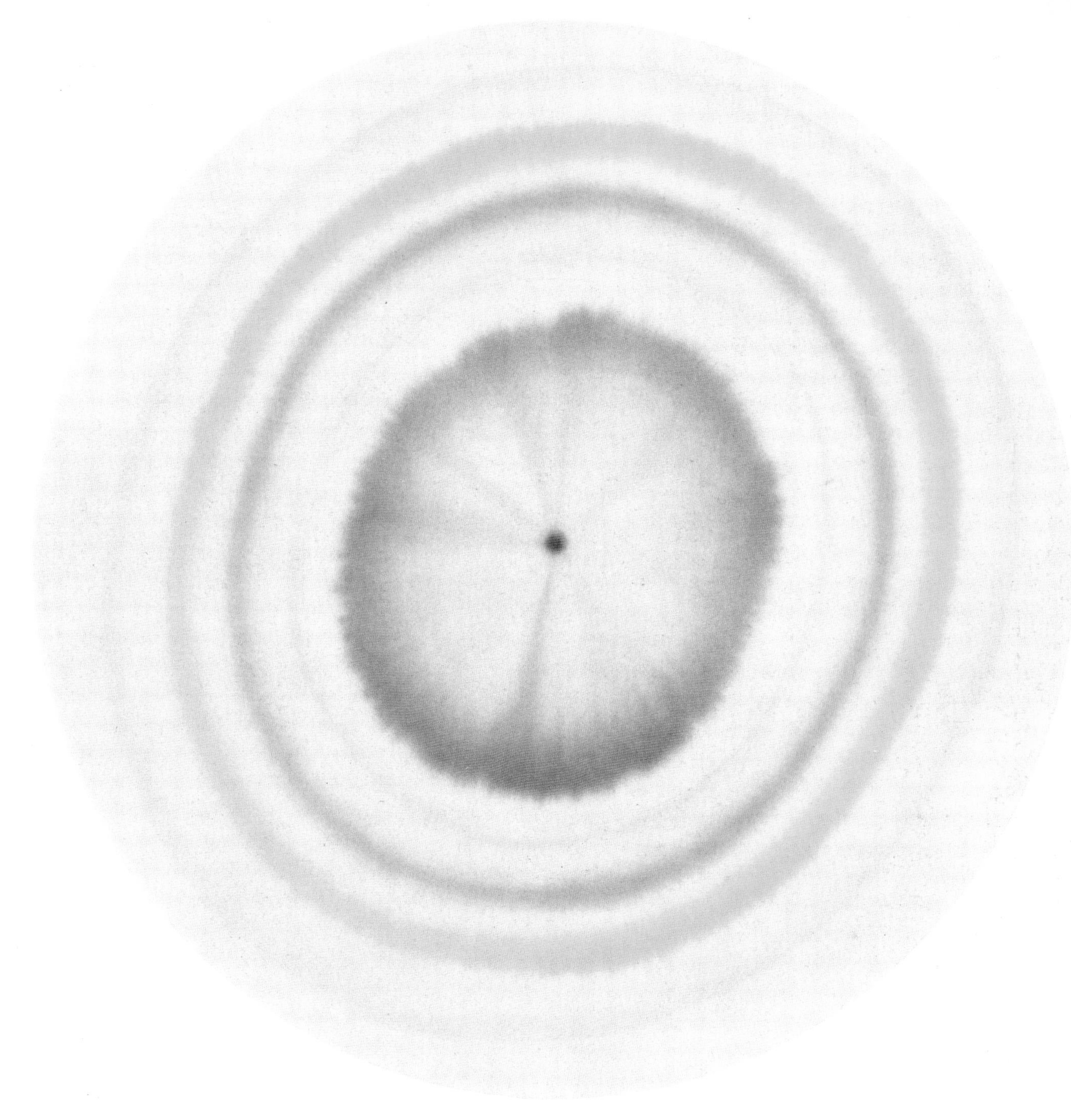

Paper chromatography
serves chemical analysis.
It is capable of sorting out
compounds that are very similar.
One drop creeps outward on
the filter paper. Clearly, one
compound travels farther than
the other. In the chromatogram
we find the constituent parts of
a solution pulled apart, "fanned
out." The method was refined
in 1944 by the Englishmen
Martin and Synge, but the
process had been used earlier
in Germany and Russia.

The electron microscope is over a hundred times more powerful than the light microscope. In the electron microscope the specimen is introduced through an air lock into a vacuum tube. The telescopic lense in the eyepiece is trained on the fluorescent screen.

Carl Zeiss, Oberkochen

Here a specimen is being prepared for the electron microscope. In a high vacuum tube it is being coated, by vaporization, with a platinum layer of 1/200,000 millimeter. This helps to bring out the surface structures of, for instance, a virus.

Carl Zeiss, Oberkochen

CIBA AG

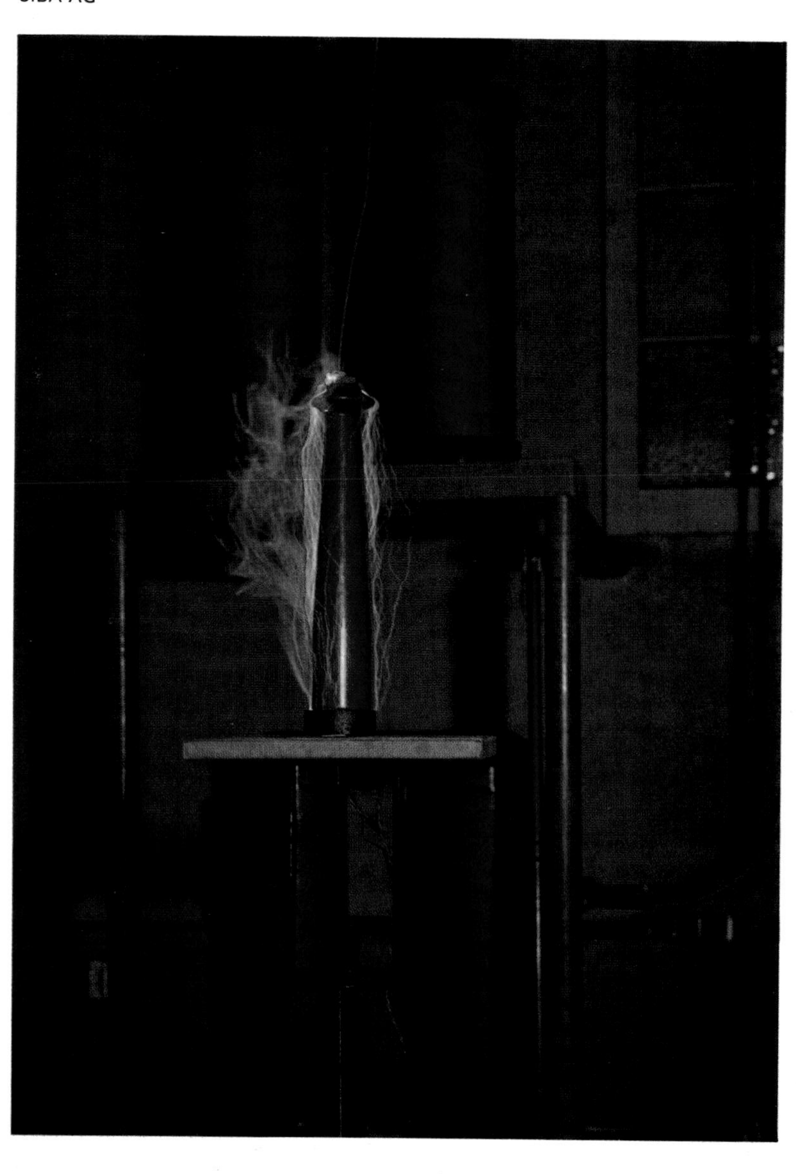

◀ Along the sides of an insulator, made of a synthetic material, electricity flows. The "instant" of spark-discharge has been stretched by stop-motion photography to such an extent that we can observe the "runnels" of the current's flow. High-voltage electricity has to be safely transported and distributed across continents: it should not "escape" anywhere along the way. Experiments in high-tension laboratories ascertain the safety limit for such transmission.

The meshing of two gears. ▶ Test gears of synthetic material are being operated under polarized light. The transmission of power, the passing on of energy, can be observed in the color play of moving shapes. Material and design show themselves to be satisfactory or in need of improvement.

Fritz Brill, Hofgeismar

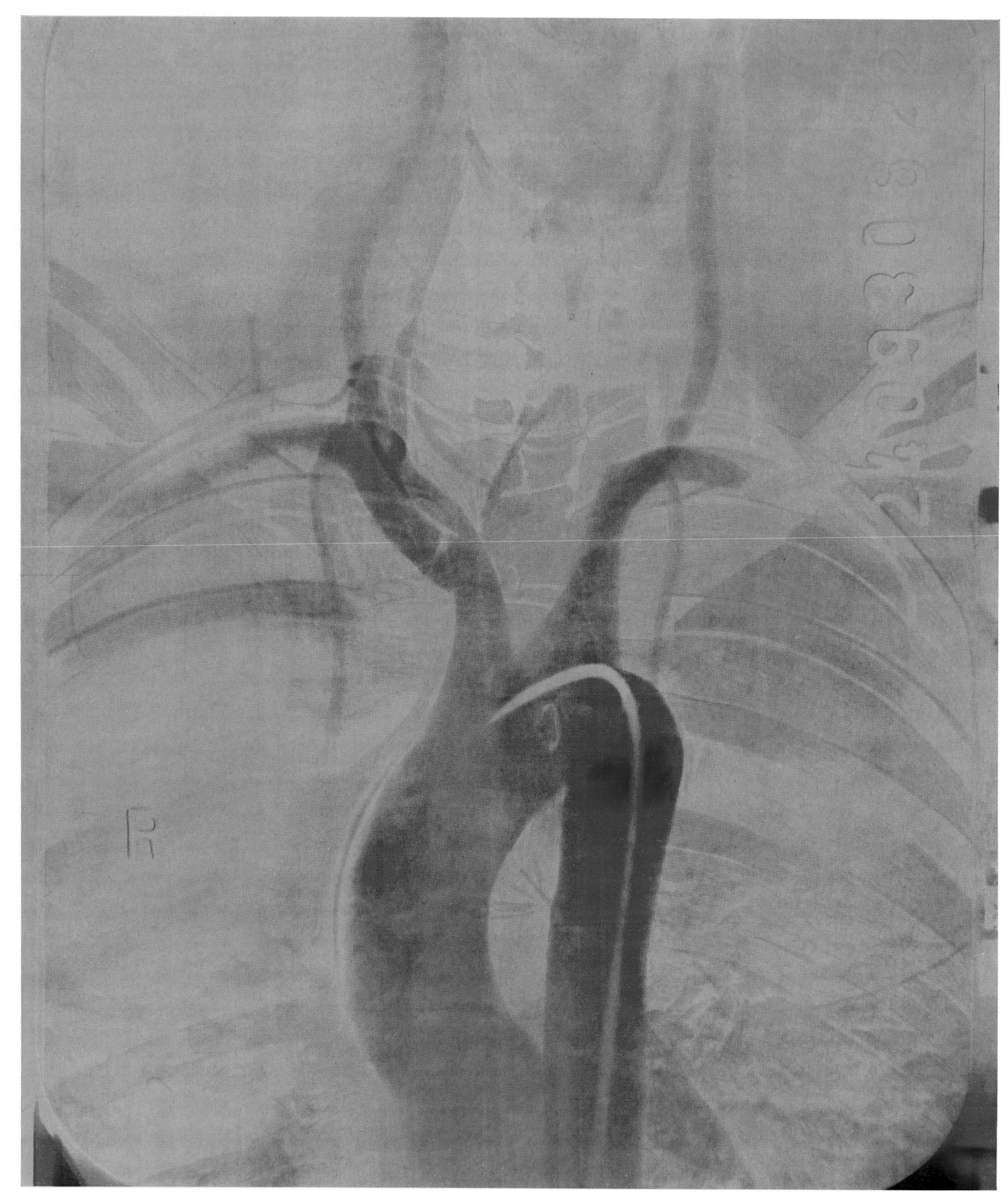
R

Modern physicists have discovered invisible "relatives" of light. They have penetrated our environment with artificially created waves and radiations, and have forced nature to reveal the previously unobservable.

For seventy years X-rays could show only "shadows," grays ranging from black to white. Now, colored X-ray pictures have become feasible. Our picture shows the aortic arch coming out of the left ventricle of the heart, and the blood vessels leading toward the brain. The picture-taking technique, somewhat simplified, is the following: two X-ray pictures are taken from the exact same position, one a "normal" one, the other one taken after a contrast fluid has been injected through a catheter into the aorta. Using a color TV-camera, both pictures can be "colored." Electronically superimposed, they produce a two-color picture on the screen.

"I did not think;
I investigated.... It was
clearly something new,
something unrecorded."
(W. C. Röntgen)

How Röntgen investigated X-rays he described in a report to the *Physikalisch-Medizinische Gesellschaft* of Würzburg. How he discovered them has remained a secret. The hand of the Würzburg professor of anatomy Köllicker that Röntgen transilluminated created a worldwide sensation.

Deutsches Röntgen-Museum, Remscheid-Lennep

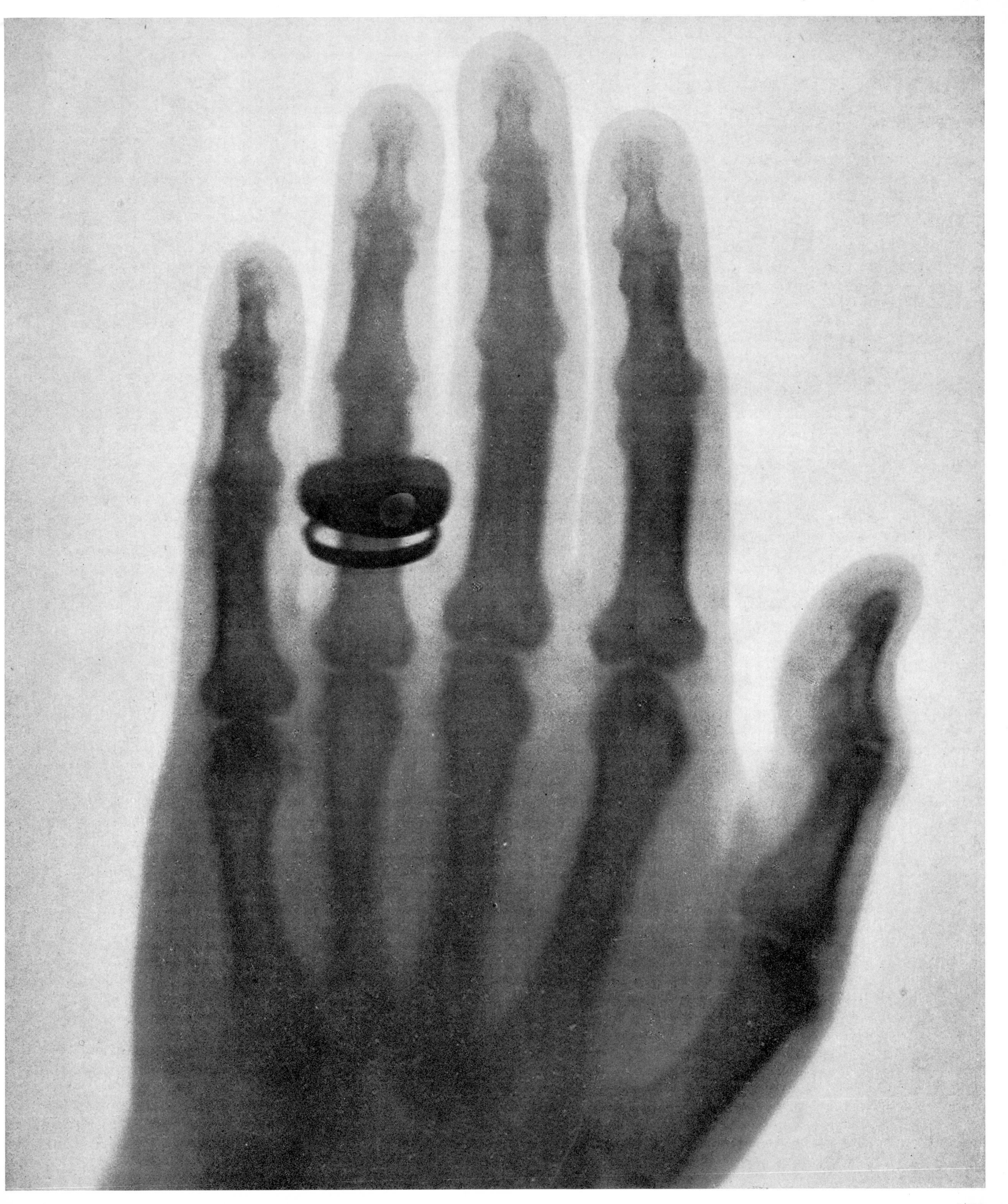

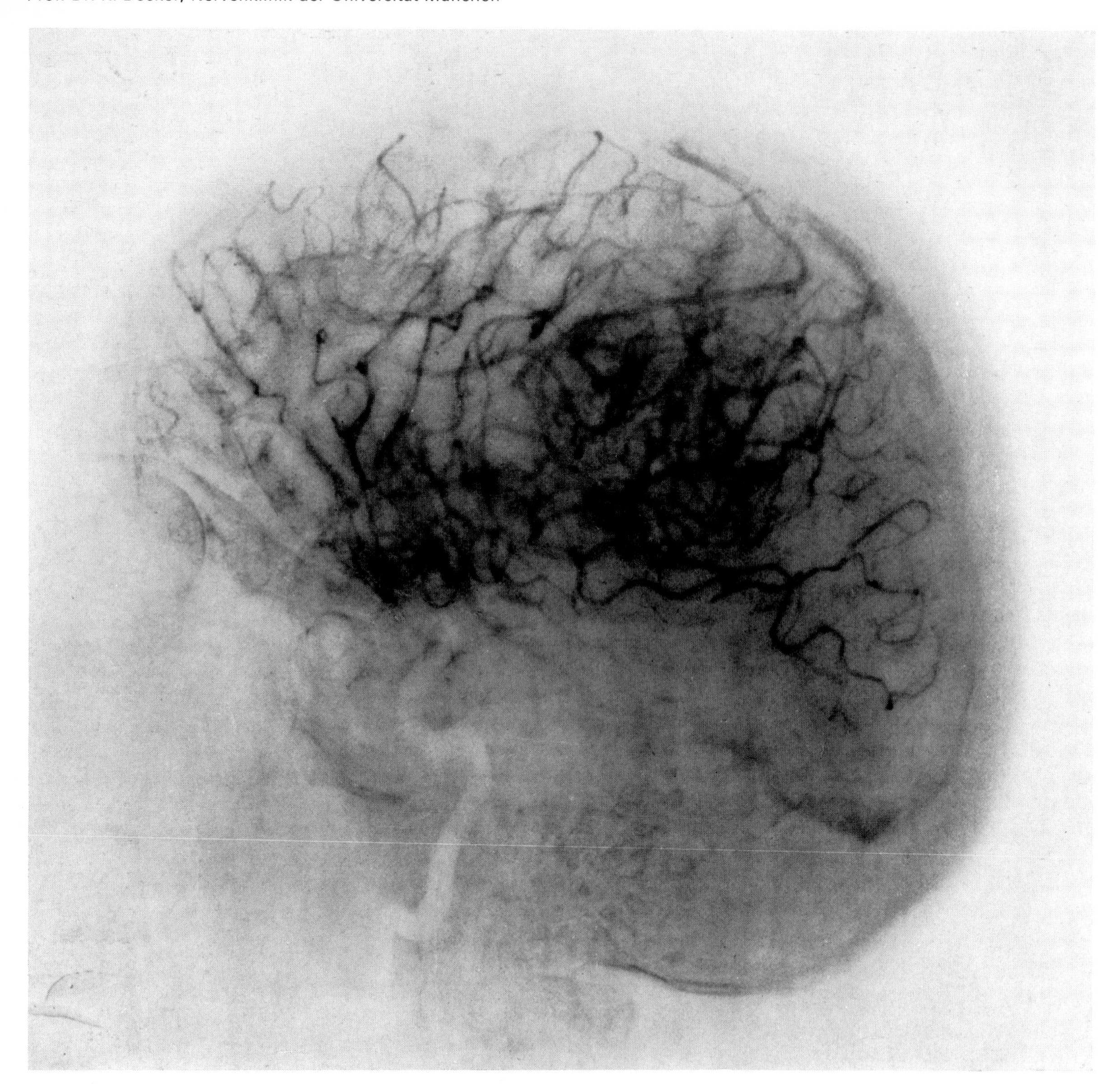

◄ Through a modern technique, the skull is made transparent and the vessels and a brain tumor have become visible.

The X-ray picture of this skull shows how the technique has been refined during the more than seventy years of its use. In front and in back of the skull bones, tissues of varying ability to absorb radiation show up as finely graduated shadows. There exist numerous methods by which X-ray techniques can give prominence to one or the other of these shadows. ►

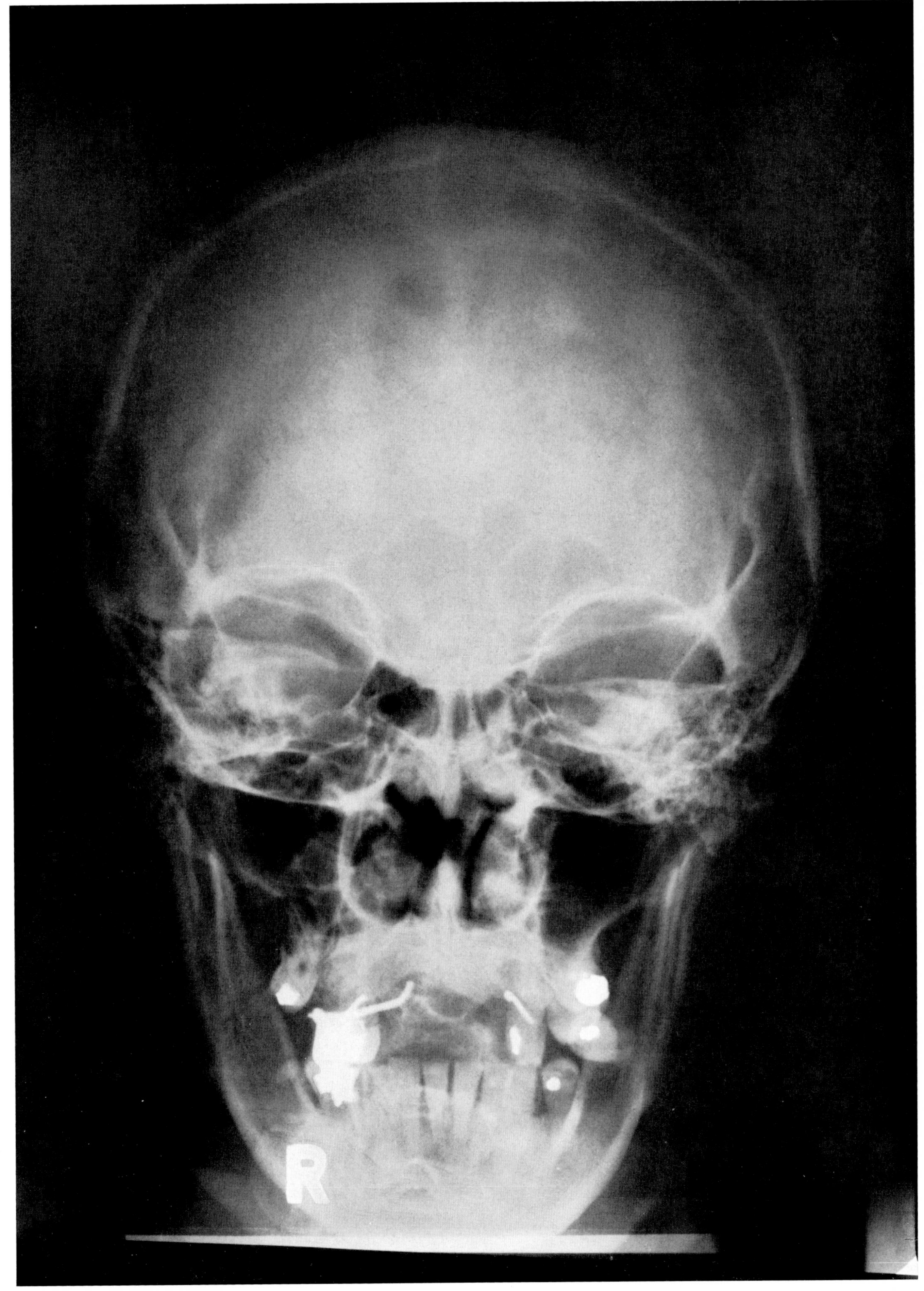
R

Haus Siemens

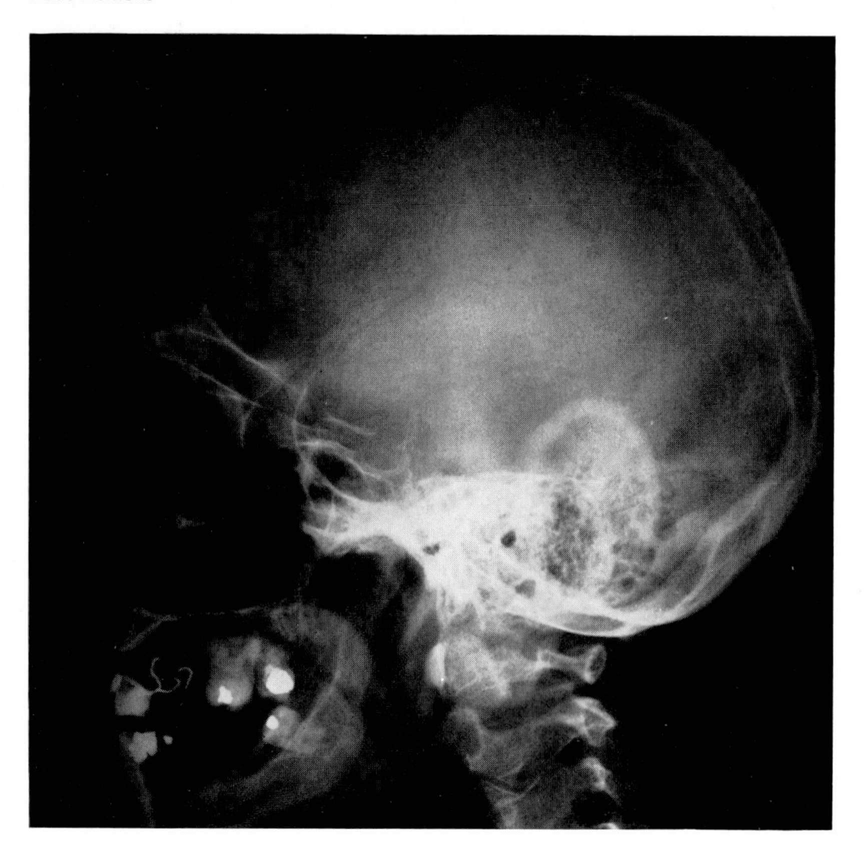

The number of lives saved by X-rays surpasses the number of lives lost in both World Wars.

Prof. Dr. med. F. Loogen, Düsseldorf

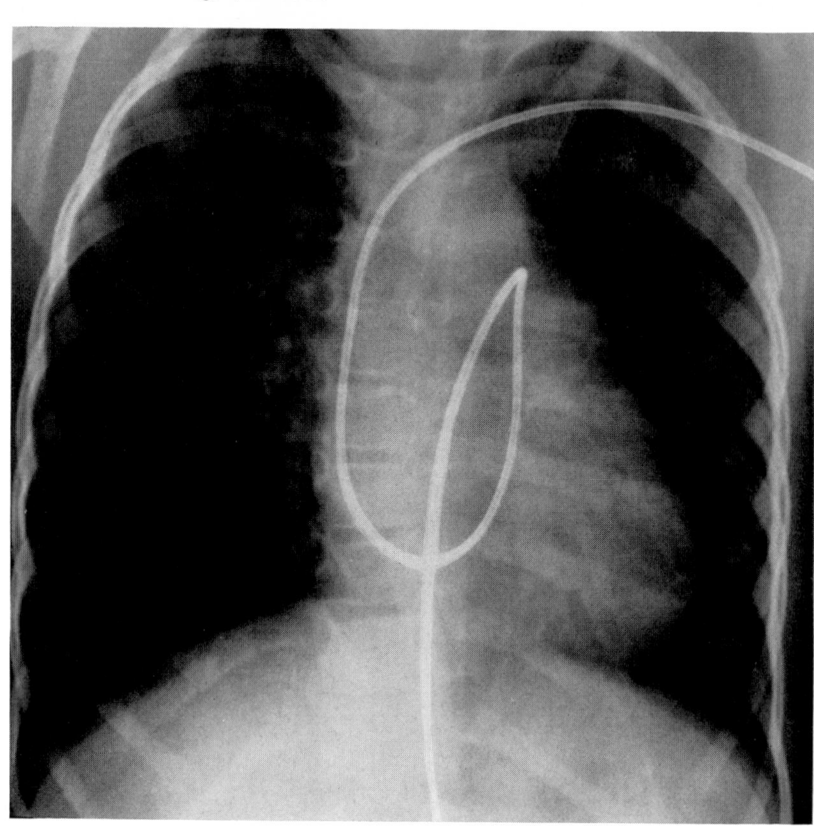

A catheter has been inserted through the arm vein twenty-nine inches deep into the heart. Werner Forssmann was able to check on the X-ray screen the results of this experiment made on his own body in 1929.

Haus Siemens

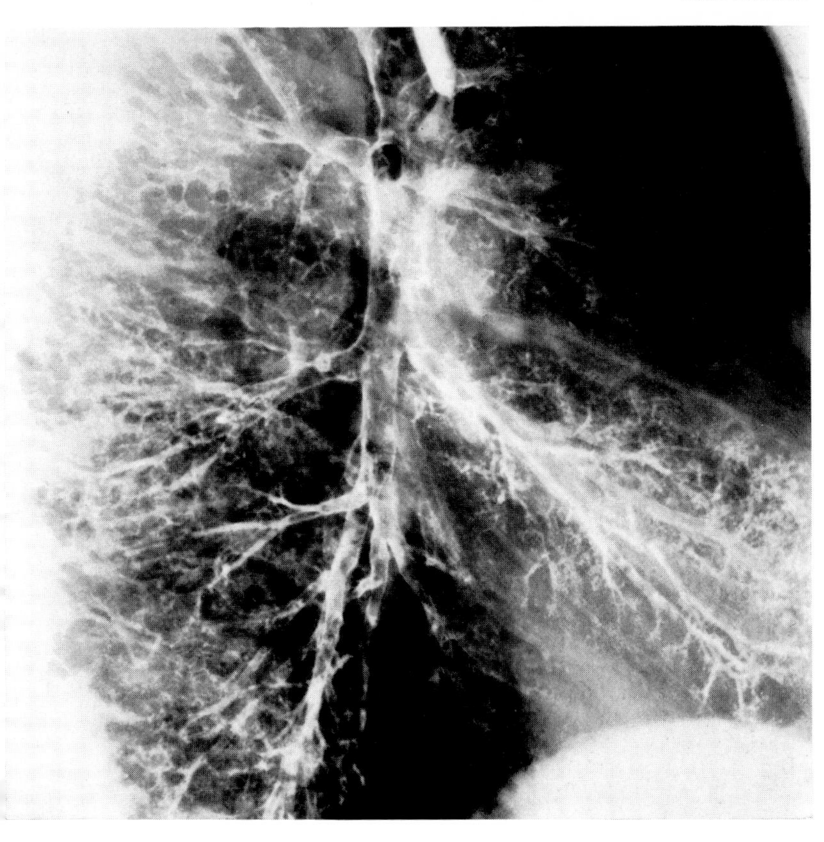

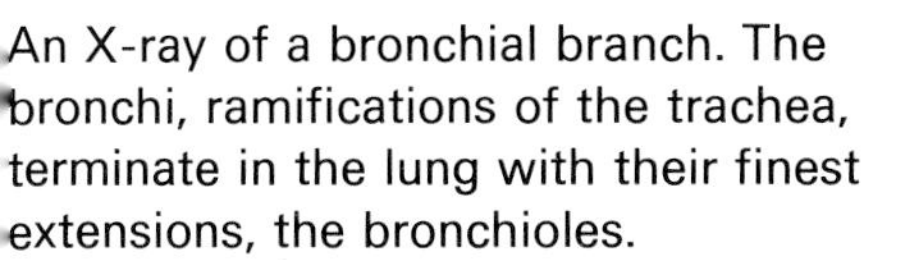

An X-ray of a bronchial branch. The bronchi, ramifications of the trachea, terminate in the lung with their finest extensions, the bronchioles.

Haus Siemens

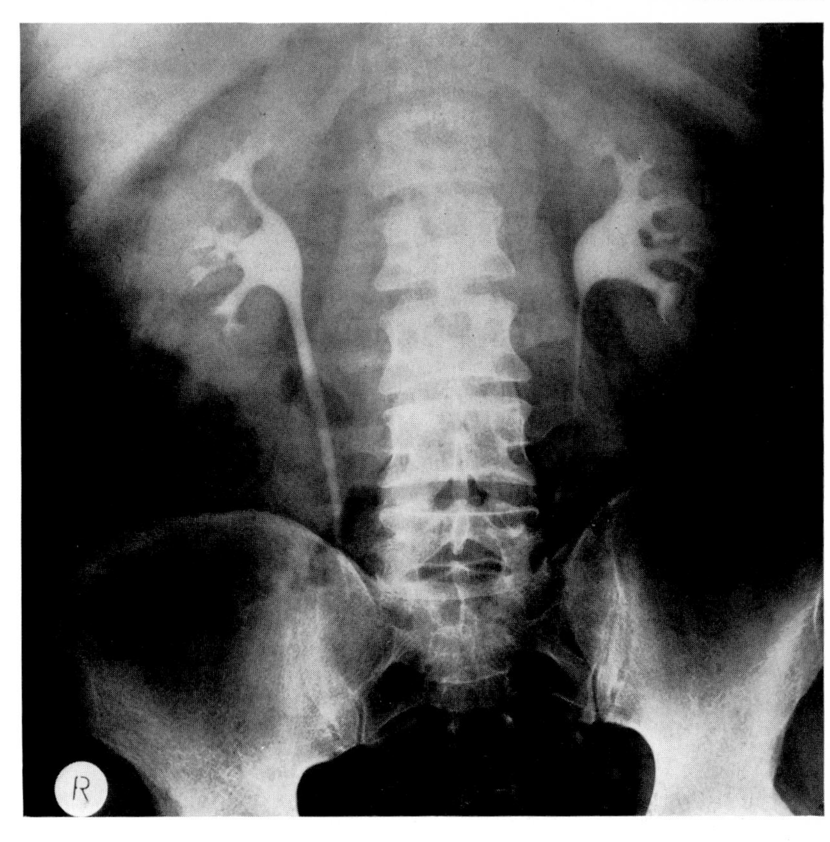

In this picture, the contrast material brings out the kidneys and the ureter. The X-ray diagnosis provides internists and surgeons with valuable information.

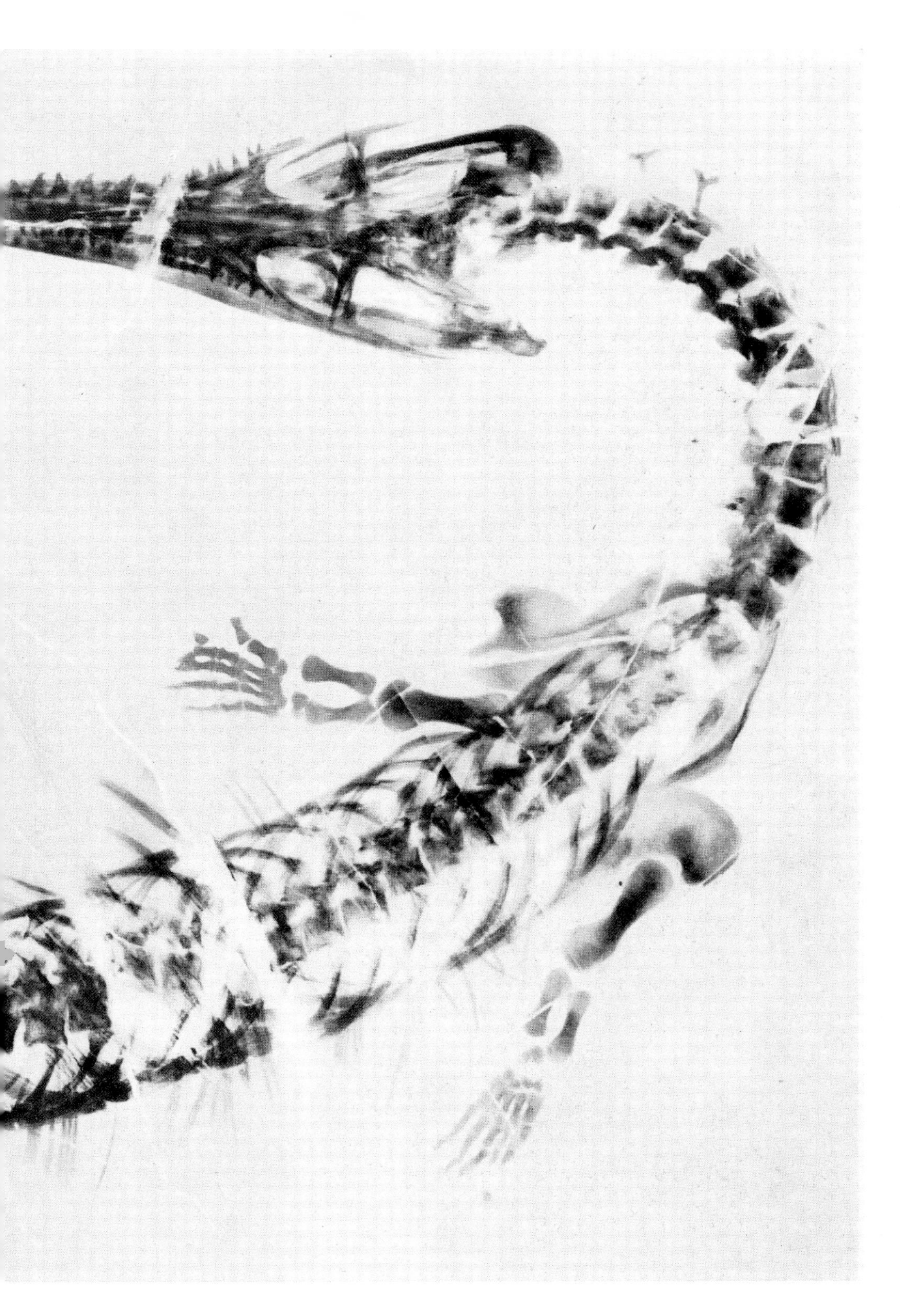

▲
A "living fossil," the chambered nautilus *(nautilus pompilius)* is a descendant of the extinct large ammonites. The full-grown squidlike animal lives in the outermost chamber; as the animal grows, successive chambers become too small, are closed off, and the gas-filled cells give the tenant buoyancy. The X-ray picture reveals the shell's remarkable construction.

◄ X–rays help to decipher the history of the earth. Here they delineate the shape of a reptile, preserved in petrified mud, that lived twenty-five million years ago.

C. H. F. Müller, Röntgenwerk

Glanzstoff AG, Wuppertal-Elberfeld

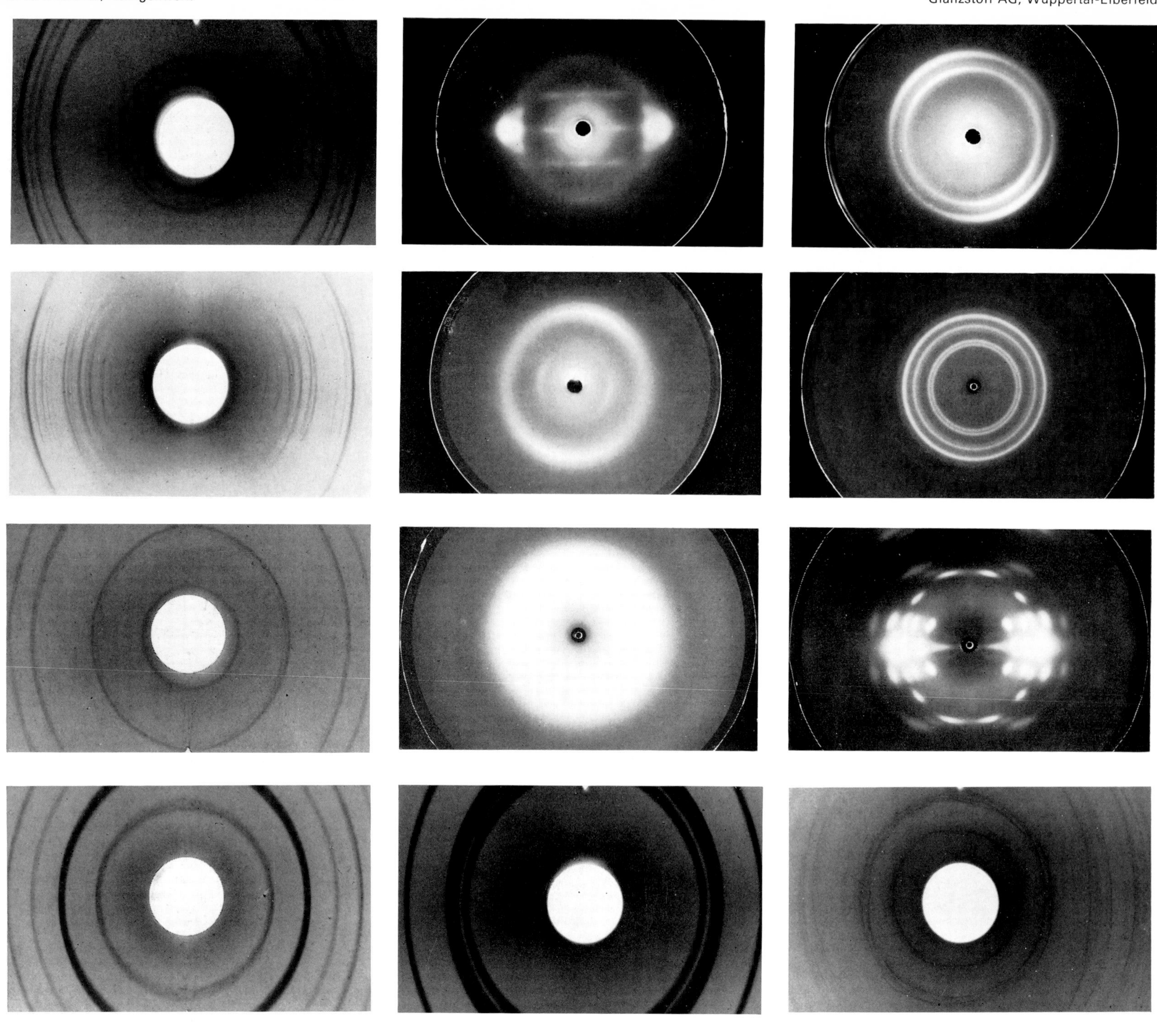

Most forms of solid matter on earth are crystalline. X-rays have the capacity to reveal the inner structure of crystals. When X-rays are directed at an assemblage of crystals, those of an artificial fiber, for instance, the rays are diffracted into patterns of concentric rings. This sort of X-ray picture provides hints on how to strengthen the fiber.

When X-rays penetrate single crystals they diffract into symmetrical patterns of dots. With similar, although not quite that clear, patterns Max von Laue proved that X-rays are "light of very short wavelength" (1912). Apparently, the grating-constant of crystals is narrow enough to diffract X-rays' light.

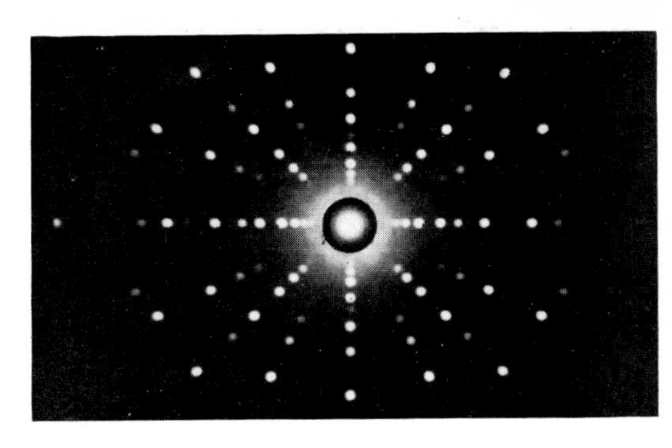

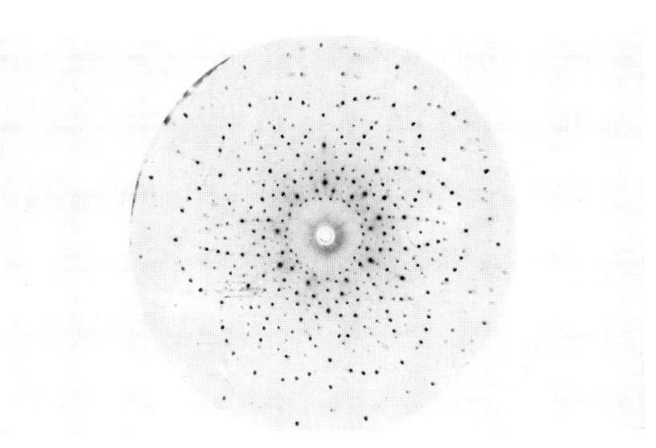

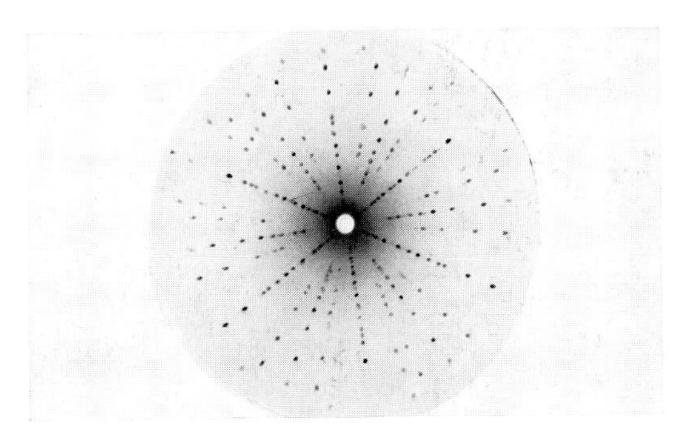

Badische Anilin & Soda-Fabrik AG (BASF), Ludwigshafen
Herbert Maas, Bad Soden (Taunus)

Dr. Hermann Kühn, Doerner-Institut, Munich

▲
◄ In painting, white lead is the equivalent of light. X-ray analysis shows the delineation of the underlying structure of a painting; areas covered with white lead stand out. The "handwriting" of the artist shows up; composition, changes, and areas painted over appear on the radiograph. When the center panel of the St. Bartholomew Altar (16th century, Alte Pinakothek, Munich) (photo above), was X-rayed, the figure of a kneeling man that had been painted over was revealed (photo at left). The surface paint was removed; the "bald man" turned out to have black hair—the X-ray had passed through the black paint. The kneeling man may have been the original donor of the altar who may have died without having paid for the work. Someone else may have been willing to defray the costs on the condition that the late lamented be painted out of the scene.
▼

Dr. Hermann Kühn, Doerner-Institut, Munich

This painting of Christ belongs to the Bavarian State Collection of Paintings in Munich. Old varnish becomes fluorescent under ultraviolet light; new varnish does not become fluorescent—it appears black under ultraviolet light. Therefore recent repairs stand out. The photo on the right shows such retouching.

Titian's portrait of a woman, known as *Vanity of the Mundane,* is now at the Alte Pinakothek in Munich. The radiograph (page 197) provides insight into the creative process: the artist corrected himself. He wanted to reinforce the expression of the woman by a more pronounced inclination of the head toward the left.

Dr. Hermann Kühn, Doerner-Institut, Munich

Vetus Latina Institut, Erzabtei Beuron

In monasteries, lettering was often removed from parchment so that the expensive material could be reused. Ultraviolet light, with a shorter

Vetus Latina Institut, Erzabtei Beuron

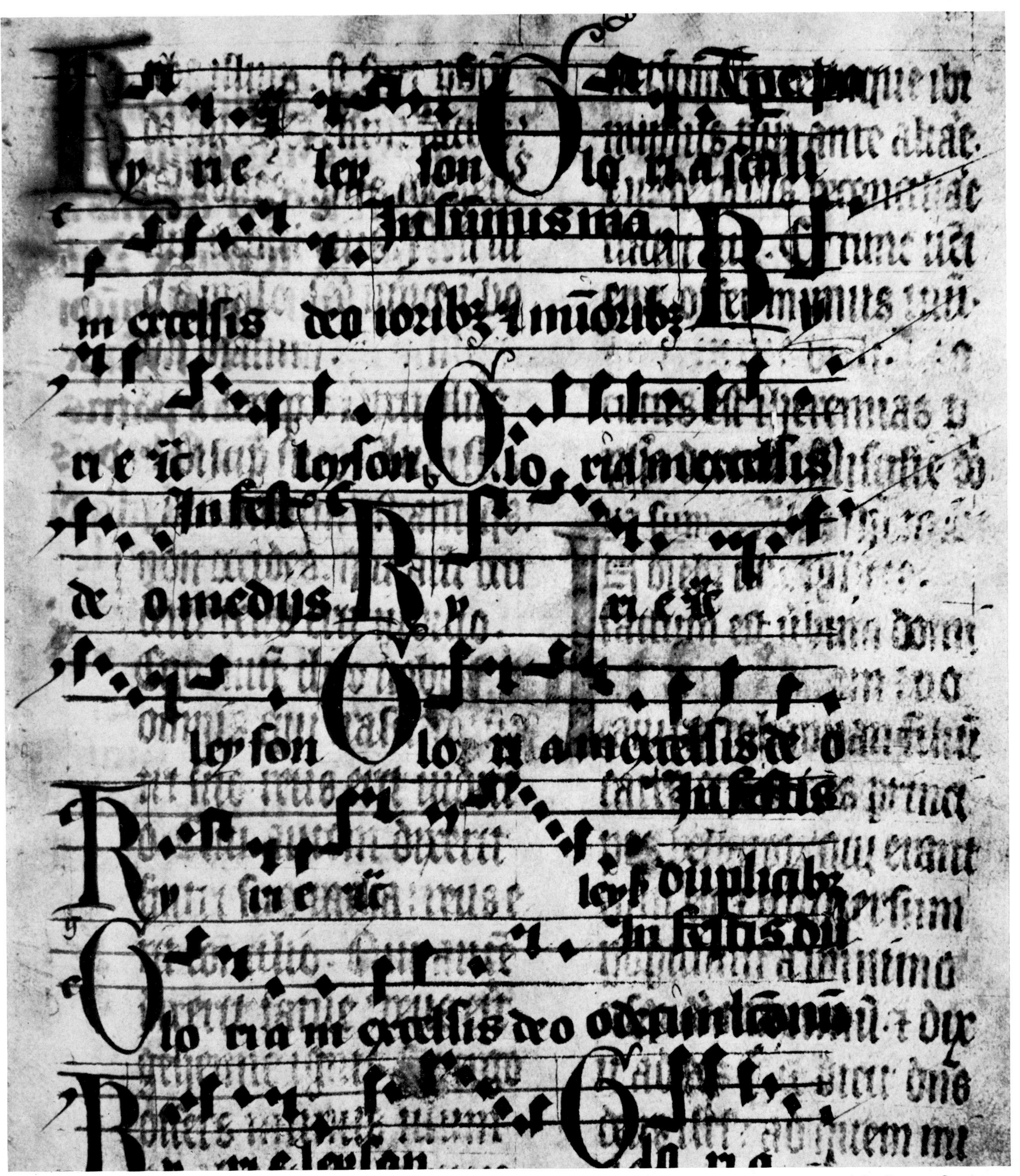

wavelength than visible light, artificially created with quartz lamps, makes the older, previously illegible text on a page of the Beuron Canon stand out clearly.

Herbert Maas, Bad Soden, Taunus

Ultraviolet light also helps in crime detection. The fingerprint on the rough paper of a dirty old envelope could not be made visible by the usual method of dusting. The criminal got traces of glue on his finger while closing the envelope; this was his undoing. The glue becomes fluorescent in ultraviolet light and the fingerprint is made to stand out clearly.

Herbert Maas, Bad Soden (Taunus)

The cancellation of a passport has been removed, the ink has been bleached out by a chemical: a forgery not recognizable by the eye or the camera (left picture). Under shortwave ultraviolet rays, the previously invisible writing shows up in bright fluorescence.

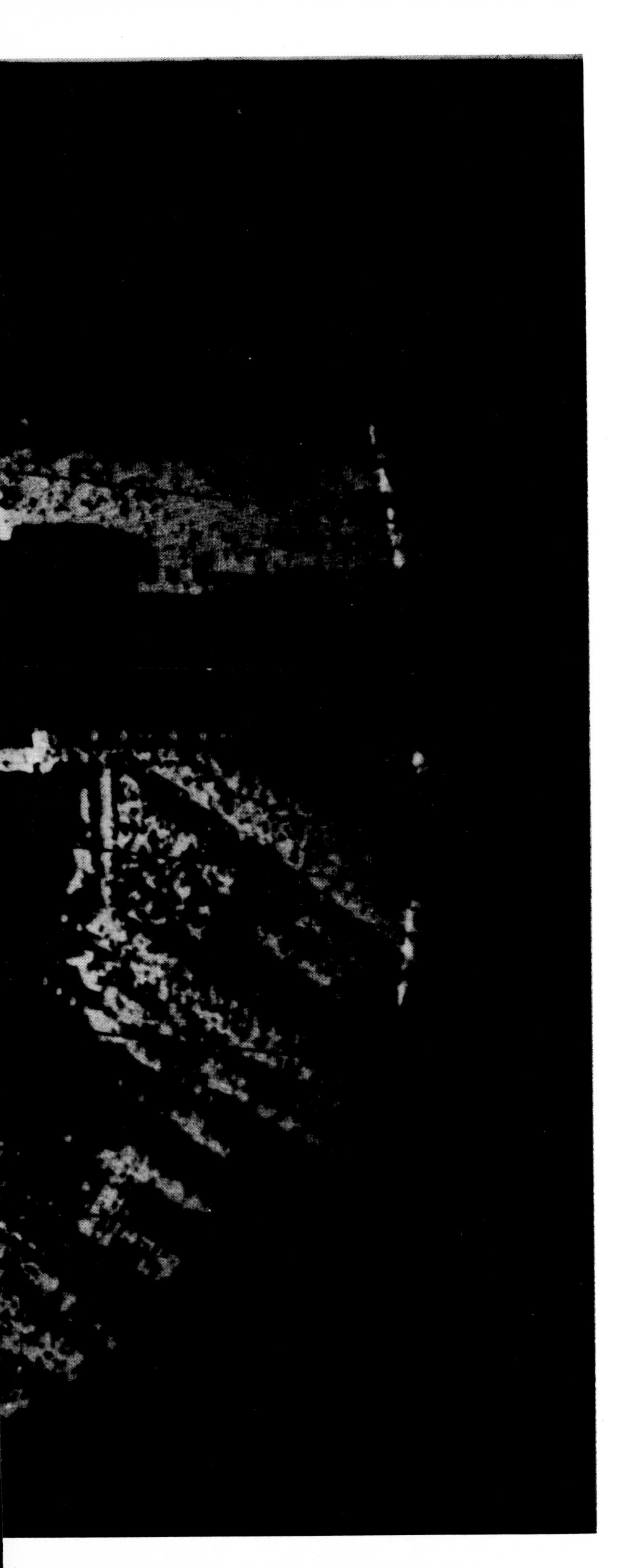

Radar cuts through night and fog.

Electromagnetic waves of half-inch length and a frequency of three billion per second can be closely concentrated into a beam that is reflected by obstacles. Radar is based on this fact. Radar transmitters send out the short radio waves and their "echo" is received. The scanned objects become visible on the round fluorescent screen of a cathode-ray tube. Our picture gives a detail of the port of Hamburg as it was picked up, through night clouds, by an airplane's radar.

Dr. Hannes, Farbenfabriken Bayer

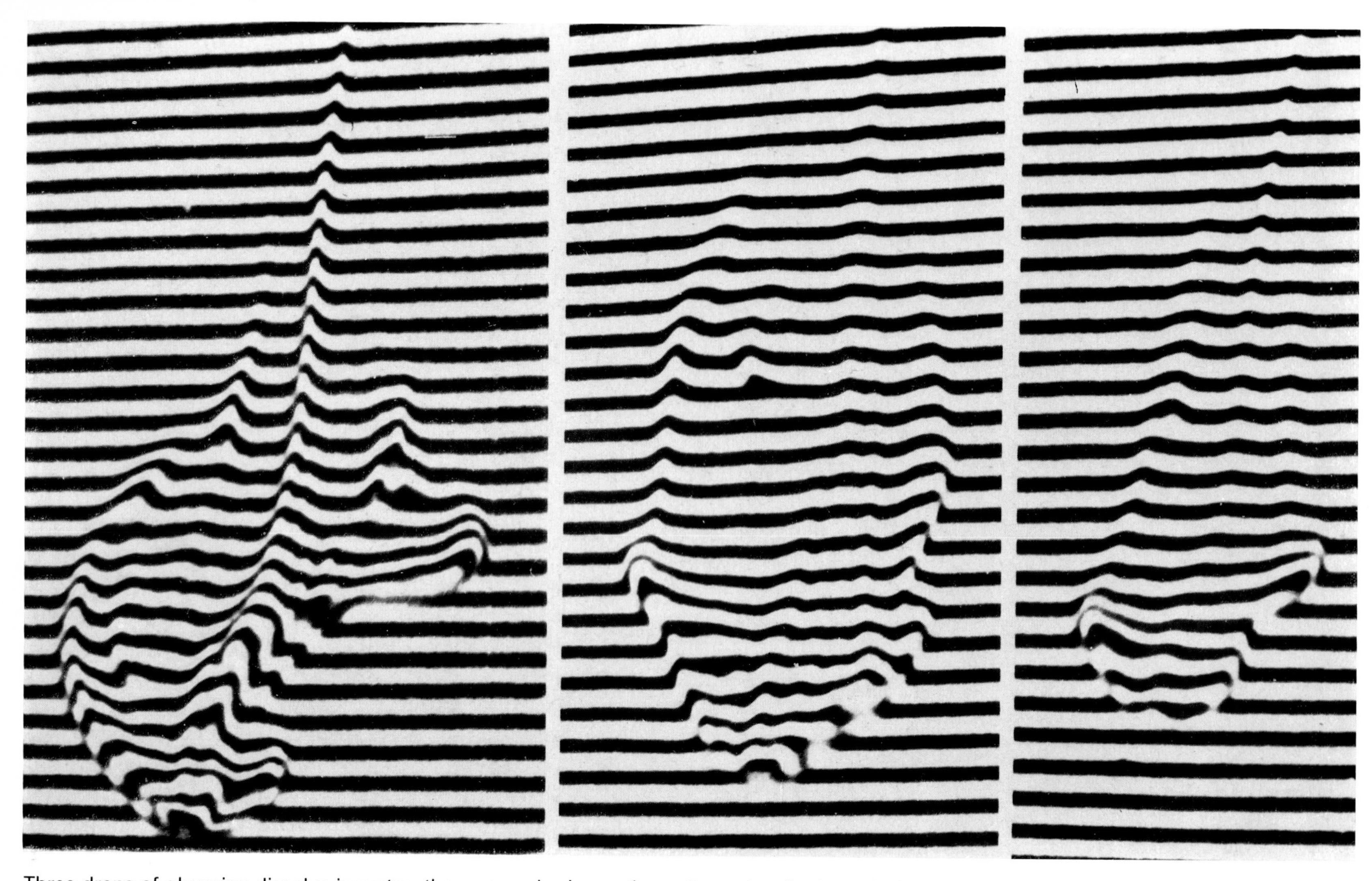

Three drops of glycerine dissolve in water; the process is shown three-dimensionally through the interferometer.

In an interferometer divided beams of light are guided along different paths and then reunited. The overlapping in part cancels, in part reinforces, the light. The resulting interference lines make visible variations in the density of liquids and gases: shown here is the turbulence of hot air after the discharge of a spark across the gap between ball and table. From the configuration of the stripes, experts can calculate how much heat was generated. ▸

Dr. Hannes, Farbenfabriken Bayer

Fritz Brill, Hofgeismar

Fritz Brill, Hofgeismar

The photo on the left shows the residue of a watery oil used in printing ink that barely sticks. The figures on the photo above were formed by a strongly adhering oil. Oil that is too sticky causes the paper to pull or even tear in printing.

Photoanalysis serves to "visually prove interrelations." Here the adhesive power of oils used in printing is investigated and demonstrated. A drop of linseed oil that held two glass plates together, leaves a strange tracing, characteristic for the degree of its stickiness, when the glass plates are pulled apart.

Technische Hochschule München

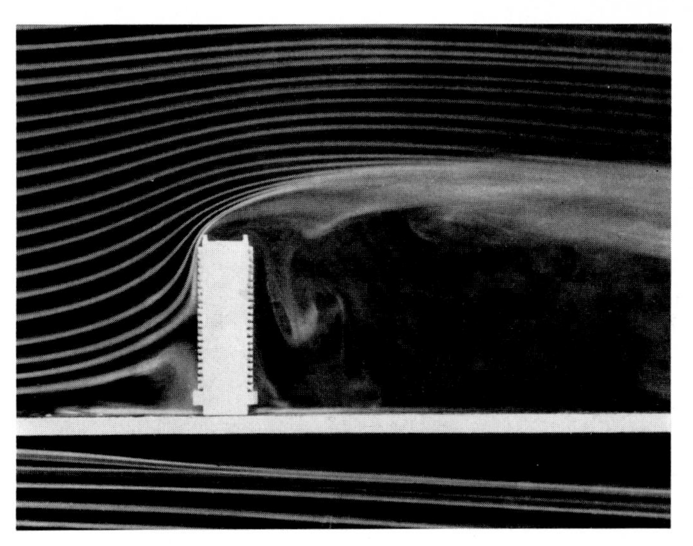

▲
The model of a high-rise building in a wind channel; measuring devices give a true-to-scale picture of the forces involved. Injected smoke allows a visual check of the air currents; the resultant information permits reliable analysis.

◄ Acoustic patterns in liquids become visible through the use of linear screens, and can be photographed. Near the end of the eighteenth century, acoustical expert Ernst Chladni already knew how to make sounds visible: with a stroke of his violin bow he caused a metal plate covered with fine sand to vibrate. The sand collected in areas of lesser vibration and strange figures formed, each one characteristic for a specific frequency (number of oscillations per second).

Manfred Kage, Winnenden

Standard Elektrik Lorenz AG

The magnetic field of a relay. Electricity can generate magnetism and vice versa.

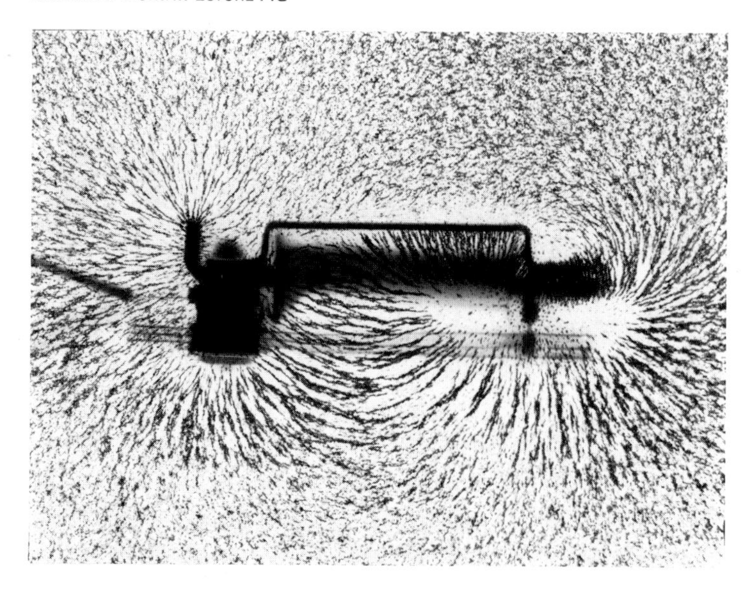

Lines of force around the pole of a magnetic bar. Relationship between magnetic iron-ore and the earth's force field has intrigued scientists for centuries.

Echter, Bavaria-Verlag

A classroom experiment: iron filings reveal the magnetic lines of force around the poles of a horseshoe magnet.

Echter, Bavaria-Verlag

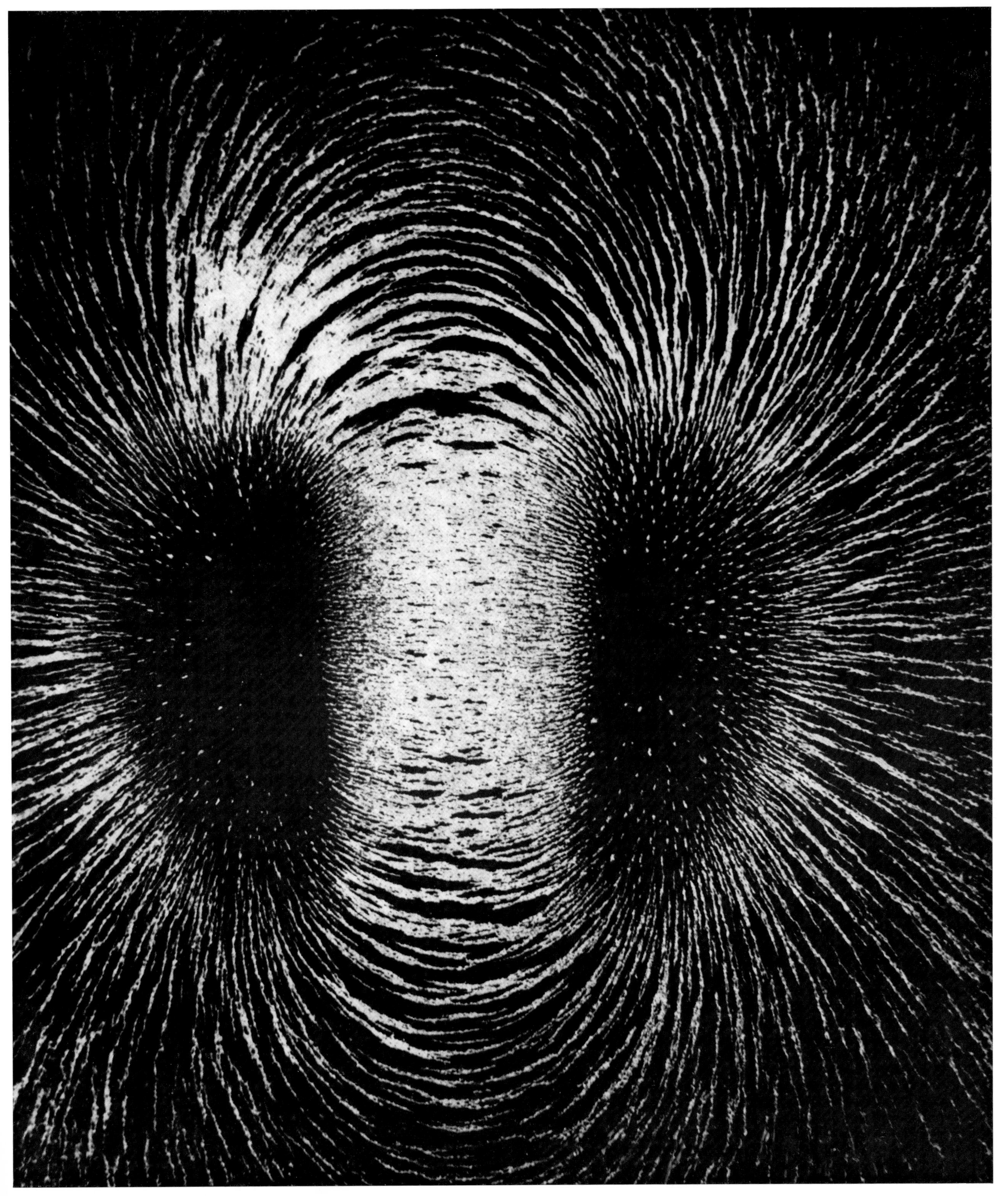

"The electron is the true hero of the century."
(Arthur Eddington, British astronomer and physicist)

AEG

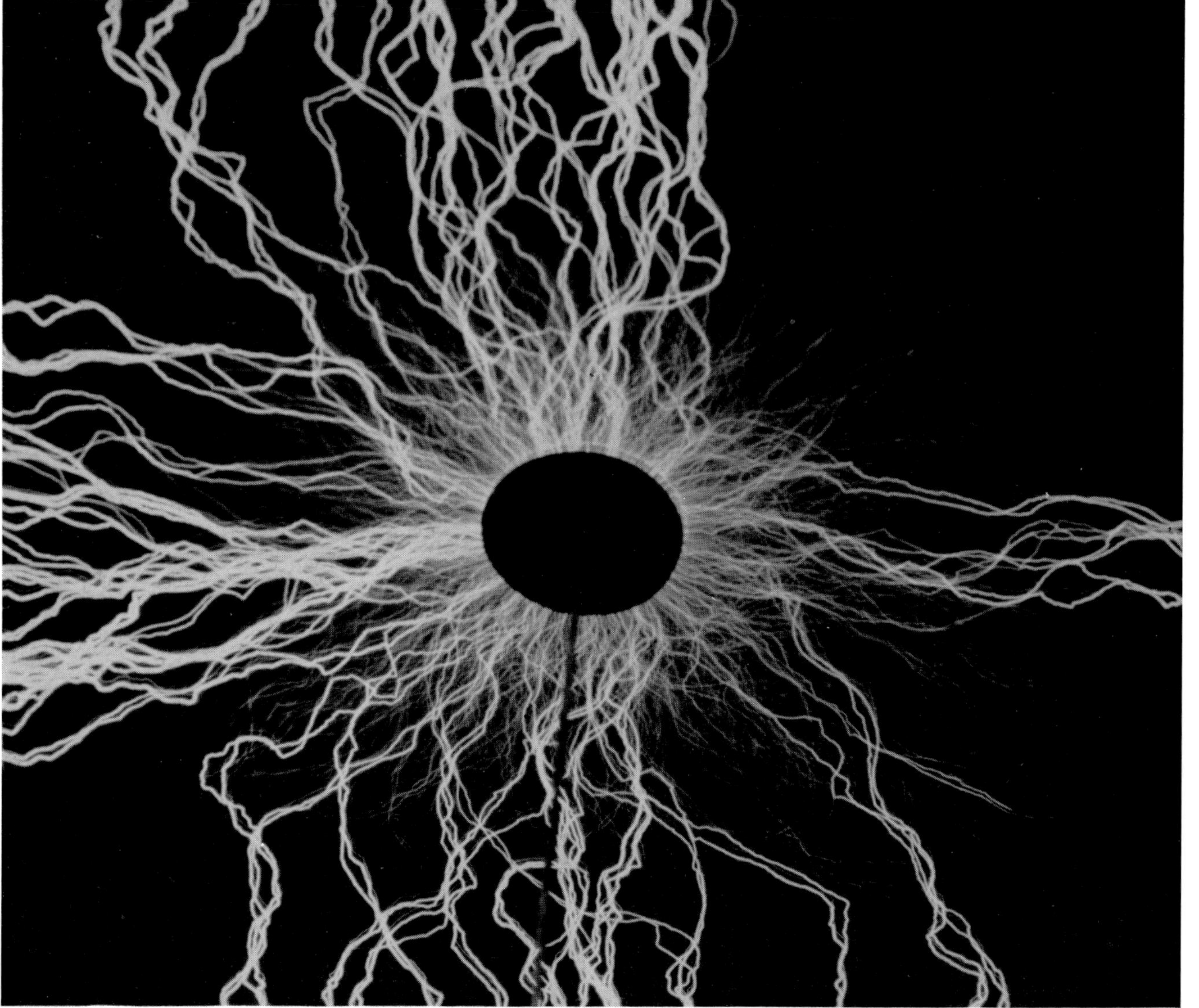

◄ In an experiment, the primordial force of electricity performs fascinating tricks: an insulating glass plate is brought between electric poles. Threads of electric current course across the plate and over its edges from the positive to the negative pole. The discharge, looking like twitching spiders' legs, creates the "Lichtenberg figures" named after an eighteenth-century physicist who was famous mainly for his literary output.

The picture shows a spark-discharge of 200,000 volts leaping along the porcelain ribs of a rain-splashed insulator. In our technological society, each one of us has the use of an electrical work force equivalent to thirty slaves. ►

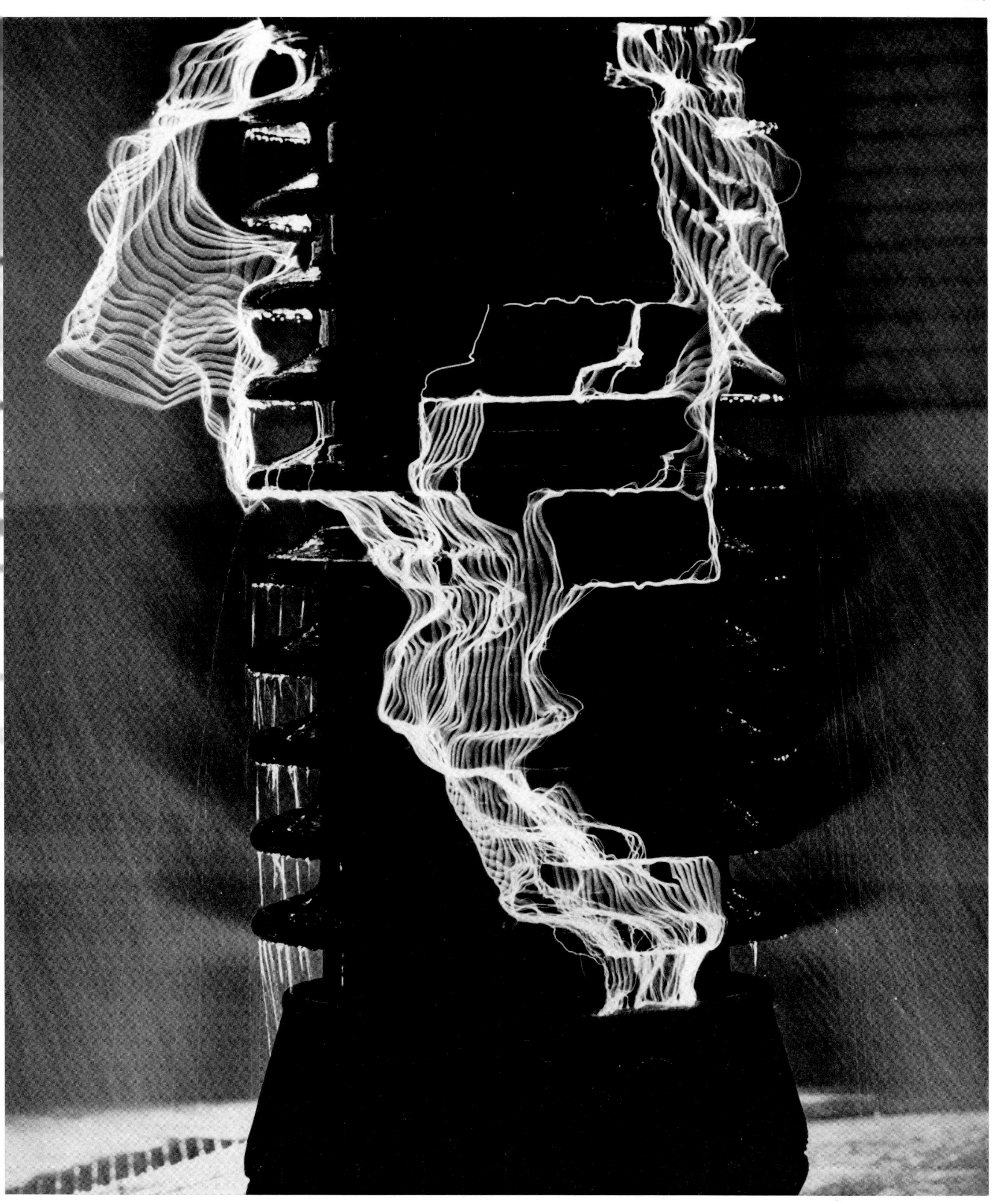

Gerhard Hille, Holzkirchen

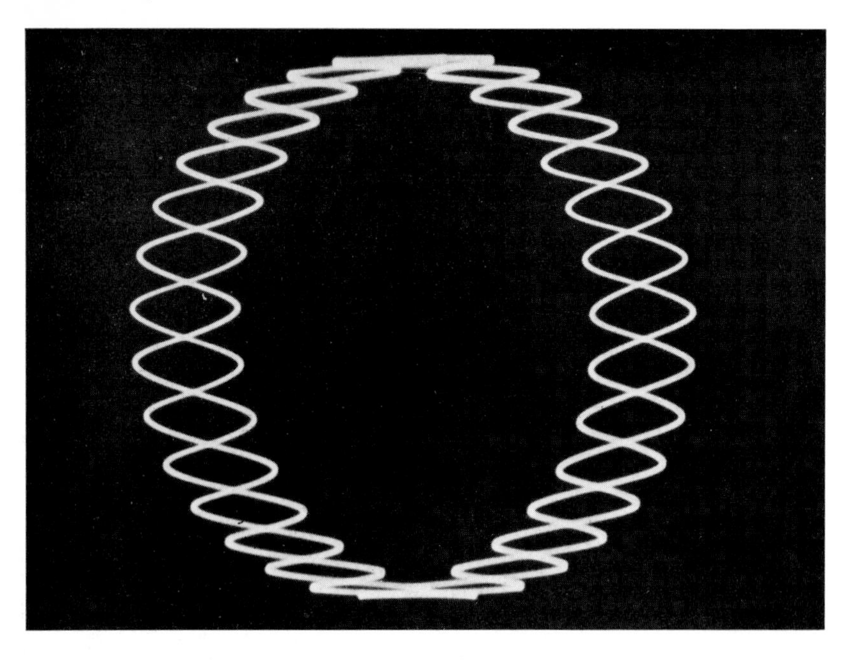

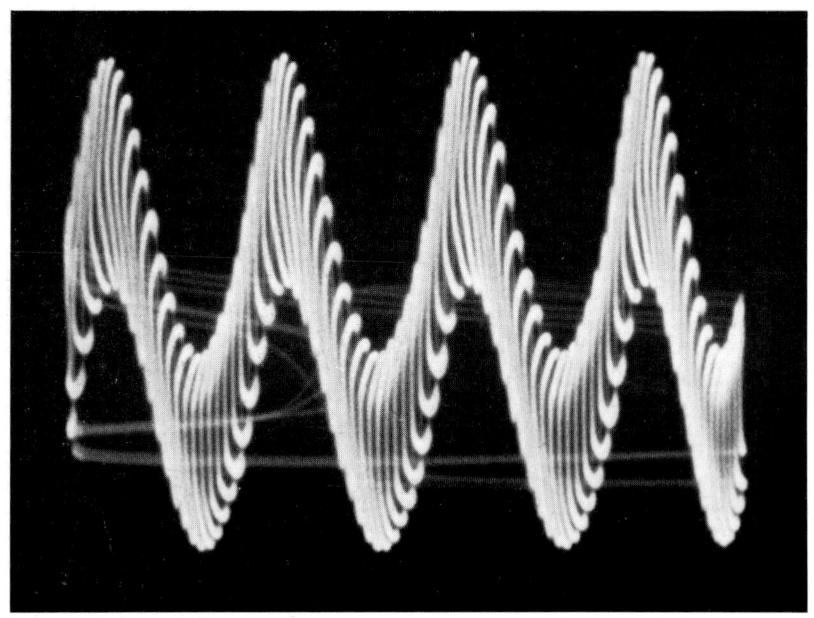

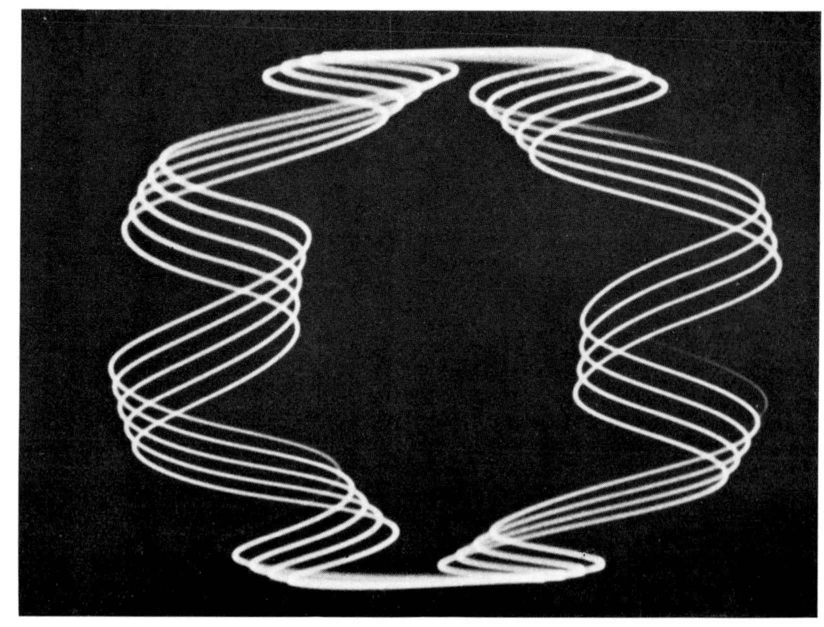

An electronic measuring device, the oscillograph, drew these oscillograms, pictures of electric oscillations, onto the screen of a Braun tube. The electrical engineer knows how to "read" these oscillograms the same way a chemical engineer reads his formulas. Oscillograms can have aesthetic attractions, as these pictures show; the technician explains the harmony of such designs with the "constancy of mathematical function underlying these patterns."

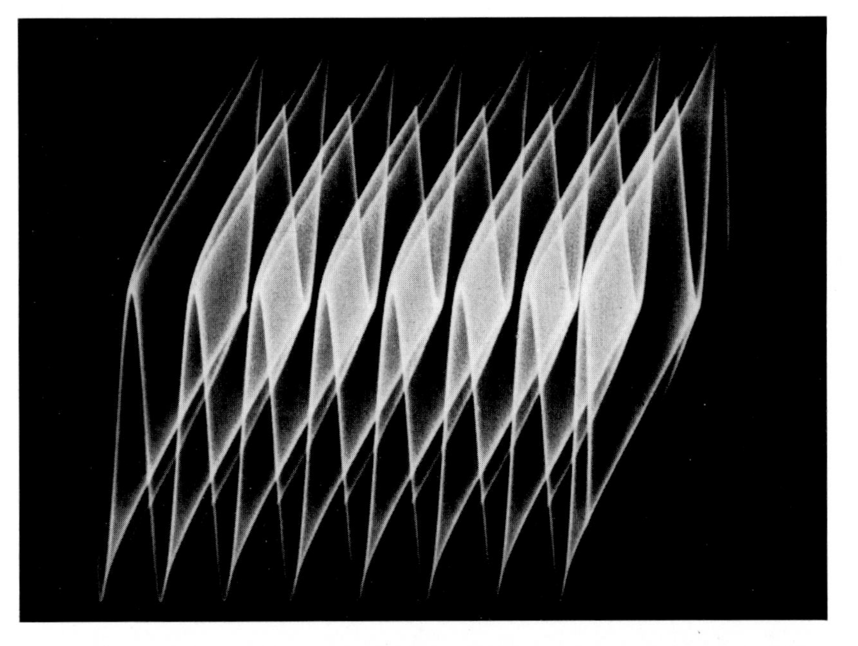

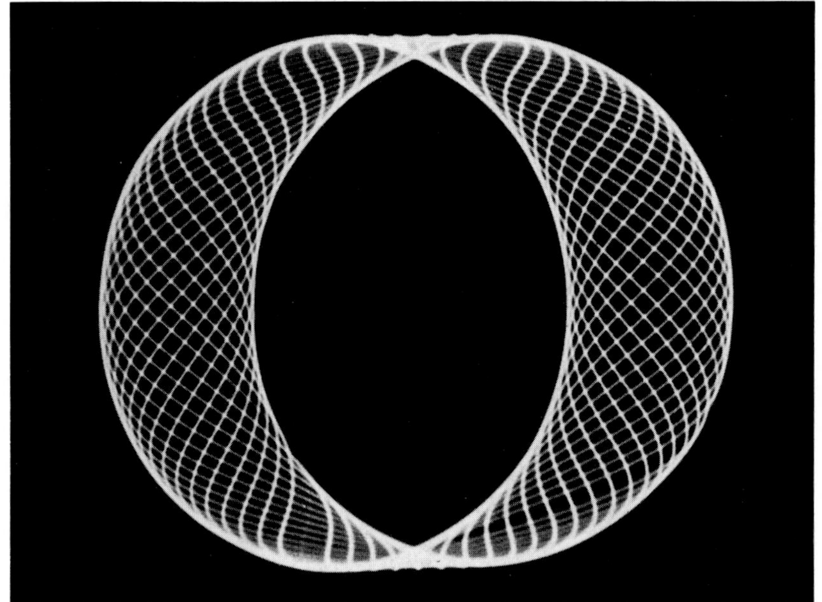

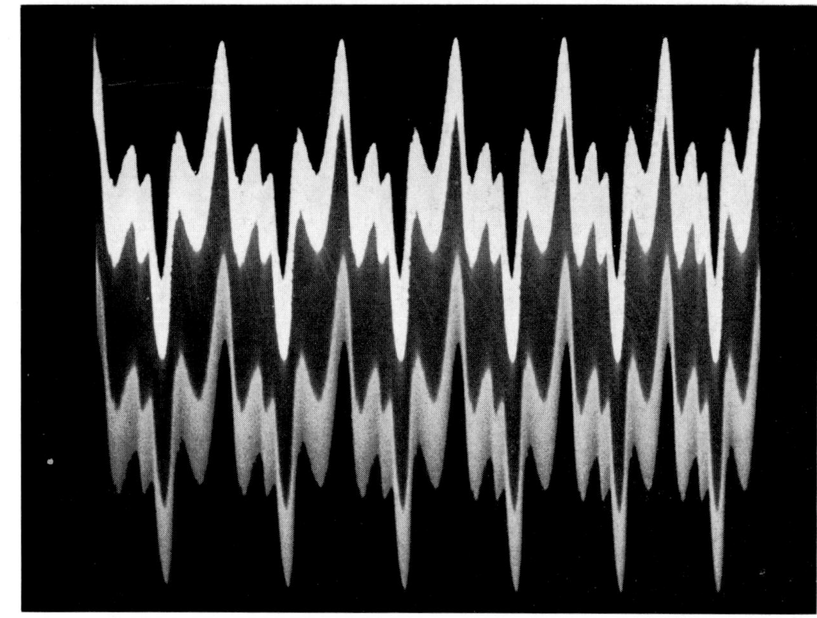

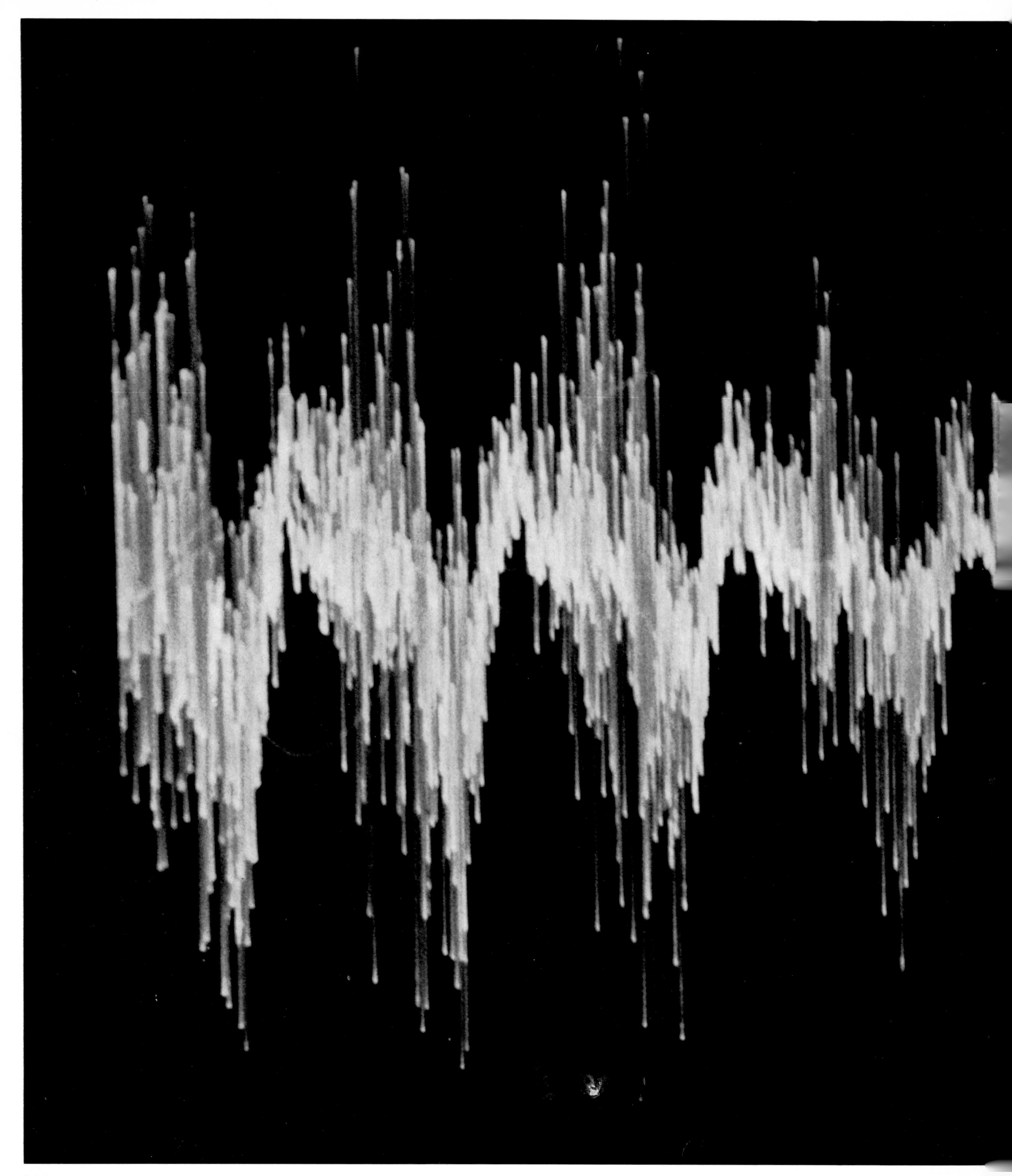

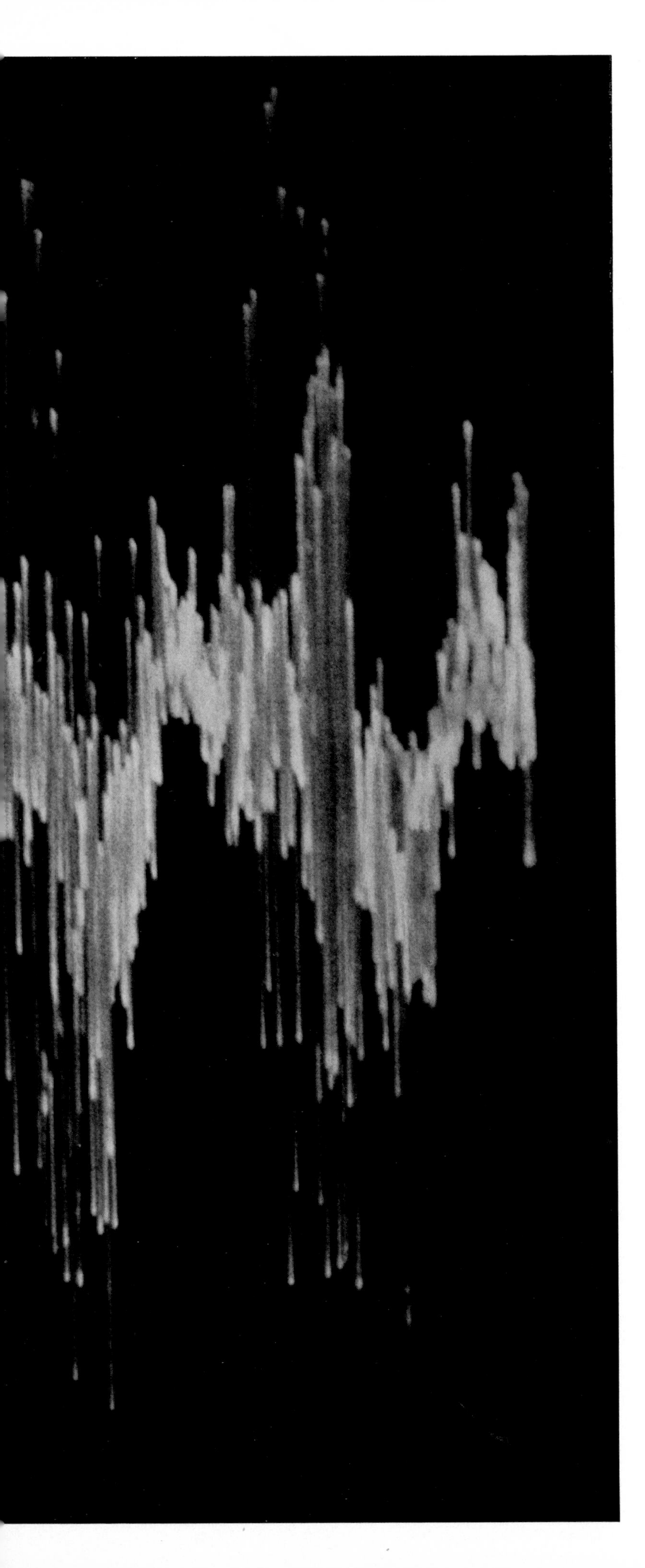

◄ Speech and music, tones and sounds, may be analyzed—via the microphone—with an oscillograph. The picture on the left shows a pronunciation check in an English-language class in Germany: the student tries to pronounce "th"; he doesn't succeed too well. The teacher once more pronounces it for him. The picture below shows how the curve should look.
▼

Prof. Dr. Ing. Ernst Brüche, Mosbach-Baden

▲ Above: a gold crystal magnified 40,000 times.
◀ At left: the surface of a human hair magnified 10,000 times.

The electron microscope surpasses the light microscope in magnifying power a hundred times. The electron microscope does not use the comparatively coarse light waves; it makes use of the very much shorter waves of electrons.

Ernst Leitz GmbH, Wetzlar

Platinum on a heat table observed through the microscope: the temperature rises from 1,400° (lower picture) through 1,430° (center photo) to 1,450° (upper photo). Within this relatively narrow temperature range of 50°, remarkable changes become visible. The metallurgist observing the changed structure of the metal is able to assign to it correspondingly changed properties.

The scanning electron microscope (SEM) works with a fine, slanting beam of electrons. The released "secondary" electrons show the specimen's surface in all its three-dimensional detail. The picture on the right shows etched stainless steel; strongly slanting electron beams (long shadows!) show the surface in 1,200-fold magnification.

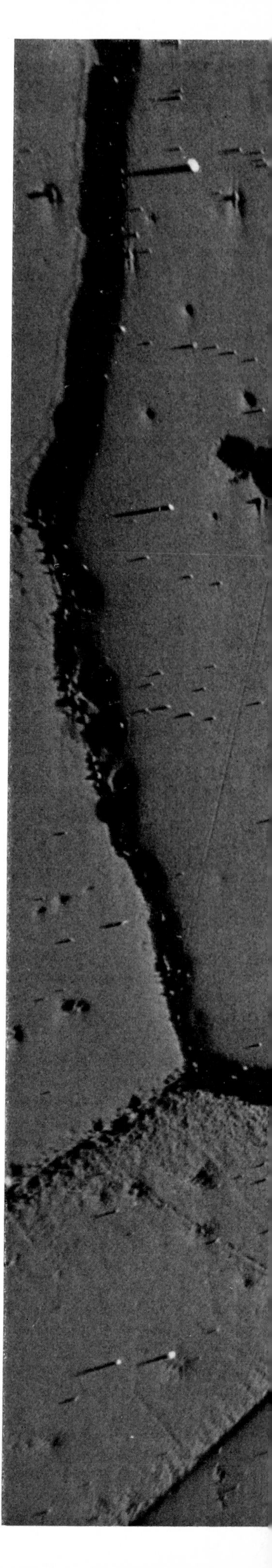

Ernst Leitz GmbH, Wetzlar

Like the bird's-eye view ► of a quarry: the surface of aluminum magnified 13,000 times.

◄ This "thicket of thorns" is zinc oxide needles magnified 14,000 times.

Prof. Dr. Ing. Ernst Brüche, Mosbach (Baden)

Prof. Dr. E. Klein, Agfa-Gevaert Ag

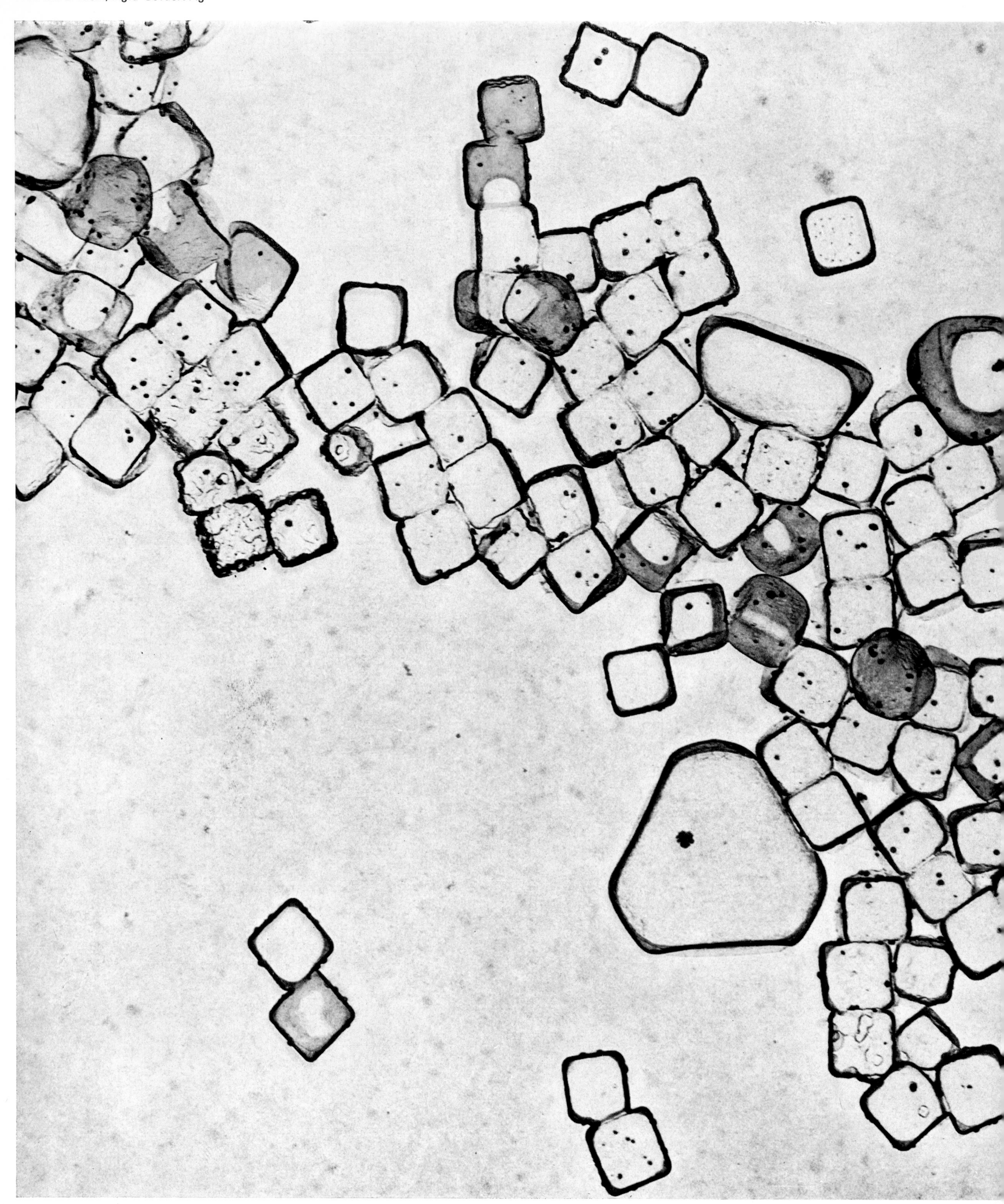

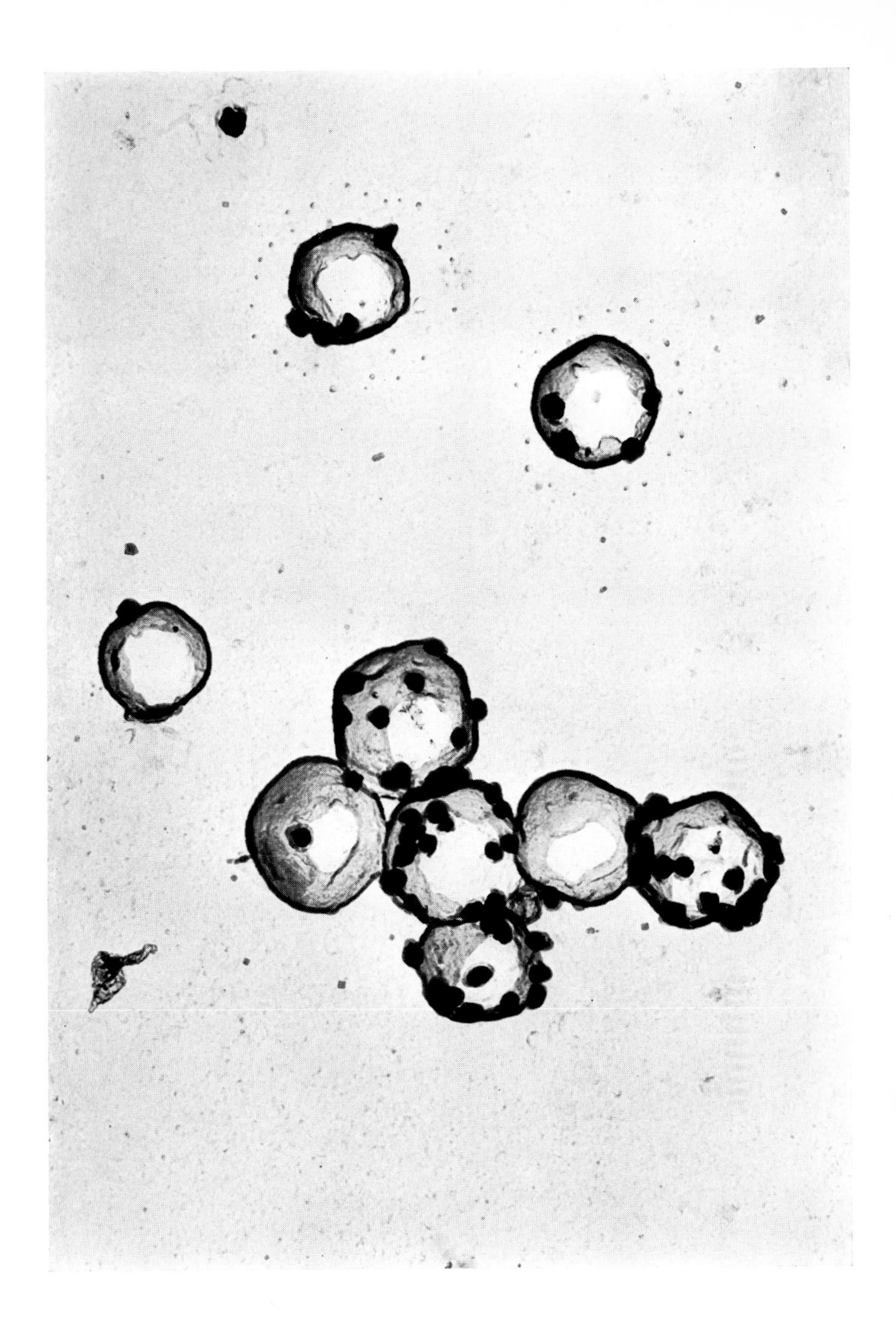

The quality of a photographic film emulsion depends on how fine and even the grain is.

The pictures at left shows silver salt crystals in an emulsion, magnified 25,000 times.

In the picture above, one can observe, in 40,000-fold magnification, the seeds of the latent image as black dots. Where silver-salt crystals are struck by a photon of light, electrons are freed, leading to chemical changes in the crystals. These changes constitute an as-yet invisible image. The development leads to further chemical changes that make the picture visible.

Seemingly an abstract ► graphic design, this is a picture of cigarette smoke through an electron microscope. The droplets floating in the haze are impaled on copper oxide needles. Magnification 1:10,000

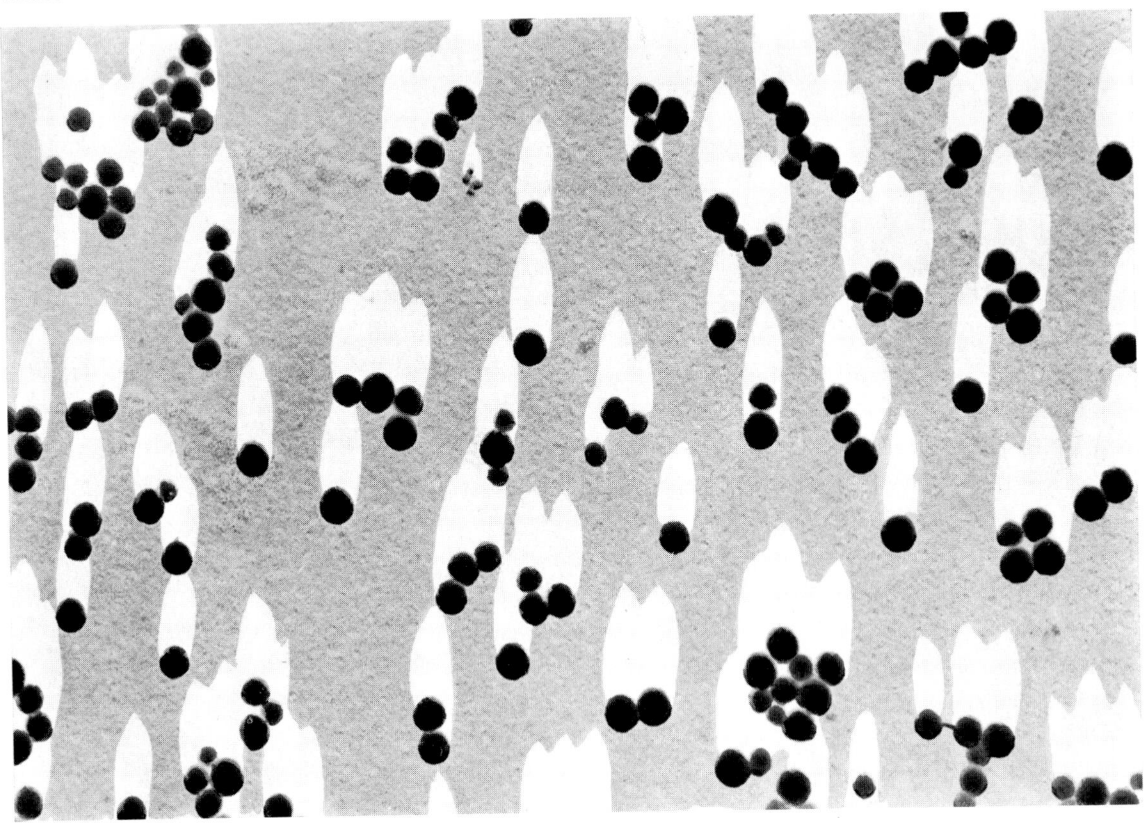

Rubber molecules about 50,000 times enlarged. Obliquely coated with vaporized platinum or gold, they seem to throw long, bright shadows and appear, therefore, more three-dimensional. Synthetic rubber is not an imitation of natural rubber but an original creation of chemistry.

Polystyrene molecules magnified more than 100,000 times. Polystyrene, a synthetic suited for extrusion, is made from a glass-clear, colorless liquid.

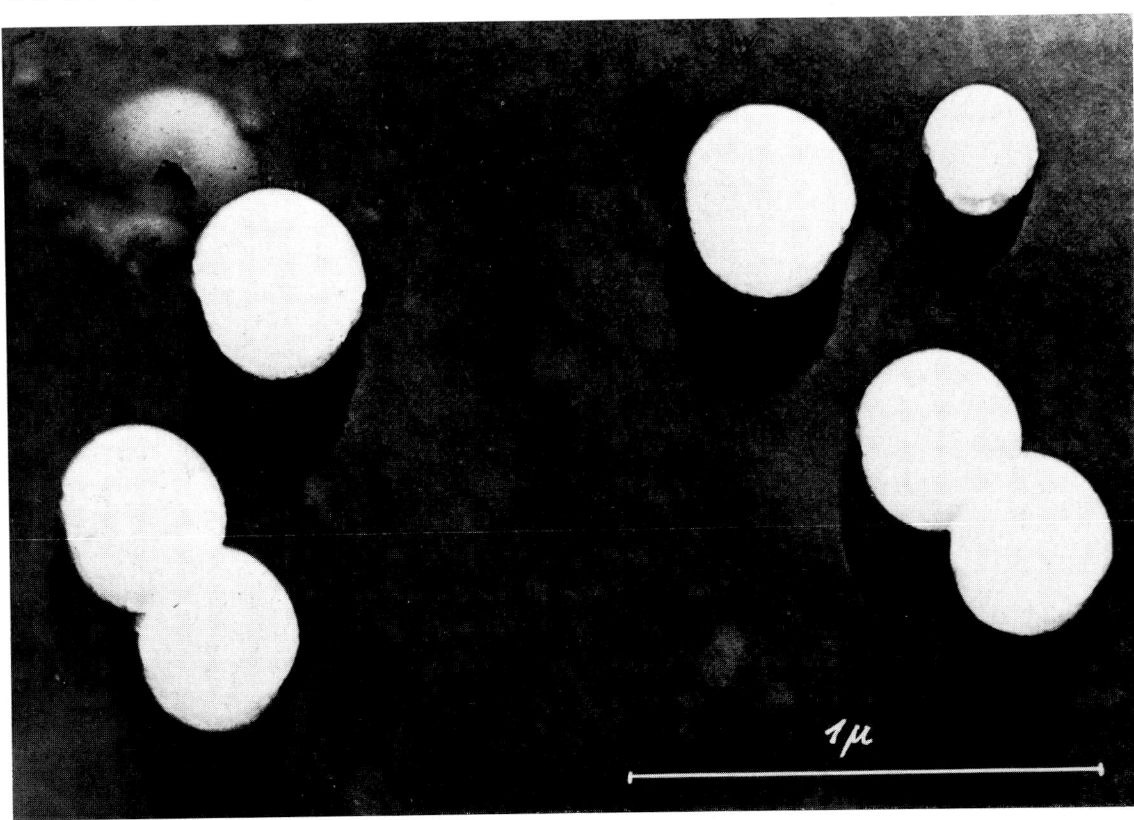

Badische Anilin & Soda-Fabrik AG (BASF), Ludwigshafen

Synthetics: substances "between liquids and steel."

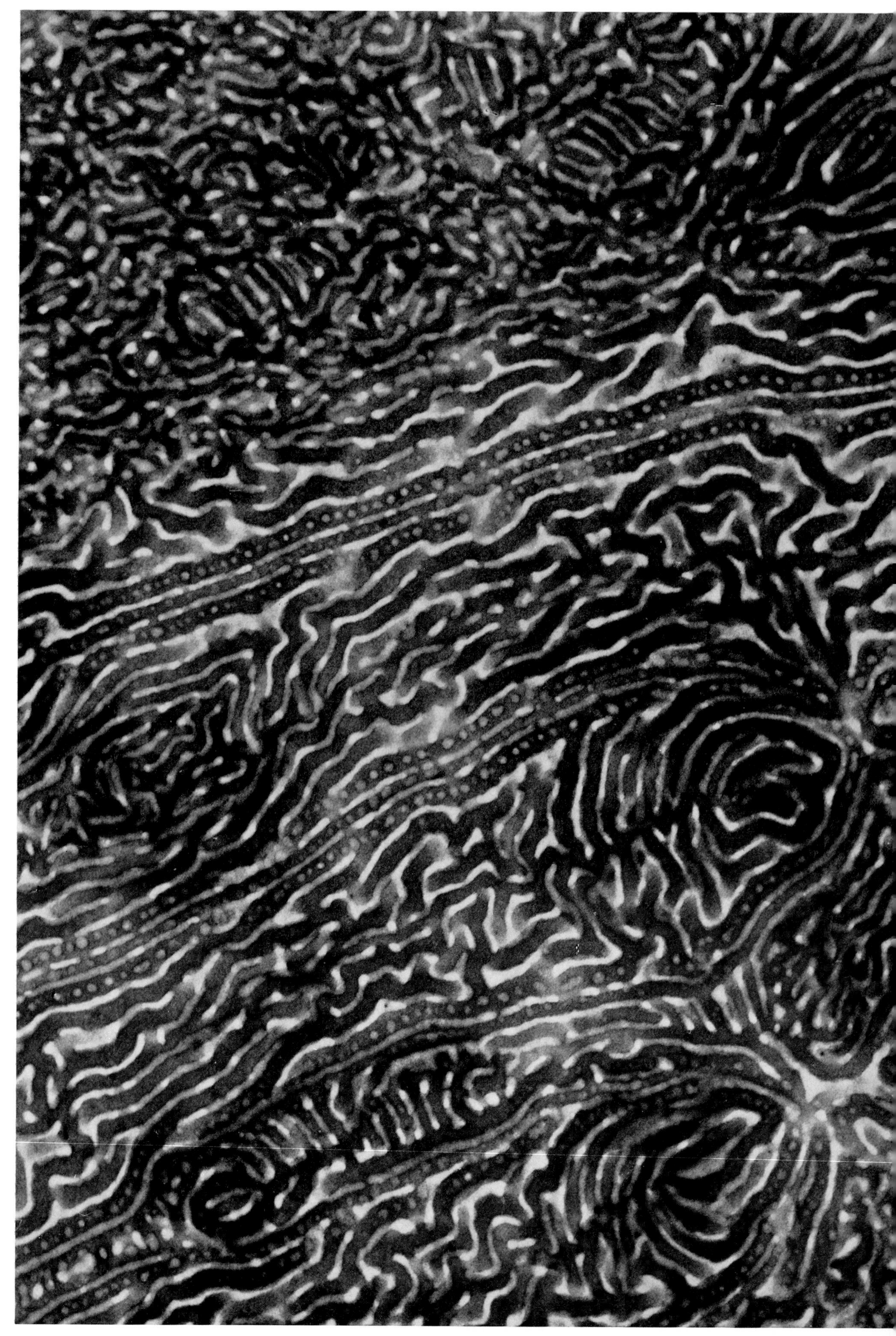

Electron-optical enlargement (65,000-fold) of butadiene-styrene block copolymers. Polymerization, linking of molecules into chain molecules, is planned here so that two different kinds of molecules are not linked "statistically" —that is, in random, irregularly constituted chains but rather in regularly alternating blocks of molecules. Block polymers have properties that are different from those of statistically arranged chain polymers.

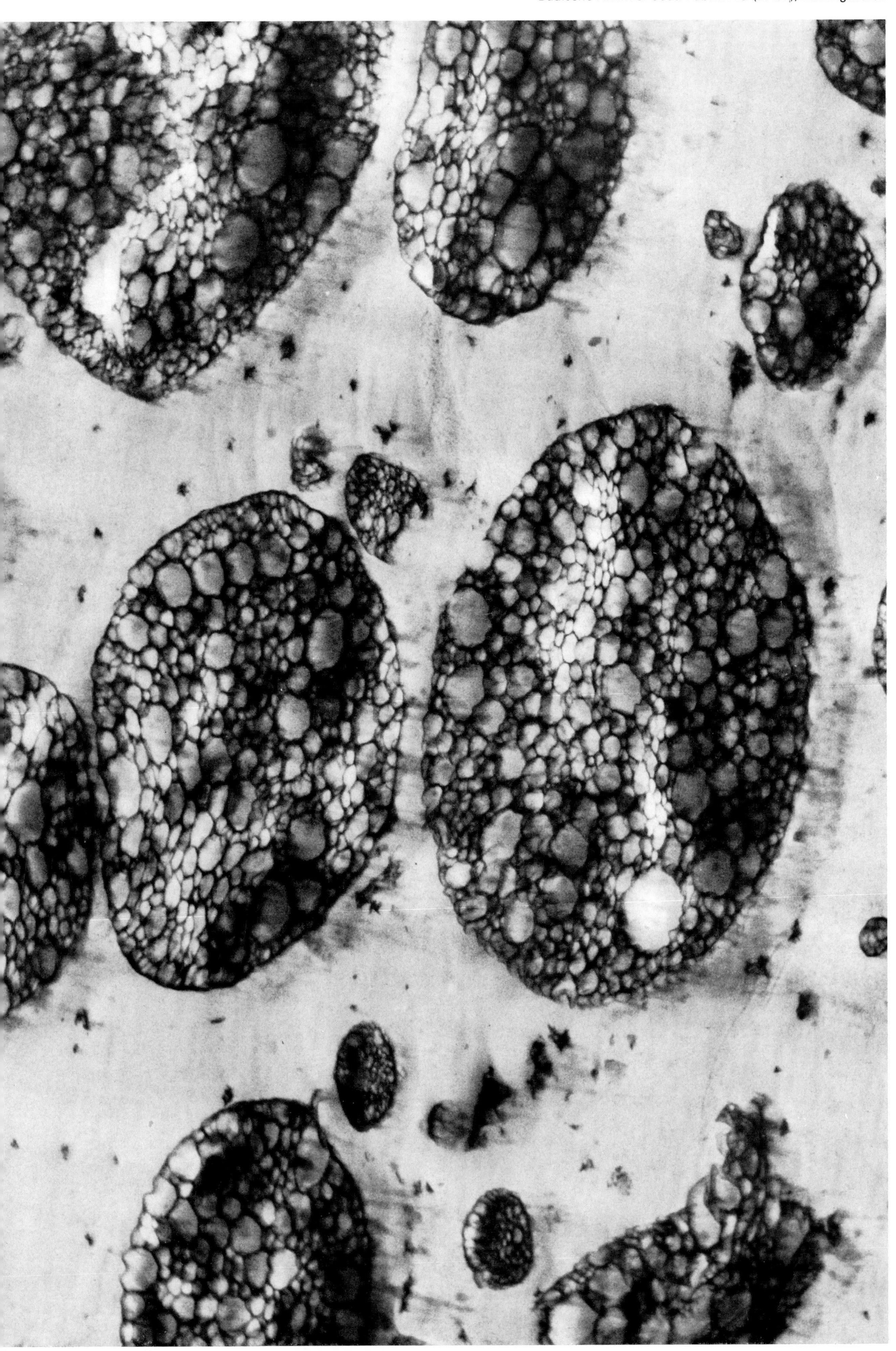

Thin slice through a shock-resistant polystyrene under the electron microscope; in this instance, a 9,000-fold magnification. Butadiene-rubber molecules are anchored within the brittle polystyrene; they can absorb shock and disperse it.

"Forms from the twilight zone of life."

Tobacco mosaic virus magnified 150,000 times. In 1935, the biochemist Wendell M. Stanley became the first to isolate a virus, the tobacco mosaic disease virus, so named for the mosaiclike discoloration of the tobacco leaf. The rod-shaped crystals are able to reproduce, and are therefore *alive*.

Robert Bosch GmbH

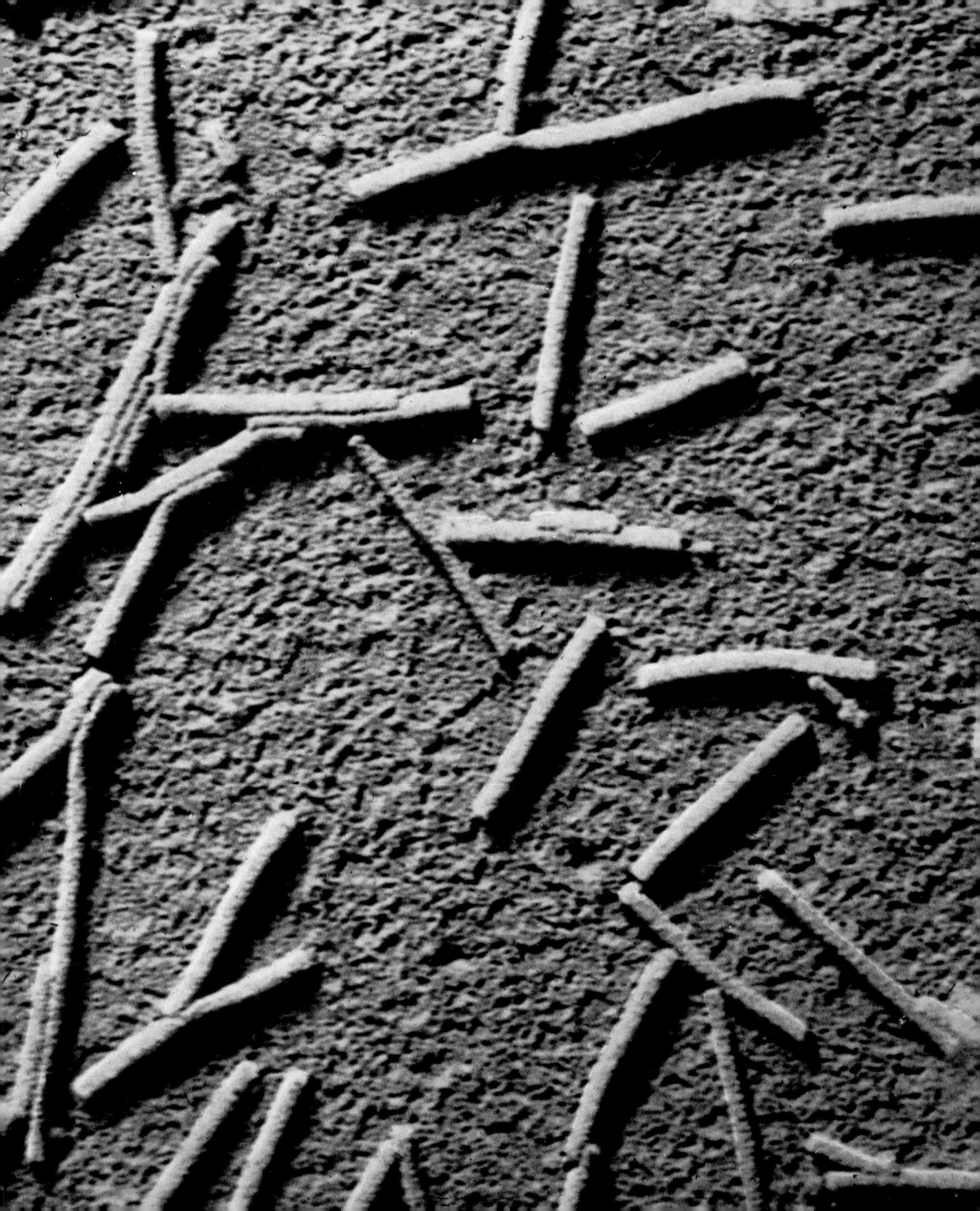

Dr. Brandes, Biologische Bundesanstalt, Braunschweig

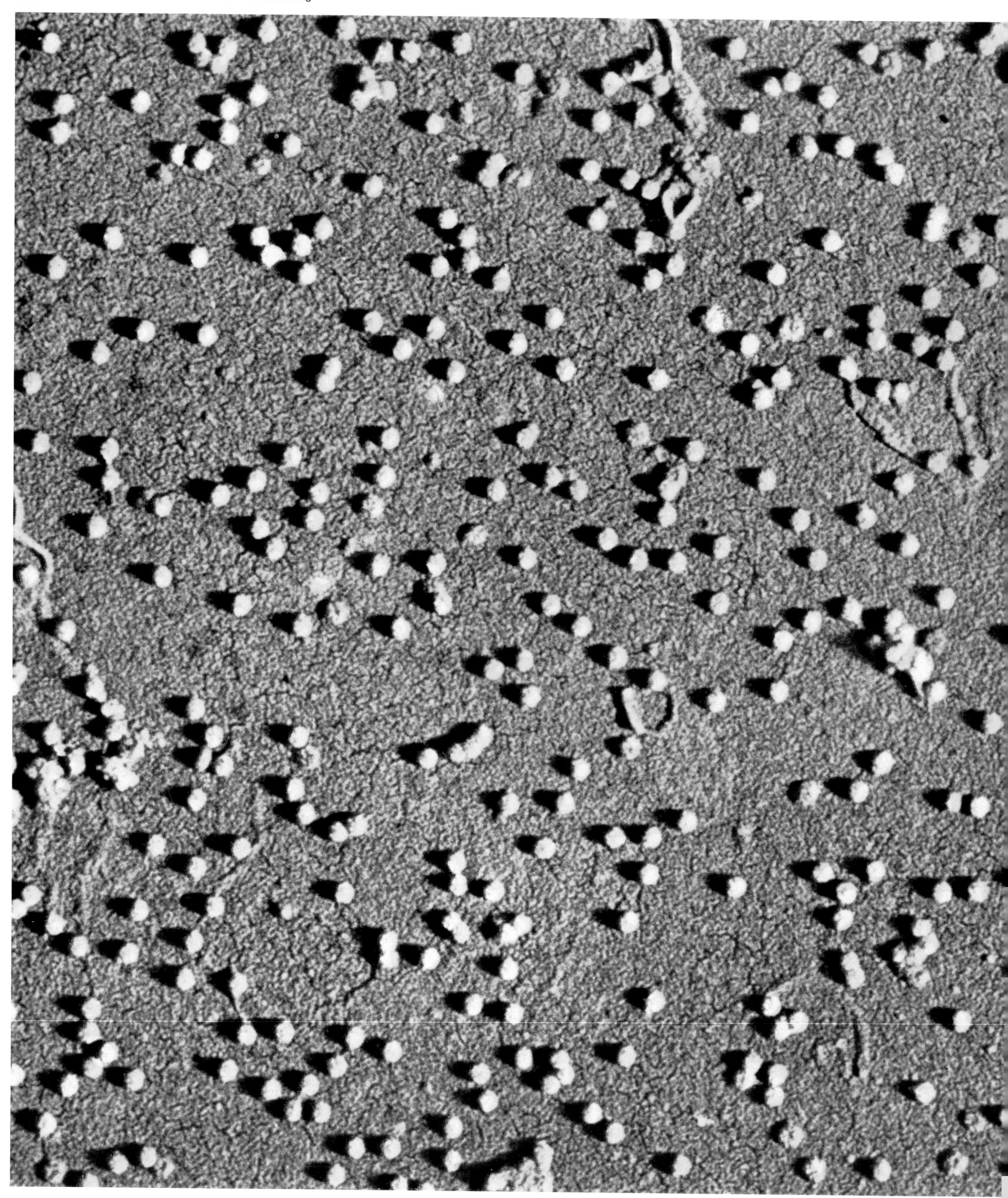

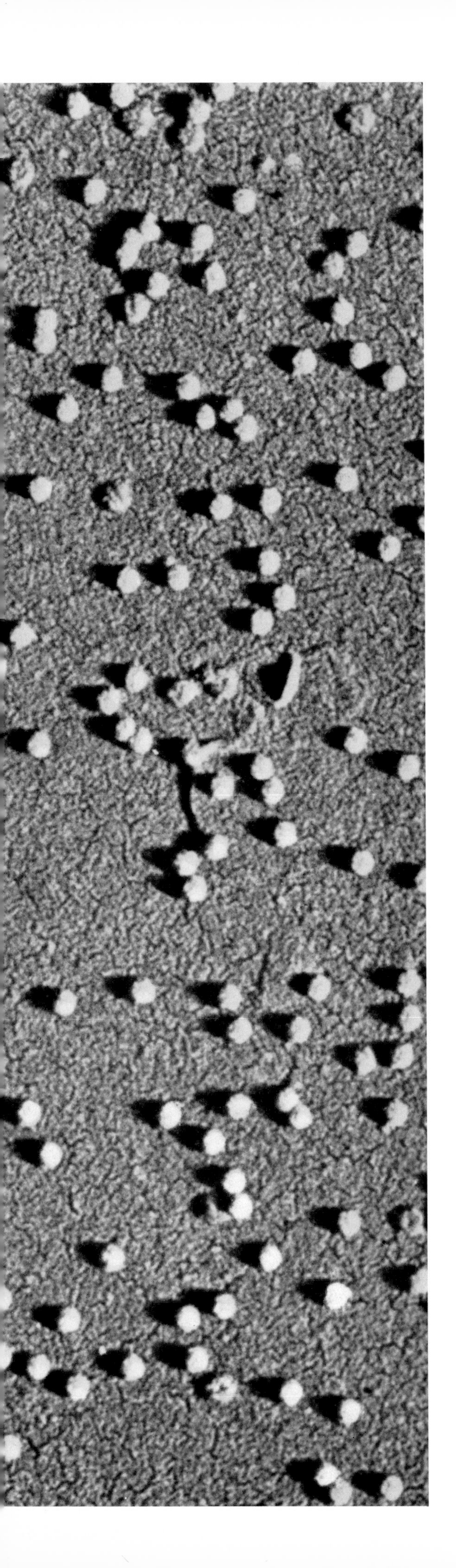

The turnip yellow mosaic virus (magnified 100,000 times), like the tobacco mosaic virus, causes a specific plant disease. The shell of these globular bodies consists of protein; in the hollow center of the ball lies the key to its life processes. An inconceivably thin layer of gold, platinum, or palladium is applied at a slant to specimens such as the one shown here; the result is an extraordinarily three-dimensional picture.

The virus of the bean mosaic disease measures only 25 to 30 thousandths of a millimeter.

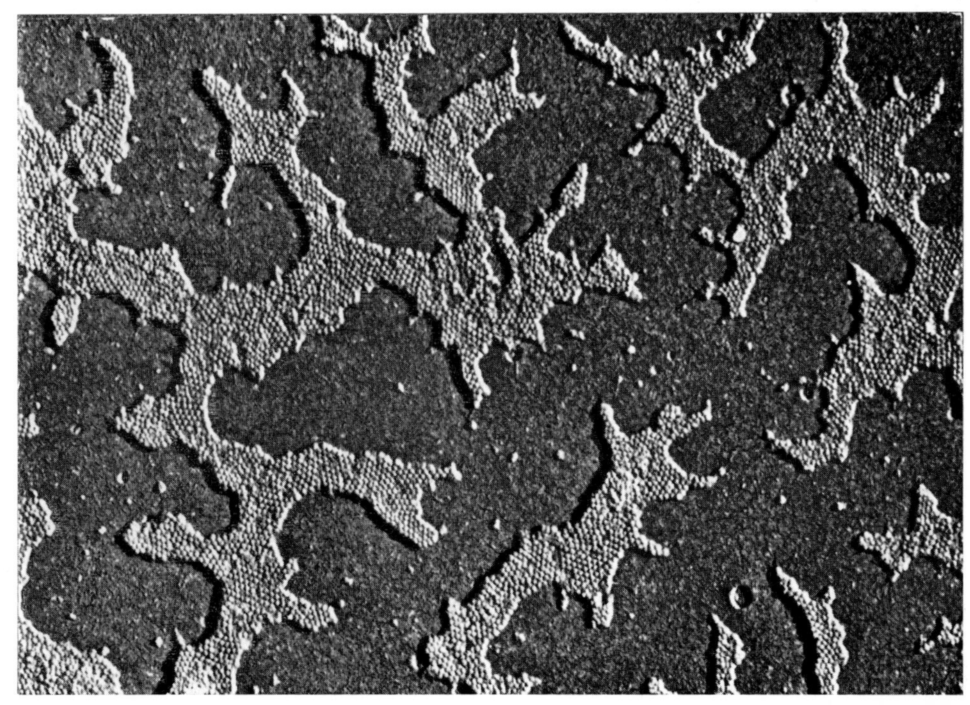

Vaccine seen through the electron microscope.

G. Müller and H. Peters, Tropeninstitut Hamburg

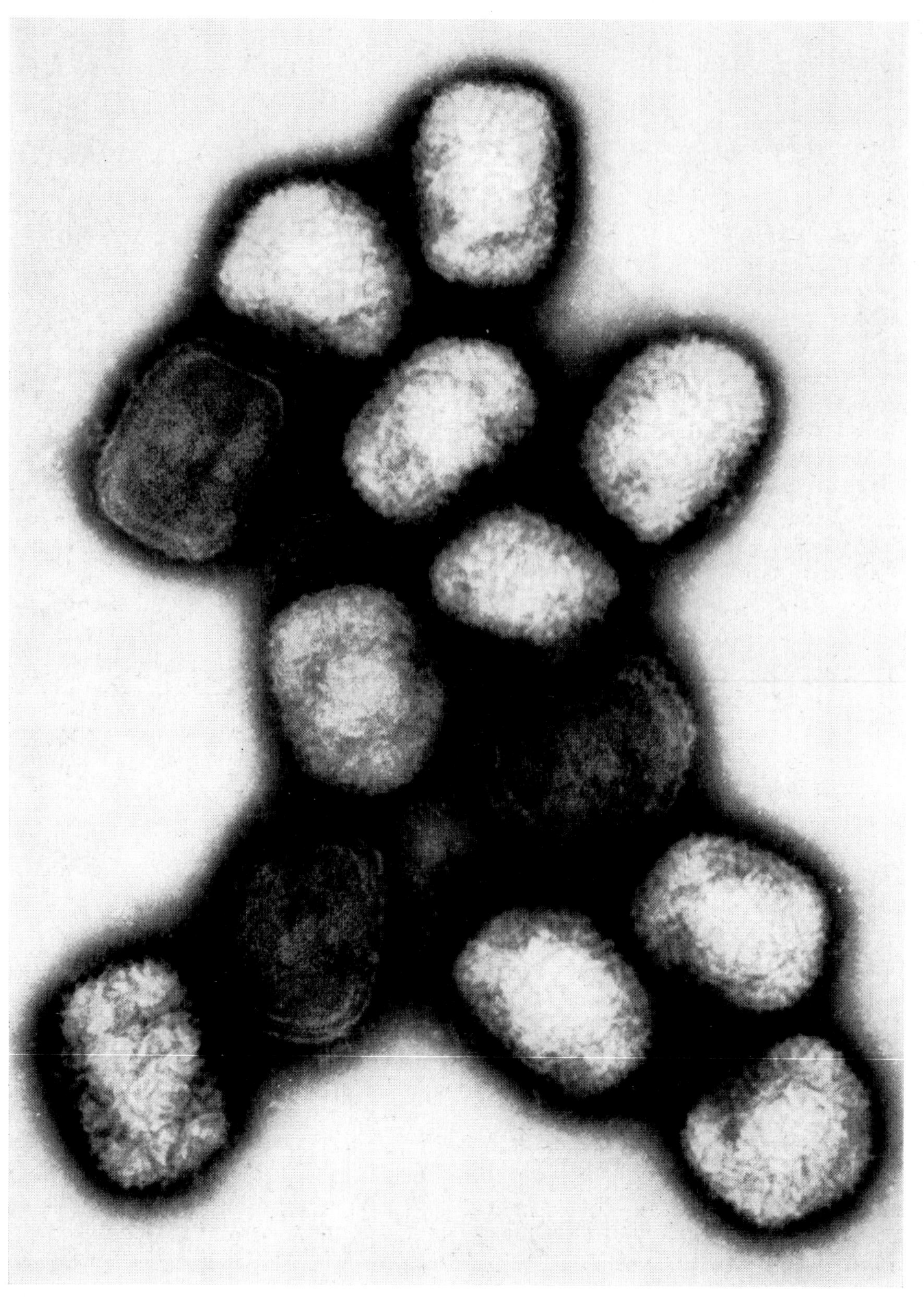

Smallpox viruses look like rounded-off paving stones. Shown here are smallpox vaccine viruses magnified 100,000 times. Introduction of a contrast material has brought out the structure more sharply. Insight into the composition of viruses means one more step toward the unraveling of life's secrets.

Viruses of smallpox vaccine magnified 60,000 times. An enzyme was made to act on the viruses, initiating a sort of digestion that partly obliterated the viruses. The molecule strands of DNA (deoxyribonucleic acid) were thereby exposed. Here too, the three-dimensional effect was achieved by obliquely applied vaporized metal. It is known now how the substance of life looks even though the coated specimen in the vacuum tube is dead.

Dr. H. Peters, Tropeninstitut Hamburg

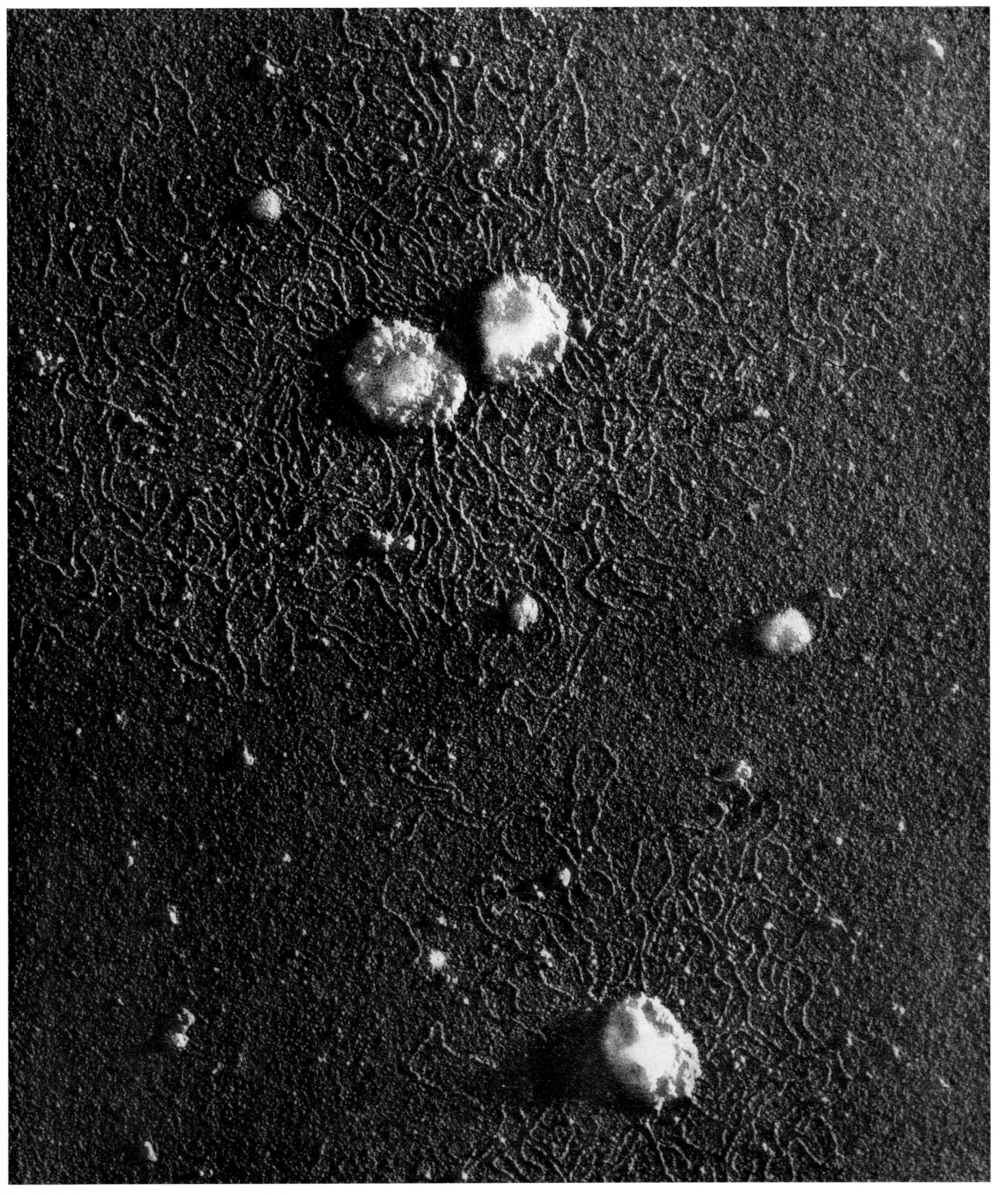

An electron-optical picture of the giant molecules that compose the blood pigment of a snail. A 200,000-fold magnification reveals "artistic" elementary patterns of living matter.

Carl Zeiss, Oberkochen

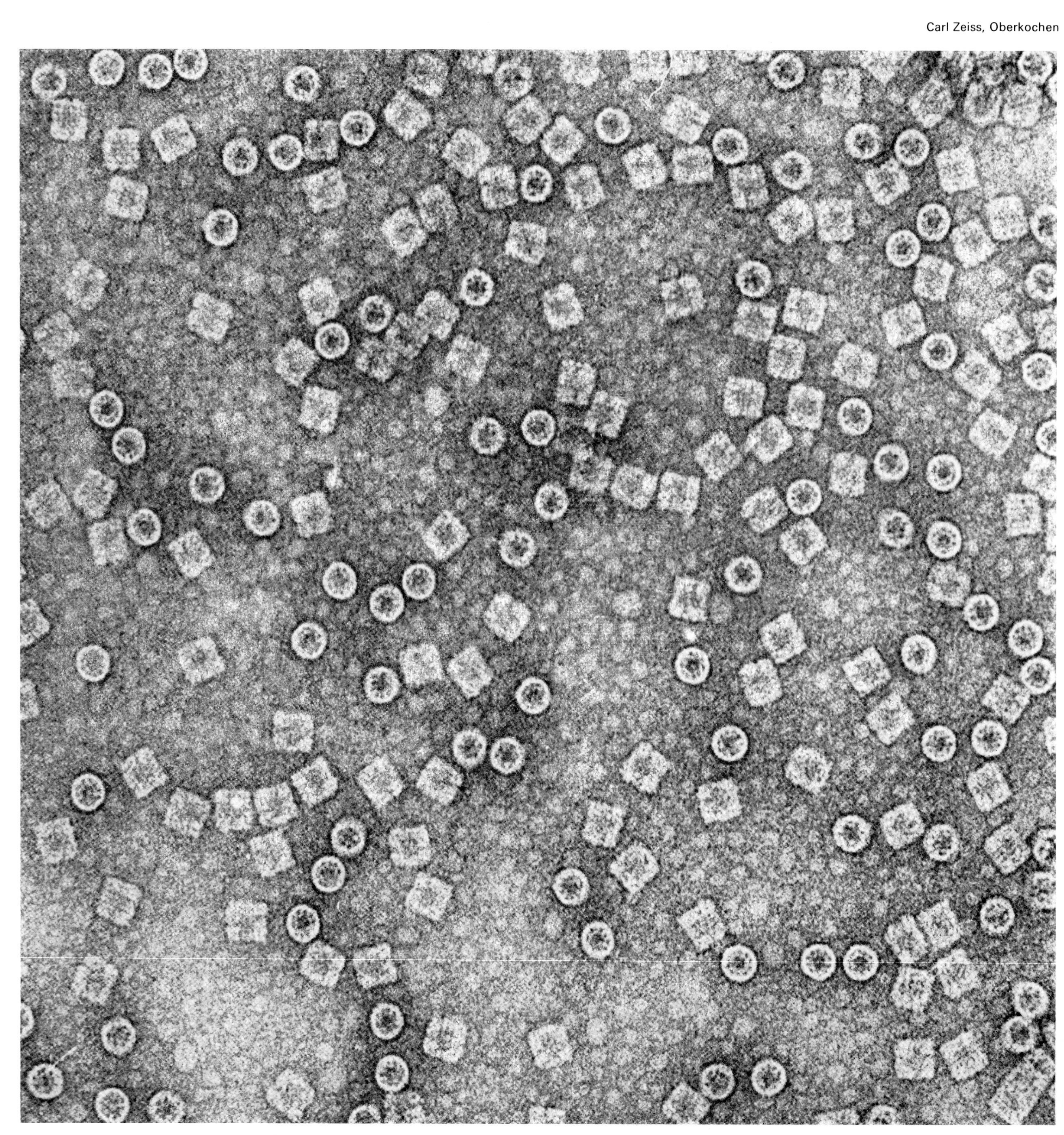

Cross-cut through mature sperm of the silverfish, enlarged 100,000 times by electron-optical means. The forms that life adopts in the realm of the minute and the most minute resemble the works of modern graphic artists.

H. W. Frickinger in the Naturwissenschaftlichen Rundschau

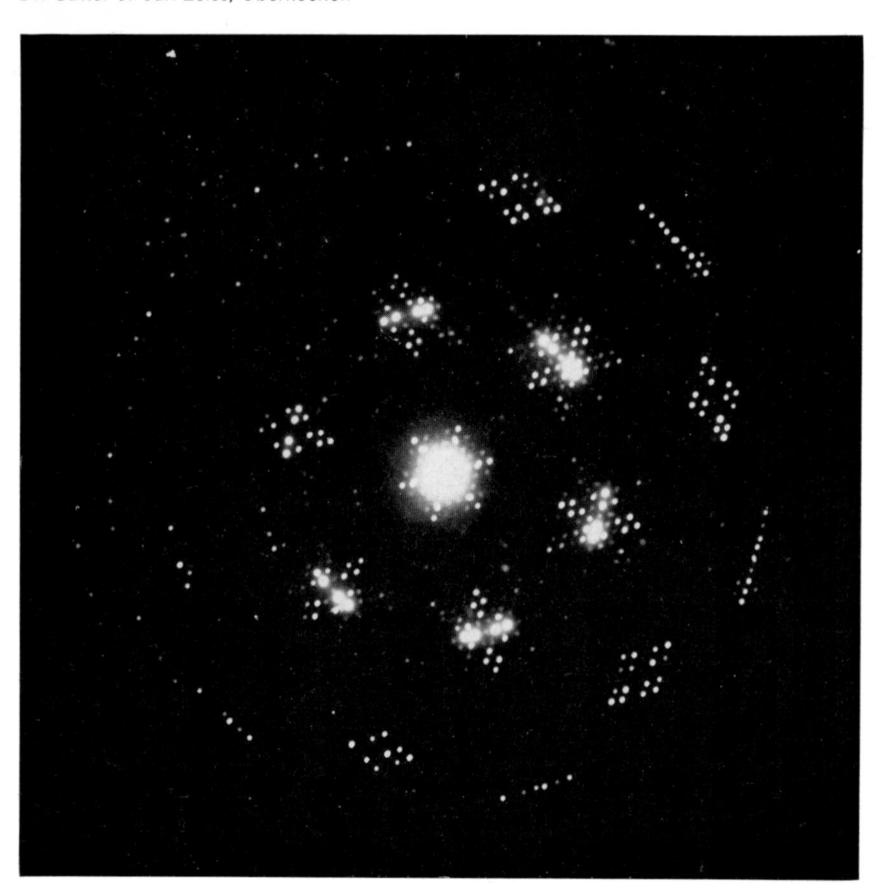

"All things are numbers."

(Pythagoras, 6th Century B.C.)

An electron beam can be diffracted like sound or light waves. Like X-rays, it is diffracted by the atomic latticework of a crystal; the resulting pattern shows the atomic structure of the crystal. Electron diffraction patterns of lead hydroxide salt (above) and of kaolin coated with vaporized platinum (below).

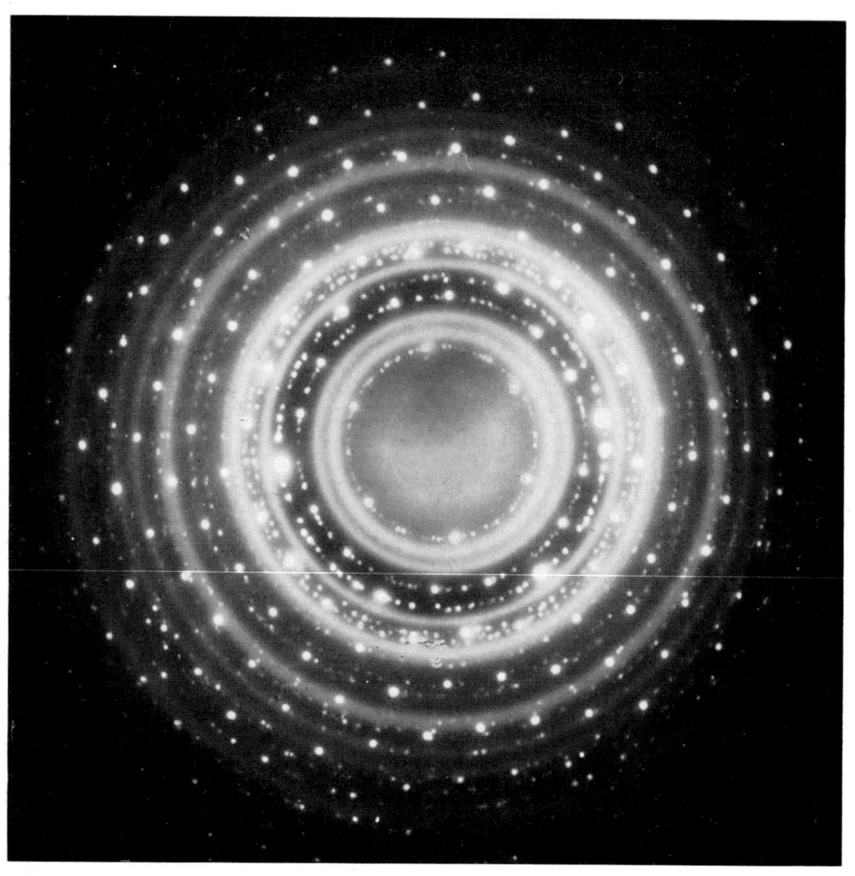

Electron-optical diffraction pattern of polyethylene, a synthetic that today is "condensed" from ethylene gas without application of pressure. A simple arrangement in the electron microscope (the fluorescent screen is located at the rear focal plane) permits corresponding diffraction patterns: the crystal structure then can be deduced.

Dr. Gütter of Carl Zeiss, Oberkochen

Prof. Erwin W. Müller

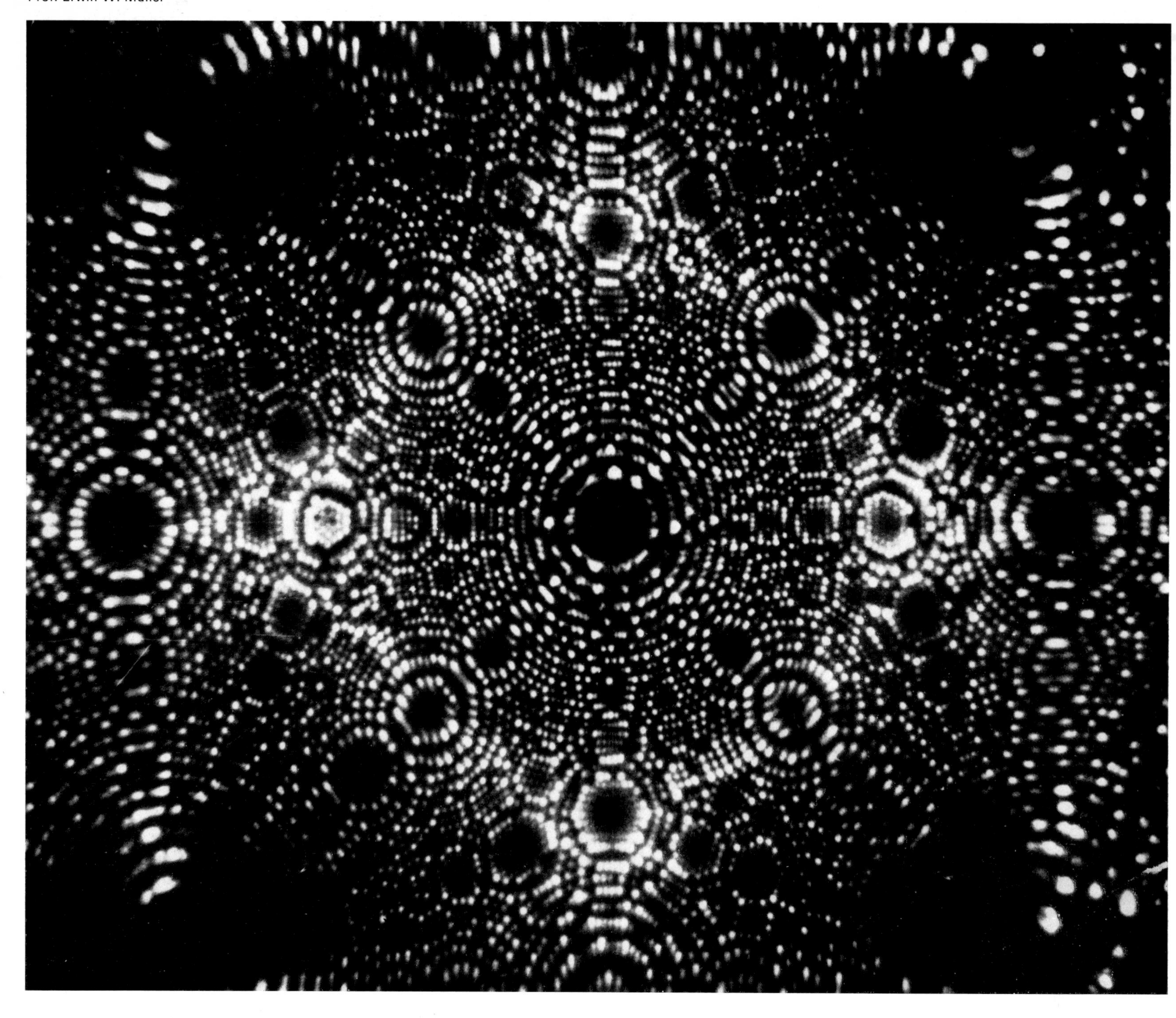

The field electron microscope makes possible magnification up to about one million times; the field ion microscope, up to about five million times. Both are the invention of the German-born Erwin W. Müller, who became a professor at Pennsylvania State University in 1952. Atoms were made visible for the first time, and truly magnificent pictures revealed themselves that permitted a glimpse into the inner structure of the universe.

Erwin W. Müller calls the principle of his instrument a "surprisingly simple" one. The apparatus is essentially only a vacuum glass tube: traces of a vaporized metal coat a point so fine that it consists of only one crystal. When high voltage is applied, ions break out of the point and–widely spread out–picture the atoms of the metal on the fluorescent screen of the picture tube.

A rustling in the texture of the universe.

Ernest Rutherford, the discoverer of the atomic nucleus, called the cloud chamber, invented by C. T. R. Wilson in 1912, "a most original and wonderful instrument." It consists of a glass vessel filled with air that is almost saturated with water vapor, like the air of a bathroom just before the mirror clouds up. In this atmosphere, the trajectories of atomic nuclei and of their component particles leave hairline traces of droplets that quickly disappear. After each photograph, it is necessary to wait until the vaporous atmosphere has been restored. Professor H. Klumb of Mainz, Germany, has constructed a permanent cloud chamber that works with a cooling plate and alcohol vapors. Tracks of radiation particles emitted by radioactive specimens can be filmed in this improved cloud chamber. The radiation of radioactive substances resembles Fourth of July sparklers.

Prof. Dr. Dr. E. H. Graul and Dr. H. Hundeshagen

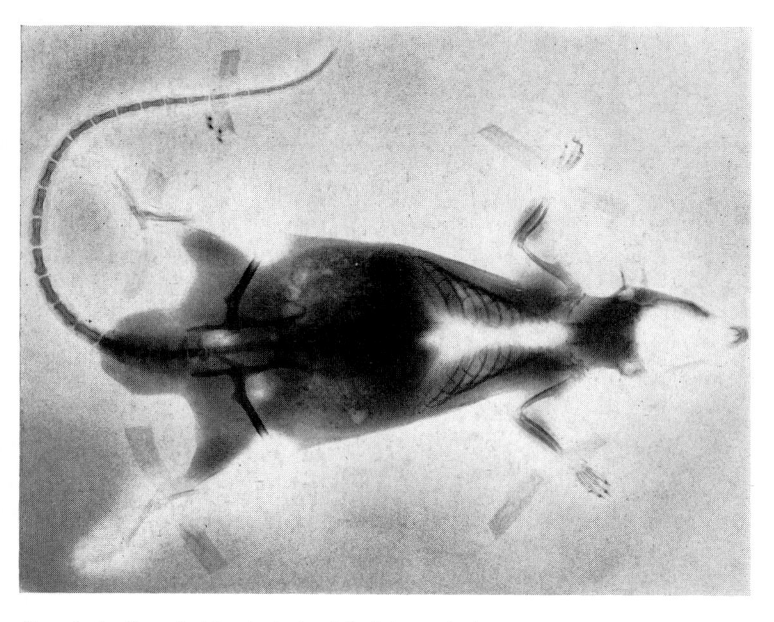

◄ The radiation picture of a rat that was placed on an X-ray film twenty-six days after it had been fed radioactive strontium. The autoradiograph shows that the radiating strontium most of all penetrated the growth zones in the bones. Radioactive tracer elements can be observed within an organism with a Geiger counter; by their radiation they give away their paths and location, and life processes are made "transparent."

Bayerische Braunkohlen-Industry AG, Schwandorf

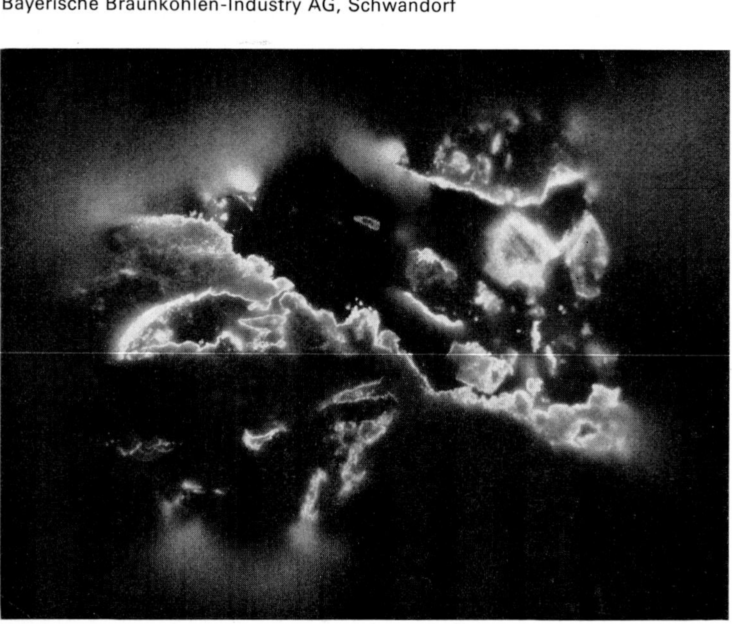

Rock containing pitchblende "exposed" for sixty-seven hours on wrapped X-ray film. The autoradiograph shows the radioactivity of the ◄ uranium vein.

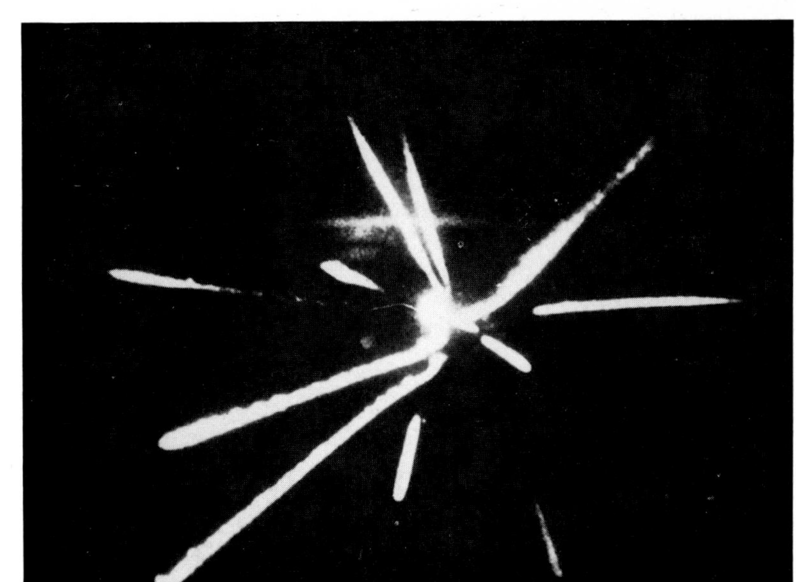

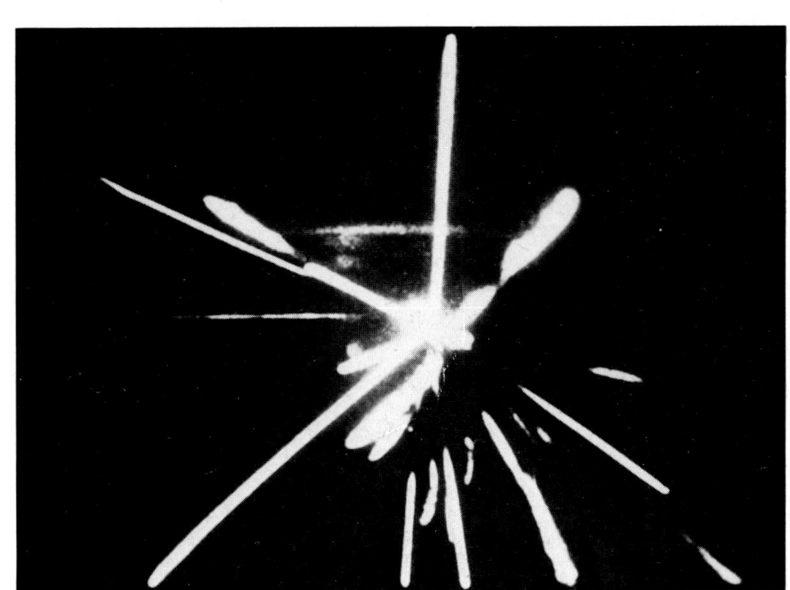

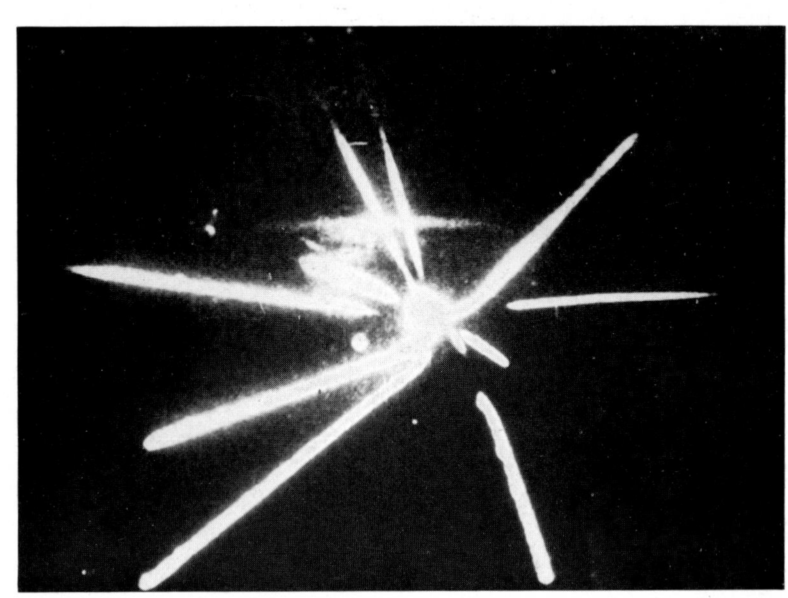

DESY–Deutsches Elektronen-Synchrotron, Hamburg

Nuclear physicists unleash storms of energy to study the behavior of matter under the onslaught. In this picture, electrons have been aimed at a target; it is hoped that the photographs of the paths of the target's fragments in the bubble chamber will help in understanding the blueprint of the universe.

The bubble chamber was devised in 1952 by the U.S. physicist Donald A. Glaser. Its principle: in an overheated liquid (generally hydrogen), colliding charged particles create ions whose paths are traced by the chains of tiny bubbles they create.

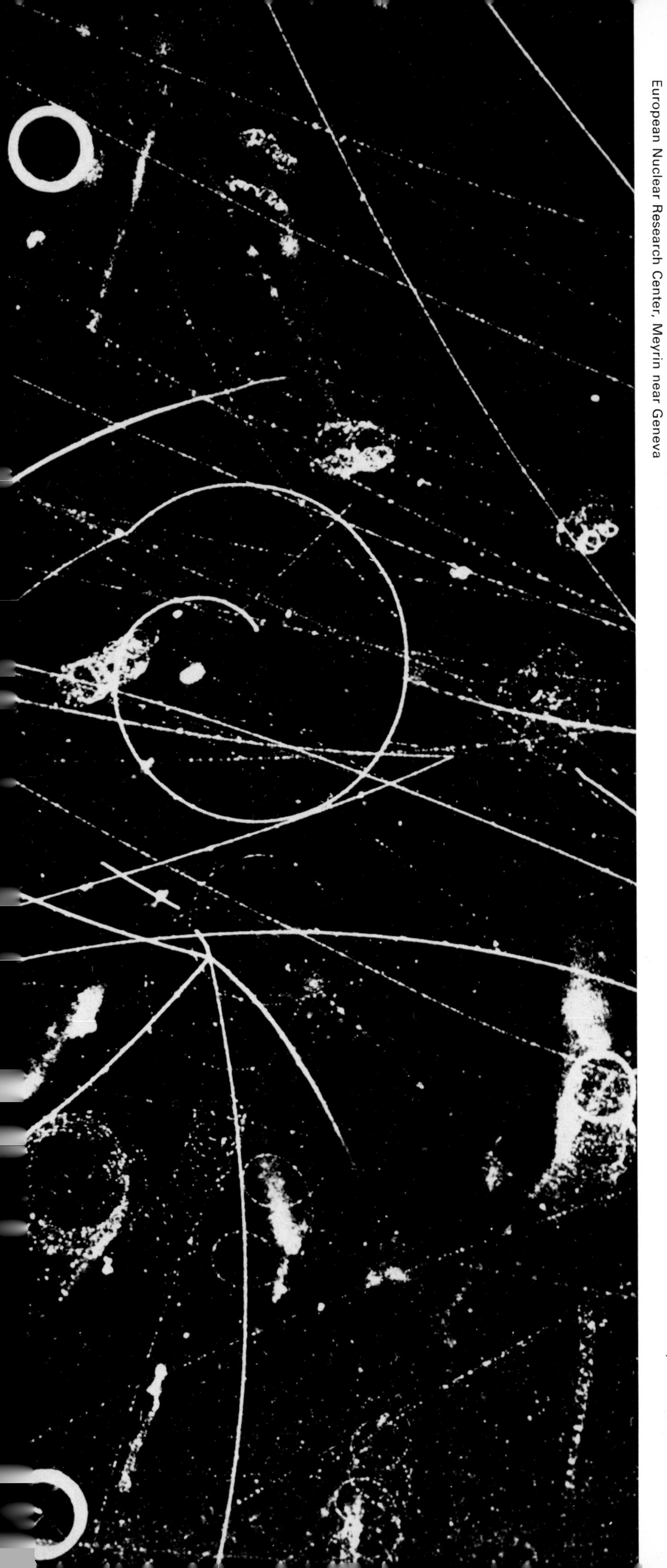

European Nuclear Research Center, Meyrin near Geneva

Institut für Kernphysik der Universität Kiel

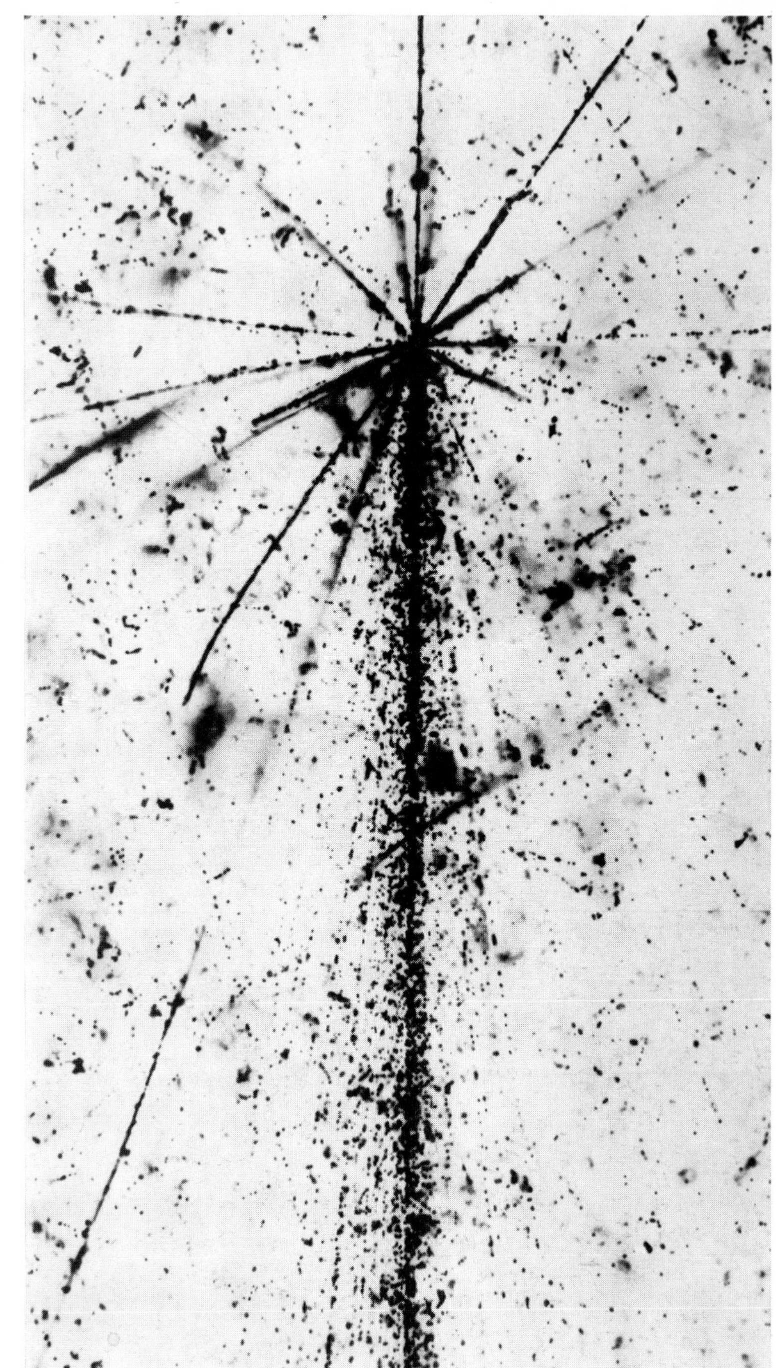

Photographic plates, carried in light-safe packs high above the earth in balloons, registered traces of cosmic radiation. The rays' transition through the film is recorded. It happens at times that a radiation particle will hit an atomic nucleus within the film emulsion and will explode it. The scattering fragments leave a star-shaped imprint that can be measured and will permit valuable conclusions. The picture shows a high-energy particle of cosmic radiation as it collides with the earth's atmosphere; it releases a shower of about 100 mesons.

◂ Building blocks of atomic nuclei have been accelerated by electric fields and powerful electromagnets to almost the speed of light. What happens when the atomic projectiles hit prepared targets, is the subject for study. The energy is transmuted into numerous, usually very short-lived elementary particles; they can be identified by the traces they leave behind in the bubble chamber.

Institut für Kernphysik der Universität Kiel

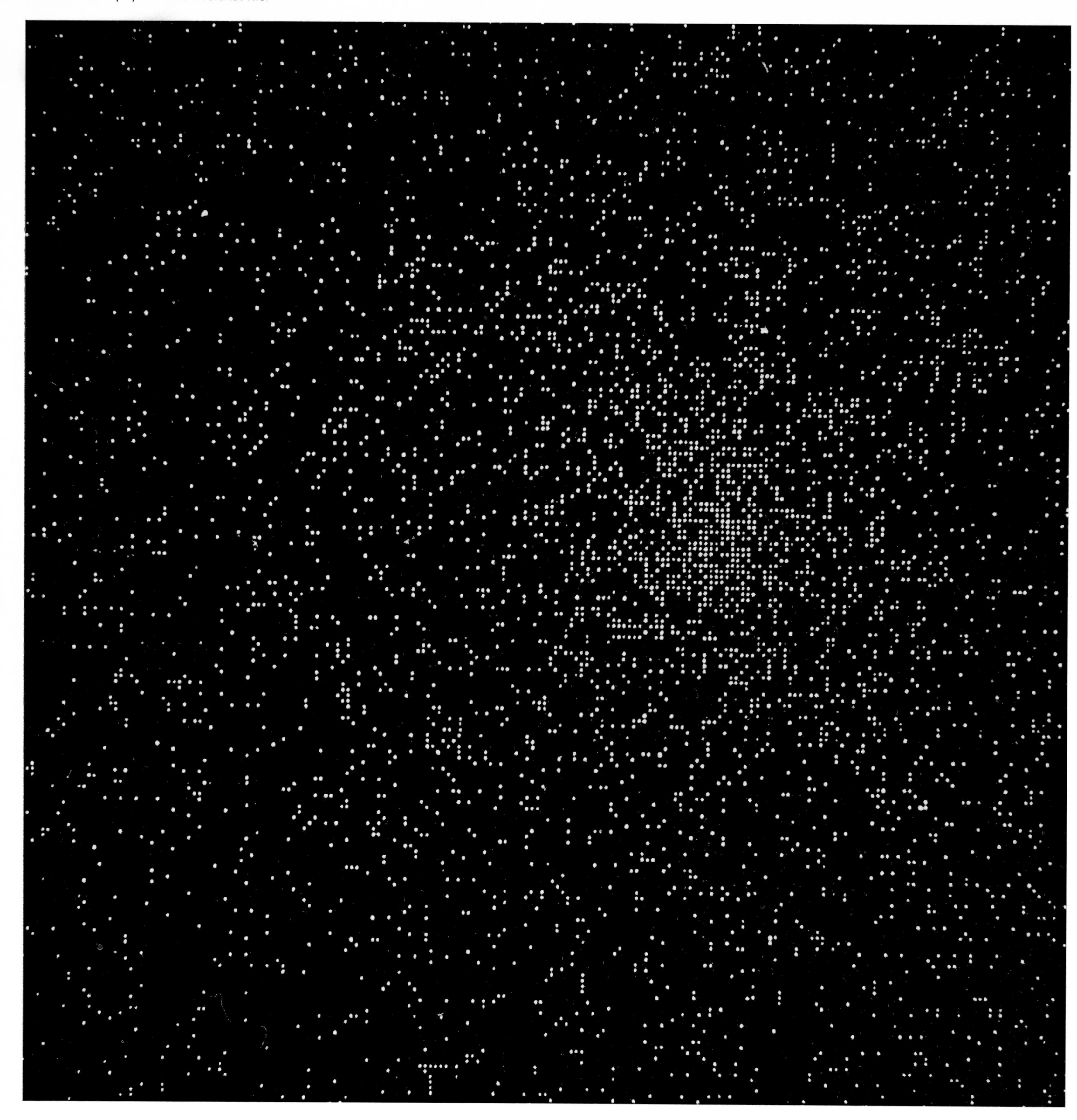

This strange picture documents a shower of elementary particles triggered in an atomic nucleus by the cosmic radiation in our atmosphere. Scintillation counters signaled the "shower nucleus"; to a 32-square-meter plate condenser containing 180,000 small neon-filled glass tubes, a high-voltage impulse was applied: tubes hit by charged particles lit up, and cameras recorded this. The only installation of this kind in Germany registered 10,000 such showers in the course of a year and a half. Our photo shows the shower core of a proton that entered the earth's atmosphere with an energy of 1.5 million GEV. It is hoped that the statistical data derived from such pictures will help explain the nature of cosmic radition. Its energy surpasses many times that produced in the laboratory.

"The universe is impelled from behind by
blind physical forces, a gigantic and chaotic
jazz dance of particles and radiations."

(Julian Huxley)

"Threefold is the step of time / Haltingly approaches the future / Quick as an arrow flies the present / Forever silent recedes the past," wrote the German poet Friedrich Schiller.

But time measurement, astronomy's child, does not want to acknowledge this threefold gait of time. It considers past, present, and future time to be the same. All time is identical—whether hours of boredom that drag or moments of happiness that seem to fly.

Translating the yearly circuit of the earth around the sun into an accurate calendar presents a complicated problem not fully solved even today. Yet the need to measure "time" was already felt thousands of years ago.

Egyptians, Greeks, and Romans had their sun clocks; they measured time by the sun's shadow. In Egypt and Babylon, water clocks were used; these were jars out of which water slowly dripped or into which it trickled, the level sinking or rising from one hour mark to the next.

Sundial, water clock, and the hourglass, whose history also must go far back in time, were all in use long before the Middle Ages. In the hourglass, time can be seen "running out." To "set" the clock may originally have meant to set the hourglass by turning it over.

Time Machines

The invention of mechanical clocks dates from the thirteenth century. The inventor is not known; perhaps he was a monk who hoped to improve the timing of his monastery's round of activities. Already the clocks of the fourteenth century have a device that attests to the desire to subdivide that valuable commodity, time, into small helpings and to dole these out to us. This device is called the escapement. To keep the clock from reacting without any check to the pull of the weight drive, to prevent it from running down faster and faster, a "brake" was installed to slow down the clockwork and force it to revolve at a steady pace. A bar with small weights oscillates back and forth; it governs the gait by alternately blocking and unblocking the crownwheel, tooth by tooth. It provides the beat that regulates the movement and forces the clock to progress in small separate steps that are as evenly spaced as possible.

It is a great experience to encounter one of those mighty mechanical clocks of the Middle Ages in an old church tower, or in Munich's Deutsches Museum "for masterpieces of natural science and technology": forged, ungainly time machines, they seem all the more impressive in our own age of high-precision instruments.

In the seventeenth century, the pendulum was added, bringing about a sorely needed improvement in accuracy. The mechanical clock did not immediately replace the older clocks, as one would have expected. It was much too inaccurate. One can well imagine how the new invention was greeted by the abbots of the monasteries, the village mayors and the burgomasters of the free imperial cities: What an expense! What a complicated mechanism! And to make matters worse, these early, expensive, noisy mechanical clocks perennially needed to be adjusted by checking them against the noiseless, much more accurate sundials!

When Galileo was already blind, he had his son Vincenzio make a drawing of the best way to fit a pendulum into a clock. The result was as expected: with a pendulum, swinging back and forth exactly sixty times per minute, the minute had been divided into seconds. The comparatively accurate pendulum became part of that era's artfully worked clocks that showed in beautiful designs the relation of the hours of the day and night to the sun's rising and setting, the relation to the stars, the phases of the moon, and the signs of the zodiac. Soon clocks were showing not only the hours and quarter hours, but also the minutes.

The splendid table clocks of the Renaissance incorporated an innovation devised by the great Dutch astronomer Christiaan Huygens—the balance spring. Another improvement, based on an invention by the Dane Ole Römer, was the transmission of the pendulum's swing to little wheels whose teeth were shaped in a special manner. This type of toothed gears is still commonly used today; they assure that the wheels work with minimum loss of momentum through friction.

In 1510, Peter Henlein, a locksmith, built the famous "Nuremberg egg," a pocket watch with an interesting type of escapement. The springs that caused the small oscillating bar to swing back and forth consisted of pork bristles. These pork bristle springs were later replaced by a wound iron spring. What a long way yet to the extravagant "time machines" of our own day: there exists a clock that "lives on air," that is kept going by the most minute changes of temperature, so sensitive that it functions even in an air-conditioned room; there is another clock that runs on light–solar cells convert light into electric energy. The most accurate time-pieces are extremely sober-looking instruments like the softly humming quartz crystal clock that keeps time to within a millionth of a second a day. The atomic clock is capable of registering even the slightest deviation in the earth's rotation. The very newest clocks close the circle: since they utilize vibration and radiation to measure time, they are related to the many thousand-year-old sun clocks.

Limits of Perception

What is the fastest action that eye and brain can perceive? Can the speed of human perception ever compete with the stopwatch–measuring one tenth of a second–of a judge at a sporting event? The late E. R. Dietze, a radio sportscaster who specialized in horse races, was famous for his unerring eye. Once, at the English Derby, when four horses crossed the finish line neck to neck as one fantastic sixteen-legend creature, Dietze not only instantly announced the winner, but also numbers two, three, and four. But a preliminary official announcement listed the horses in a different order. "Let's wait for the photographs of the finish," was the reporter's unconcerned reaction. As it turned, out, the judges had to reverse themselves–E. R. Dietze had been right. Eye and brain had taken in minute details, had registered them like snapshots within a fraction of a second, had evaluated them critically and confidently, before he even had completed his sentence. An amazing feat that illustrates the potential of human perception.

Yet vision does have narrow "time limits." There are phenomena that inherently do not require enlargement or special lighting to make them visible, events that take place in "full view" and yet still cannot be seen by the human eye because they happen much too fast or much too slowly. Only photography can expand to an unimaginable degree the limits of vision.

Nature likes to take its time. On windless days, clouds seem to stand still. It becomes evident after a while that they have changed, but how they shift can be observed only with some difficulty. It is an event close to the limit of the perceivable.

Folded mountains have risen up over millions of years. The history of man is too brief to have registered such changes from firsthand observation. Even the growth of plants is hidden from our view. Observation after an interval of time will show the cumulative effect, yet we cannot really see the change happening. Within a few days a tree bursts into bloom. We may admire the progress; the progression we cannot see.

Condensing Time

Excessively slow motion can be made visible only by the motion picture camera, a camera that can condense the happenings of days into a brief reportage, that can contract time into a manageable quantity. Picture sequences arrived at in this way are astonishing. What takes place in the course of days and hours, unrolls in thirty seconds: the flax-dodder, a climber, winds itself around a morning glory which in turn begins to entwine its rival. This slowly progressing reciprocal embrace "condensed" in time, turns into a dramatic scene.

Fritz Brill of Hofgeismar, near Kassel in Germany, recorded with time-lapse photography how a little silkworm emerges from its pinhead-size egg, an event that after all takes two minutes. Photographing such a tiny object in precise focus and properly three-dimensional lighting is by no means easy. Very strong light is required, especially so since the silhouetted movement inside the egg also has to be made visible. Cooling agents and ventilation must be used to counteract the heat from the lighting. "The preparatory biological work kept us busy, all right. Time and again we weighed the eggs to make sure there was life within. For months they had to be stored in a cool place; then, for the decisive days, they were placed in an incubator and exposed to rising temperatures. By the time there were signs of life indicating that emergence from the eggs was imminent, the egg had to be in position, perfectly lighted and with all the auxiliary apparatus in working order. All in all, this was an undertaking that required meticulous planning of every detail."

The assignment seems a simple one: an egg, the size of a pinhead, from which a silkworm will emerge, a two-minute action to be recorded in close-up and time-lapse photography. It sounds so easy, and yet, it turned into a rather exciting undertaking.

At some other time, Fritz Brill filmed a silkworm spinning its cocoon. In eight days, a mile-and-a-half-long silk thread is produced by the worm. The film shows the whole sequence in one minute. Within the transilluminated cocoon, the worm appears as a moving shadow. Its leisurely work becomes a hectic struggle, a frenetic action, when seen in time-lapse photography. The spinning design, passed on by nature over hundreds of thousands of years, can thus be demonstrated in an instant.

Fritz Brill has called his field of work "photoanalysis," a concept, concerned with the "optical documentation of interrelations," that has become widely accepted. Photoanalysis is able to point the naturalist into new directions and to provide him with new insights. The germination of a bean, a process that extends over four days, can be studied on a three-yard-long filmstrip. Documentation that "condenses" a sequence of events to manageable size provides reference material that can be studied as often as desired, frame by frame, or as a whole interrelation of movements.

Fritz Brill has his own laboratory where he solves problems with ingenuity and scientific precision enhanced by artistic camera work. To photograph the germination of a bean in time-lapse technique Brill used photocells that electronically directed the time-lapse apparatus. The instrument was set up like a loom. "When the threads are well arranged, everything runs by itself. If necessary, the apparatus can run in this fashion for weeks, taking one picture every minute or every hour." At the showing of the film it became evident that everything had been taken into consideration: the sprout did not come up jerkily, but it slithered out of the ground like a snake and on toward the light. The first leaf unfolded as if by magic.

Photoanalysis may take on some very pedestrian tasks. Fritz Brill has filmed how a cake "grows": how the baking powder does its job, how the cake rises inside the heat-resistant glass pan, how the gases of the leavening agent escape, how the cake turns brown, how it cracks. The camera had a ¾ inch depth of field. The dough rose 2¾ inches. Brill had installed a rig that caused the cake mold to descend as the dough was rising. Moisture vapors had to be eliminated. The camera had to function without a hitch even at 372 °F. Shifts of no more than $^1/_{24}$ inch would have caused the action to unroll unevenly. It is obvious that this kind of work requires much thought and advance preparation, much patience and technical know-how.

One of the trickier assignments was to photograph a Madagascar spider, the producer of a fine silk thread that has technological uses. A tiny spot on the hind part of the spider had to be brightly lighted without actually putting the "spotlight" on the animal; the spider will not spin when it feels threatened.

Hans A. Traber, from Zurich, has filmed the maturing of sperm cells in the testicles of a grasshopper to show the behavior of the chromosomes responsible for the still puzzling workings of heredity. One can see how the chromosomes, small filaments, align themselves in the center of the cell; how they suddenly move apart, as if by command; how the cell becomes pinched-in at the middle and finally divides. Twelve hours, telescoped into thirty seconds. With a microscope and time-lapse photography we get close to the threshold of the ultimate secret of life.

Time-lapse sequences are among the most exciting experiences made possible by photography: by compressing time, one can show a mushroom literally "shooting" up; a drop of acid attacking metal; crystal needles growing in a solution; oil drops drifting like comets in an emulsion, trailing their tails of droplets after them; water, whose surface tension has been reduced with the aid of soap or detergents, carrying off a grease spot. Time-lapse photography changes the trivial occurrence into absorbing adventure. The researcher and the scientist gain valuable clarifications through it, while the layman receives the gift of fascinating pictures.

Expanding Time

And now for the very opposite: whenever events occur in fractions of a second, in much too brief an instant, then it is slow-motion that makes the invisible visible. Sports-lovers know it well: the sight of the goalkeeper at a soccer game who stretches sleepily toward the slowly drifting ball, who dreamily follows with his eyes the ball floating past his fingertips. A twelve-yard penalty kick or the decisive goal, repeated in slow-motion, are by now familiar feasts for the eye. A jump from the high diving-board, a swing on the parallel bars, a full circle on the horizontal bar, the apparently weightless leaps on the spring-board—these become—with as little as two or four times "expanded time"—experiences of dreamlike beauty. A flight of pigeons alighting on Venice's St. Mark's Square reveals itself, when seen in slow-motion, as a most graceful spectacle.

To give our eyes the perfect illusion of movement, motion pictures are reeled off at the rate of twenty-four single frames per second. When a scene has been shot with two or four times that number of frames per second, and this film is then projected at the normal rate of twenty-four frames a second, the action appears slowed down, dreamlike, unreal. In this way, the legerdemain of the card juggler becomes a phlegmatically performed "how-to" lesson.

How long have we had slow-motion photography? Toward the end of the last century, French scientists filmed the fall of a cat safely landing on its paws, and a celluloid ball dancing on a water-jet in a shooting gallery. In 1894, motion pictures were taken with a "photographic shotgun"; at one hundred shots per second, motion was slowed down four times, and one could see a greyhound floating along a racetrack.

In the 1920's, the German engineer Thun constructed a camera capable of shooting 2,000 pictures per second. This camera was subsequently perfected at the Allgemeine Elektrizitäts Gesellschaft, a research institute, first under Dr. Werner Ende's direction, later under Dr. Hehlgans's. By 1932, a rate of 80,000 pictures per second had been achieved. Two years earlier, Werner Ende had filmed falling drops of water, flying bullets, electric sparks, the burning out of a fuse.

At about the same time, Harold E. Edgerton, professor at the Massachusetts Institute of Technology, worked on the problems of high-speed photography. Edgerton is also one of the great pioneers of "high-speed cinematography," the filming with thousands of frames per second. Edgerton's work has caused sensations: the multiple exposure shots of the stroke of a golf player, his motion fanned out in fifty and more positions, taken at one hundredth of a second intervals, and with a shutter speed of one hundred thousandth of a second (speed of the ball = 240 feet per second); the whirling flight of a hummingbird whose wings flap up and down more than sixty times a second, dissolved into razor-sharp pictures; the vibrating tip of a fencer's foil; soda water squirting out of a siphon; the roll of drumsticks; the lightning-fast strike of a poisonous snake; all this shown as "frozen" motion that can be analyzed frame by frame.

The Edgerton Crown

Professor Edgerton has discovered beauty that had never been seen before: the fall of a drop of milk filmed at 2,000 exposures per second—just a drop falling into a shallow bowl, but what a surprise! One sees how the drop detaches itself, how the narrow neck of liquid breaks, rolls itself into a tiny ball, and follows the larger ball; how the drop, falling into the milk, raises a circular wall, how the liquid rises out of the center, how a column grows out of the crater, forming on its summit another ball, which slowly follows the collapsing column downward. The spectacle of thousands of dancing, glistening splashes during a rainstorm is magically slowed down in the professor's photographs, and an exciting interplay of surface tension and gravity is revealed.

When a drop falls into a shallow dish of milk, it forms a crown whose points each bear a "pearl." Slow-motion makes this structure last for a moment, until gravity destroys it. The Museum of Modern Art in New York has displayed such an Edgerton crown.

Scientists and technicians never tire of photographing the falling drop. There is no institute active in the field that has not experimented with this attractive theme. The falling drop of milk creating a twenty-five-point crown, the falling drop of water forming a crown of glass—these crowns are the symbols of international congresses for high-speed photography and high-speed cinematography, where experts meet every two or three years. In 1965, I was present at a lecture by Edgarton at the Zurich congress. A slender sexagenarian, with the relaxed manner of the typical American university professor, he set up a few instruments with cheerful calm. During his lecture, he lightened the dryly scientific with vivid demonstrations and an occasional joke.

Frozen Motion

For the filming of rotating motions at constant speed, as for instance the turning of a propeller or a turbine, one uses a lamp that flashes a light that is stroboscopically controlled. "Stroboscopy is in a way an optical illusion."

Edgerton started a small desk fan; he turned a flickering light toward it. Sure enough: from the blurred circle of the rotating fan emerged clearly the shape of the fan-blade. If a flash-gun shoots as many flashes per second as needed to illuminate the blades of a turning propeller always in the same position, then the blades seem to stand still. An optical trick causes the very rapid rotation to "freeze." One is reminded of a phenomenon sometimes observed at the movies: the spokes of a wheel seem to come to a stand-still or even to turn backward. This effect appears when the frequency of picture frames per second and the rate of the object's rotation are in a certain relation to each other.

Harold E. Edgerton caused a disk with a black and white pattern to rotate. A rapid succession of flashes synchronized with the speed of rotation lighted up the disk in the same position at each rotation. Out of the gray blur of the swirling disk emerged the black and white pattern, which moved forward a little, then back, and then came to a halt. If it had not been for the whirring sound of the moving disk, one would have believed that, indeed, it was standing still. Of course, said Edgerton, switching off the machine, this only works with a turning wheel, a turbine, or a propeller rotating at a constant speed.

During the early 1930's, Edgerton was constantly on the lookout for fast motion. Whenever he came upon an action too fast for the eye, he went to work: card tricks in a variety show, splintering glass, the flight of a dragonfly, the erratic flapping of bats, the smashing serve of the tennis player, the springy leap of the ballerina.

Edgerton was curious to know how a cat manages to lap up her milk, how it "shovels it in" just as an elephant takes in his food with his trunk. He has been interested in something as commonplace as the peeling of an orange. Shot at 2,000 frames per second and projected at normal speed, the pictures show the volatile oils squirting from the skin pores as if from many tiny fountains. Technology assumes here a quality of playfulness that is almost poetic. And, playfully, high-speed motion pictures show what they can achieve, how motions too fast for human vision are slowed down for observation and study. Motion pictures of the serve of a tennis player, slowed down eighty times, show how the impact o the racquet flattens the ball and indents the strings.

The Captured Bullet

Demonstrating to his students how to catch a bullet in flight with high-speed photography, presents no problem for the professor. In his Boston laboratory he had a shooting gallery set up. There, a rifle is firmly fixed in place; a playing card is the target. The card is positioned in the direction of the shot so that the bullet will hit it edgewise. The professor wants to catch this very moment. He places a small microphone in front of the rifle barrel, slightly off the trajectory. Its task is to record the sound wave that accompanies the bullet and to trigger the flash–half a millionth of a second!–with an electric impulse, at the precise moment the shot will tear the playing card apart. It would also be possible to trigger the flash by means of a photoelectric cell that would react to reflected light or to shadow.

The shot cracks! The playing card is torn in two and Edgerton has photographed this very event that happened too fast for the eye to see. The flash was perfectly timed! The professor is a shrewd psychologist–he arouses his students' interest with games. Aspiring technologists are–and remain–big children; it is fun to shoot away at the king of hearts. The playing card in the process of being torn by the bullet is a classic masterpiece of high-speed photography, and Professor Edgerton likes to present to his many friends and admirers all over the world a postcard with this photograph as his signature.

Another Reality

"Dr. Edgerton's photographs provide a unique and literal transcript of that time world beyond the threshold of our eyes. In these pictures are not only facts to help us in seeing and doing, but new aesthetic experiences, new horizons in observation, the stimulation of penetrating a new world of time where the 'visual impact of the split second' gives a fresh aspect to the commonplace." (J. R. Killian)

Making flying bullets visible may be an excellent example of the wide range of possibilities of ultra-high-speed photography, it is by no means the ultimate goal of extreme stop-action photography. Modern technology works with dizzying speeds, far exceeding that of a rifle bullet. Extreme stop-action photography uses increasingly shorter exposures to catch up with the ever faster motions of events unrolling at speeds of several miles per second, such as explosions, the air-flow in faster-than-sound wind tunnels, the rotation of ultra-centrifuges.

Walter Thorwart, who works in Hamburg together with Dr. Frank Früngel in the field of "impulse physics," in the course of testing high-speed cameras, has made visible and understandable on film some fast-moving everyday technological processes: at 3,000 frames per second, that is, with "only" 120 times slowed motion, one can "leisurely" watch what happens inside a vacuum flashbulb during 1/125th of a second–how the metal strands inside the bulb catch fire, burst into flame, fade out in an afterglow following the "flash."

A thousandfold time-retardation creates a dramatic and instructive event out of the burning-out of a defective radio fuse during a short circuit. The wire spirals melt into pellets that form a chain of tiny black pearls: the soldered connection that was meant to open, did so too late.

The arclight of a welding torch, precluding close scrutiny because of its excessive brightness, discloses its workings when filmed at around 6,500 frames per second–how drops form at the tip of the electrode and how these blend into the welding seam. In this way, calculations and experiences can be checked against visual evidence. The research engineer gains valuable documentation; the layman, a look into something resembling the crater of a volcano.

A planing machine paring off metal shavings may seem a commonplace piece of machinery to the expert. It ceases to be that, once it fails to function properly without any apparent cause. When, by slow-motion, the machine's processes are slowed down as little as ten times, the two seconds it takes to pare off a metal shaving become one third of a minute. The film can be projected again and again, endlessly, and this opportunity of seeing the process over and over may possibly bring the "insight."

That a window pane breaks we are all aware of since childhood. How it breaks, we only know since there are high-speed motion-picture cameras, ultra-slow-motion projection. At 2,000 pictures per second–that is, 80 times slowed down projection–sudden cracking becomes a comprehensible happening that gives the technologist important information. The way a pane of glass breaks, whether into sharp daggers or into blunt crumbs, may mean the difference between life and death. Very often substantial improvement of a product depends on whether it has been possible to observe, step by step and in detail, promising intermediary experiments.

Let us turn once more from technical considerations to the poetry of the falling drop, to Edgerton's crown, which is, and will remain, a hallmark of high-speed cinematography. It can happen that within the crown, the drop falling next, forms a second smaller crown or that the liquid column growing out of the center of the crown breaks up into a chain of descending globules.

One may combine the falling drop with another structure of miraculous beauty, an opalescent soap bubble. The drops generally fall straight through the bubble without damaging it. It takes a surprisingly long time before the delicate soapy skin, only a few molecules thick, falls victim to the barrage of drops.

Soap bubbles already played a part in the first successful experiments with extreme slow-motion. The pictures were taken in 1903, with the help of a very fast succession of sparks that cast shadows on the film and imprinted it with a silhouette of the object. With a series of, let us say, 1,200 sparks per second, an equal number of silhouette pictures of the action was taken. This made it possible to record how a shot-gun pellet pierced a soap bubble, how the fine skin burst, how the exploding structure dissolved into floating particles, how the droplets rapidly combined into a twisted chain.

The conviction began to grow then, that once shorter exposure times would be possible and faster moving film available, a new and absorbing world of motion would open up.

In 1908, A. M. Worthington, professor of physics

at the Royal Naval Engineering College, Davenport, England, published *A Study of Splashes.* He wished to share, he wrote, "some of the delight that I have myself felt, in contemplating the exquisite forms that the camera has revealed... the multitude of events, compressed within the limits of a few hundredths of a second but none the less orderly and inevitable...."

Professor Worthington photographed in darkness, using electric sparks; from his pictures he gained valuable insights into the mechanism of surface tensions and into the changes of form taking place in the bounding surface of a liquid.

"It would be an immense convenience," he wrote in conclusion, "if one could use a cinematograph and watch such a splash in broad daylight. But the difficulties of contriving an exposure short enough to prevent blurring, either from the motion of the object or from that of the rapidly shifting film, are very great. Anyone who may be able to overcome them satisfactorily will find a multitude of applications awaiting his invention."

A Cinematographic Trick

We have talked of "ultra-slow-motion." Where science deals with optical magnification, it distinguishes a magnifying glass from a microscope. When we talk about time, we might compare slow-motion photography as we know it in filmed sporting events (where time is expanded two to four times) to the "magnifying glass," and the more extreme expansion of time in modern ultra-high-speed cinematography to "microscopy."

"Just as the magnifying glass and the light-microscope make the world of the small accessible to the eye and to the photo-camera," writes Walter Thorwart, "so do stop-motion photography and slow-motion film projection make it possible to observe fast motion. The step from the magnifying glass to the microscope, in optics, corresponds to the one from slow-motion to ultra-slow-motion projection. The factor of time expansion rises from the roughly threefold of the usual slow-motion in newsreels of sporting events, to the several thousandfold of the latest time-expanding techniques. To the microcosm there, corresponds the infinitely small time-segment here."

Obviously, terms like "time magnification" and "time microscopy," or "time expansion" and "time contraction," are no more than conceptual aids. Time is unchangeable; it can neither be expanded nor contracted. A second always lasts a second. What is possible is this: pictures of an event can be taken in large intervals in the case of time-lapse photography, or at lightning speed and with briefest possible intervals in the case of high-speed photography, and then projected at the normal rate of about twenty-four frames per second, the speed that is geared to the sluggishness of our vision.

When a moving object has been photographed at a rate of twenty-four pictures per second, and the film is projected at that same rate, the motion appears free from jolts and completely natural. The comical jerky movements of people in early motion pictures were caused by the fact that the action was filmed with fewer that twenty-four pictures per second, but was projected at that rate.

"Expanded time" is not real, it is a cinematographic trick. To be sure, it is not the kind of trick that a director of a film drama would employ to show something that never actually happened; it is rather the trick of making visible, through manipulation, a real event that in its too fast progression is beyond the horizon of human vision. Time expansion, "time microscopy," increases our comprehension and awareness, opens up barely imaginable new dimensions.

Life expresses itself in motion. For three-quarters of a century we have had motion pictures, we have been able to "store" movements and to reactivate movements. Obviously, we would like the motions on film to correspond to real-life motions. Experience has taught us to have a fairly good idea of the speed of normal movements. We are familiar with many thousands of motion sequences. We know by instinct the rate at which they progress, regardless of our good or bad sense of time. Whenever a motion seems different we are immediately alerted. If someone's movements are noticeably slow, we think of sickness or an attack of faintness and wonder whether we shouldn't give some assistance. To see a man cross the street at a time-contracted or time-expanded pace strikes us as anomalous. On film, either one appears slightly comical. Exaggerated time contraction soon makes us nervous; exaggerated time expansion quickly becomes boring—assuming always that the movements are familiar ones and that we do not wish to see them changed, as for instance in a motion picture drama. But time contraction and time expansion become interesting whenever a drive for new scientific knowledge is involved, especially where the unaided eye is not up to the task anymore, where it can follow the "much too slow" or the "much too fast" events only via manipulation of time, by time-lapse or by high-speed cinematography. What then becomes visible, is a completely different, surprisingly new reality. A bursting soap bubble is a common sight—but how it bursts can be observed only in motion pictures.

Aiming for the Trillionth of a Second

The faster the motion, the shorter the exposure has to be for a sharp picture to result; the more pictures per second, the clearer the information furnished by the film when it is projected.

Exposure times in high-speed photography begin where the fastest exposures used in ordinary photography leave off, at about the millisecond, one thousandth of a second.

They range via the microsecond into the range of the nanosecond, although the advance guard of "high-speed physicists" have begun to measure in picoseconds.

It takes an effort to conceive of such time segments: the millisecond, one thousandth of a second (0.001 sec.) can be found on the more elaborate of amateur cameras. The microsecond is one millionth of a second (0.000,001 sec.), the nanosecond is one billionth of a second (0.000,000,001 sec.) and the picosecond is one trillionth of a second (0.000,000,000,001 sec.). To register such time spans, to measure them and to devise the means for such short exposure times may seem believable to us who live in the twentieth century and are constantly made aware of technological advances. The time spans themselves remain what they always were: inconceivable.

Professor Hubert Schardin, who died in 1965, and who was one of the great pioneers of high-speed photography, tried to find some way to make such time fragments comprehensible. He tried it with a linear measure: the distance from Basle to Rome, 1,000 kilometers or 621.4 miles, he equaled to one second. The millisecond then corresponds to one kilometer, the microsecond to one meter, the nanosecond to one millimeter. The picosecond, the trillionth of a second, is not visible to either the unaided eye or the magnifying glass; it can be seen only through a good microscope. The comparison, therefore, starts with the 1,000 kilometer distance between Basle and Rome and ends in the invisible.Even if one were to equate the distance to the moon with one second, the picosecond would still only correspond to one third of one millimeter. That, at least, would be visible to someone with normal eyesight. Let us consider how far unimaginably fast light (186,000 miles = 300,000 kilometers per second) travels in one picosecond. It "creeps" three tenths of a millimeter.

Comparisons like the above define the enormous range that is observed and measured by physics, and show that the concept "time-microscopy" does not really do justice anymore to the scale involved. The light-microscope magnifies only about one thousand times. Even so remarkable an instrument as Erwin Müller's field-ion microscope that for the first time made atoms (if not their nuclei) visible, has a magnification of "only" five million.

Ultra-high-speed photography, on the other hand, operates in nanoseconds, "expands" an event one billion times.

Capturing the Split Second

At this point one could let matters rest. Many people have retained from their schooldays a horror for equations, diagrams, figures, and technical concepts. So I'll pass these by and try to describe, with a minimum of "technical jargon," the adventures of technicians who engage in "time microscopy." How is one to visualize their tools, the "microscopes" with which they expand time, million-, billion-, trillionfold?

Let us briefly recapitulate some previously mentioned points: to get sharp pictures of very fast-moving objects we need first of all correspondingly short exposure times. The common store-bought electronic flash makes exposures of around one thousandth of a second possible. For many purposes this is still much, much too slow. Also, it takes seconds for the next flash to be triggered. For a motion picture camera, the still-photographer's electronic flash is therefore eliminated. In its place, a microsecond flash has to be used. This is a source of flickering light that can emit short bursts of light in extraordinarily fast regular succession.

Since the 1950's, there exists the high-frequency electronic flash called "Strobokin" that flashes up to 50,000 times per second, and whose spark frequency can be precisely controlled. Each spark lasts for only one millionth of a second; it gives off an exceedingly bright and relatively cold light. Ways had to be found to keep the sparks from holding over and merging into an arc, and to insure that the sparks were of precisely timed duration. Dr. Frank Früngel of Hamburg, Germany, created an almost pinpoint microsecond flash in a spark-discharge tube filled with rare gas. The device that prevents hold-over and times the spark-gaps, involves a hydrogen-filled tube. With exposure times of one millionth of a second, very fast movements can be "frozen."

Incidentally, an interesting physiological fact emerges in this connection: the human eye can see a projectile in flight very well, if an appropriate flash tube lights up the scene for very brief time spans. Assuming that five bursts of light are flashed at a frequency of 20,000 flashes per second, then the projectile will be lighted five times for one millionth of a second, during one four-thousandth second. It seems unbelievable, yet it is borne out in practice: the projectile can be seen very clearly by the unaided eye. The eye's optics and the brain's conception are capable of taking in, with the help of ultra-high-speed flashes, a segment of an event taking place within a millionth of a second. Judged by such capacity, "quick as a wink" is quick, indeed.

How does one arrive at a film capable of shooting, in the required time, the required large number of phases of an event? In the ordinary motion picture camera, film advancs in steps; it stops twenty-four times per second for a brief instant to take a shot of a single phase. During the intervals of darkness, the film is advanced. With the commonly used slow-motion camera, the film can be speeded up about four times. At about one hundred pictures per second, the film material itself sets the limit; at higher speeds, the film threatens to tear from the jerky motion.

For higher speeds the film has to run in an uninterrupted motion. It is whipped along by two motors: one activiates the feed roll, the other the take-up roll. Such cameras briefly howl like sirens: thirty yards of film race through the camera within half a second and virtually shoot past the objective. A rotating prism is synchronized with the passage of the film. It equalizes the movement of the film, and, in a manner of speaking, pulls the image along in much the same way an experienced sports photographer makes his camera follow a racing car with a sweep of his arm. With a sixteen-millimeter film, one achieves in this way about

8,000 pictures per second; that is, slowing down the motion 300 times.

If the film is rolled up on a large drum in the camera, it can race past the objective with the speed of an airplane. For a slow-motion sequence, the shutter of the camera is opened for one full rotation of the drum. The flash is electronically triggered; the speed limit is 50,000 pictures per second. The brevity of the flashes results in razor-sharp pictures. It is true that the film is only five feet long.

If one aims for even more pictures per second, one has to forego moving the film. A rapidly spinning mirror sweeps successive phases of the event across a series of small objectives that are arranged in a circle around the mirror. Rotating mirror cameras are capable of shooting pictures at a rate of several million per second. The stationary film has room for only a relatively small number of pictures, but on the other hand, they are exposed for unimaginably short periods. What happens in a time progression is recorded in linear succession in a circle of from 25 to 125 frames. But haven't we been talking about millions of pictures? Yes, but "millions of pictures per second," refers to the speed of picture-taking, not to the actual number. Since such a camera can be operated for only a very short time, it has to be exactly synchonized with the event to be photographed, so as not to miss it.

"One million pictures per second" means that this number could be fitted into a second; it does not mean that that many pictures were in fact taken. A sequence of 25 pictures shot at 1/1,000,000 second means that an event has been recorded that lasted for 1/40,000th second.

It is possible to select a set-up that does away entirely with moving parts. With this technique the scene is lighted or transilluminated by a succession of sparks. The images, each imprinted on the film by a single spark, are pictured by separate objectives, side by side on one photographic plate. The practical limit of this technique is two million pictures per second. This corresponds to a 100,000-fold retardation. In many cases it is possible to gain the looked-for information from a short sequence of high-quality pictures.

Sparks and Shadows

In the late nineteenth century, the Austrian physicist Ernst Mach photographed sound waves and flying bullets with the help of electric air-gap sparks. But it was Carl Julius Cranz, who some fifty years ago became truly the master of spark photography. In 1928, he created together with Hubert Schardin, who was his student at the time, the so-called Cranz-Schardin multiple-spark camera. With 3,000-fold time-expansion, they filmed shells piercing or ricocheting from armor plate. Schardin filmed the impact of pistol shots on one-fourth-inch-thick plate glass with 8,000-fold time-expansion—that is, 200,000 pictures per second; he filmed the impact on a sheet of safety glass at 400,000 pictures per second. The Cranz-Schardin spark-flash method uses a chain of sparks igniting at unimaginably brief intervals and casting light for a tiny span of time repeatedly and so sharply, on a projectile, for instance, that its shadow can be photographed just as sharply. Shadowgraphy and schlieren photography yield excellent results. Flow and shock processes, ballistic and aerodynamic phenomena can be made accessible to the human eye.

Knowing how to deal with fast currents, with turbulence, with thrust- and shock-waves becomes increasingly important for modern technology, operating at ever greater speeds. Supersonic wind tunnels are ambitious and costly installations. Much can be explored more easily and cheaply in the so-called shock-tube, a device that permits the study of the same phenomena in miniature. The shock-tube is a wind tunnel reduced in size in which wind velocities up to twenty times the speed of sound can be simulated. Of course, everything happens much faster inside the shock-tube; but "time-microscopy" remedies this drawback.

It is now possible to observe what happens to a test object as it breaks through the sound barrier. Calculations are confirmed or refuted by the evidence. Accidents can thus be investigated and catastrophies prevented. The designer gains unimpeachable documentation; the researcher gets answers to his questions and the answers lead him to more questions. An artist might be inspired by the nearly inexhaustible wealth of shapes. The "true" nature of our surrounding world reveals itself in fantastic pictures.

At a rate of 100,000 frames a second, Hubert Schardin has filmed in the shock-tube how a supersonic airflow passes all around a wedge. One can observe a shock wave rolling over the obstacle, the play of opposing forces, reflected waves penetrating or overlapping each other, turbulences forming. As laws of nature are transformed into abstract illustrations, the pictures take on a peculiar mathematical beauty and a esthetic charm. They are nature's own designs that the scientist can read like a formula. Movements of inconceivable speed can in this way be dissected by technology.

Electro-optical Shutters

Millions of pictures per second are by no means the upper limit. But we now have to turn again to the problem of the shutter. Mechanical shutters, as we mentioned before, are far too slow. What was needed was an electric system that would allow the light to pass only for the briefest of time spans. The so-called Kerr cell provides this looked-for shutter: it consists of a glass container holding nitrobenzene and having, at each end, a sheet of polarizing material. Photographers know that, when taking pictures against the light, a polarization filter will eliminate the reflections from, for instance, a mirror of water. A polarization filter blocks reflected and thereby polarized light, light that vibrates in one plane. With two polarization filters, turned at ninety degrees against each other, the passage of all light through the nitrobenzene can be blocked. And now the so-called Kerr-effect, named after the Scottish physicist John Kerr who discovered it in 1875, comes into play. When an electric field is applied to the nitrobenzene, this substance acquires a particular quality, it becomes double-refracting—it depolarizes polarized light. The blockage is removed; light can pass! Electric fields can be applied for the briefest time spans; correspondingly brief are the periods when light can pass through the Kerr cell, resulting in a very fast shutter speed. Today, speeds of one trillionth of a second can be attained by means of Kerr cell shutters.

The image converter can also be used as a "shutter." In an image converter, a visible image is transformed into its electronic replica, which then is made visible again on a fluorescent screen. By permitting the electrons to proceed to the screen only during the desired minute time segments, the shutter is opened for just those tiny time spans. Sequences can also be produced: separate electron images are projected on adjoining sections of the screen by electric deflection.

This technique permits plasma physicists to photograph their experiments. Plasma is gas ionized through high temperatures. It has been possible to heat plasma very briefly in "magnetic bottles" to millions of degrees. The attending events can now be photographed and analyzed. The technique is also used in research of the grandest energy project of the future: controlled nuclear fusion.

Combining Techniques

It stands to reason that one should be able to combine slow-motion with other techniques employed to see the unseen, and thereby reveal the doubly invisble. For example, one can combine microscopy of space with microscopy of time. Two areas, each in its own way outside the range of our sense perception, become thus accessible to the eye.

Fritz Brill has investigated what happens to printer's ink in rotogravure printing. How does the ink behave within the gravure cells of varying depth etched into the cylinder? On a high-speed press the ink is subject to a considerable centrifugal force. On a printing cylinder of 1,100 millimeter circumference there are 7,700 gravure cells lined up in each row. The microscope is focused on one group of these cells on a space of less than one millimeter square. The color photograph is to be magnified at least 40 times. The light has to come from the side in order to produce the most three-dimensional effect possible. Color film requires a lot of light. When the cylinder rotates, at a speed of 10 yards of paper per second, the picture under the microscope changes 63,000 times every second!

In order to get a sharp picture nonetheless, a 1,200-volt spark-generator is used that shoots off sparks of less than half a millionth of a second duration. The sparks are electrically timed so as to illuminate always the same group of gravure cells and their imprint on the paper, while the cylinder rapidly spins. Photography and electronics together make the unimaginable come true. Half a millionth of a second is enough to "freeze" the motion of the rotating cylinder, to catch again and again the same carefully designated spot for the microphotography: the ink in the tiny cells, of which a millimeter can accommodate 7 times 7!

Light can be "manipulated" and combined with high-speed photography to solve stress problems that defy mathematical solutions: the fast moving parts of a machine undergo enormous stresses which the material must be able to withstand. The circle of vanes on the spindle of a steam-turbine rotates under a pressure of 200 atmospheres. The centrifugal force exerted on the vanes equals a weight comparable to that of a locomotive. Modern technology wants to achieve the utmost with the greatest economy. Transparent scale models are fashioned of plastics and exposed to proportional stresses. They are transilluminated with polarized light; high-speed photography then reveals shifting patterns of strange beauty. The technologist draws his conclusions from these stress patterns. He sees where and how they make themselves felt.

Some events in the microworld are very fast, for instance, certain crystallizations, chemical reactions, interactions of liquids, biological processes. Under the microscope, the cilia, with which protozoa swirl food into their mouth and move around, generally show up as nothing more than a halo. At hundredfold magnification, the object's speed also appears one hundred times magnified. Fast flashes with "cold" sparks that do not harm the single-celled animal, nonetheless make high-speed filming possible and clearly show the functioning of these swirling hairs in dreamlike slow-motion.

Where fast events take place in regions where light cannot penetrate, X-ray flashes or series of X-rays flashes have to do the job. The inner functions of a motor become visible in this way. In cases where self-luminosity bars observation of rapid events, again, X-ray flashes are used to photograph what the radiance of the object hides. The hampering luminosity can be overcome by simple means, a piece of black paper, for instance; X-rays pass effortlessly and record a shadow picture of the object.

Are there limits to high-speed photography? Technologists distinguish between what is technically achievable and what is reasonable and desirable. One would not use a faster shutter than necessary to picture an object sharply. One would not increase the apparent expansion of time through ultra-slow-motion to the point where our eye is no longer able to make the connection between phases of the motion. Events that occur in the shock-tube or during detonations are close to the limit of today's phototechnical capability. The world of nuclear and plasma physics may hold phenomena that are outside this limit. "New questions lead to the development of new methods. And new ways of attacking problems lead to new questions." Walter Thorwart finds this dynamism that has pushed forward scientific theory and research especially pronounced in the field of high-speed photography. Who would doubt that the ingenuity of the technologist will advance beyond today's frontiers into the no man's land of the as-yet unknown?

On the Tracks of Nuclear Particles

To photograph events in bubble chambers, like the ones at CERN, the European Atomic Research Center near Geneva, three flashlamps that for three seconds shoot synchronized flashes, are used. They catch the fine trails left behind by the elementary particles as they pass through the bubble chamber. Slow, fine-grained film is used. Considerable depth of field is required. Highest possible light-energy at shortest possible flash is essential for catching these quickly vanishing trajectory traces. Each individual flash lasts 150 microseconds; that is the seventh part of a thousandth of a second.

Some time ago, anti-protons were produced, elementary particles that in a way belong to another world, a "minus-world." The existence of the anti-proton had been predicted, but now it has been identified and it can be "produced": an act of creation, uncanny and magnificent at the same time, a feat that touches on the last mysteries of matter, on that which bonds the core of the universe.

The life-span of an anti-proton is a nanosecond. A nanosecond = a thousandth of a thousandth of the thousandth part of a second = a billionth of a second. This could be ascertained beyond a doubt through high-speed photography.

Professor W. T. Runge, head of the Telefunken Research Institute after World War II, is a scientist who has the gift of making the complicated simple. In his book *Electronics Is Not Sorcery*, he asks: "But what insights and discoveries await us, once we can measure the thousandth part of a nanosecond, the picosecond, once we can grasp the experience of the anti-proton during its life-span, during that one nanosecond? They will correspond to those gained by the increased magnifying power of the microscope. We do not know what we shall see then, but we do know that it will be something entirely new. And if one were to see nothing new, that would simply be another incentive to try to increase magnification even further."

The Motive: Curiosity

Professor Runge cannot believe that this persistent searching, this urge "to resolve the duration of time into ever smaller segments," is ultimately based on a desire for economy and profit: "Neither Philip Reis nor Graham Bell, when they invented the telephone, were primarily concerned with marketability; rather, they were fascinated by the idea of transmitting the human voice instantly over long distances. Actually the telephone was created," so claims the smart professor with a wink, "to permit separated lovers to hear each other's voices in the evening, and to engage in conversation, and the only reason why business is permitted to use the telephone during the day, and higher rates, is to finance the enormously costly intercontinental set-up and its maintenance—something the lovers, of course, could never pay for." The question about the "why?" leads one to "a basic human tendency which needs no further justification, which is its own ultimate justification. It is... the urge to push back the limits of comprehension as far as possible and even further, and it is the curiosity about the means to this end and about the degree to which we may succeed."

"Automation plays an important role in high-speed photography," says Walter Thorwart in conclusion. "The entire process that is involved in high-speed photography is pre-set for fractions of a second. The 'pushing of the button' must in most cases be left to the object itself. High-speed photography, therefore, is not a mere task for phototechnique. Its success depends on the interplay of optics, precision mechanics, and electronics as well as on the utilization of physical phenomena like the Kerr-effect." It is precisely this multiplicity of branches of knowledge that contributes to high-speed photography, that keeps the technologist in that field constantly on the lookout.

In space travel, the most adventurous technological undertaking of our time, high-speed cameras record every essential step, beginning with the moment of ignition, when the fiery stream of gases appears and the rocket lifts off the launching pad. High-speed cameras are focused on the blazing furnace of a seemingly superhuman technology. They see what happens in fractions of a millionth of a second; they make it possible to later analyze the event, regardless of the mission's success or failure. Whether something happens in a second or in a much, much shorter time, it is preserved as flawless documentation for the probing eyes of the engineers. High-speed cameras accompany the flight into space. They record the separation of the rocket's stages. They film the return voyage, the reentry into the denser atmosphere, the fiery glow of the heat-shield that protects the capsule from burning up like a shooting star. They not only bear witness to a great feat of engineering, they testify to the dynamism of our technological age. And yet, the pictures of the falling drop of water still constitute the most entrancing demonstration of the achievements of technical ingenuity: an event eclipsed in time, enters the range of human vision.

A new dimension gains in depth.

Time is unchangeable—it cannot be condensed, nor can it be expanded. One second lasts one second. Speeding up events in time or slowing them down can only be done via the motion picture camera. Extreme slow-motion, "microscopy of time," opens up new dimensions.

"Dr. Edgerton's photographs provide a unique and literal transcript of that time-world beyond the threshold of our eyes. In these pictures are not only facts to help us in seeing and doing, but new aesthetic experiences."
(J. R. Killian)

Professor Harold E. Edgerton, Massachusetts Institute of Technology

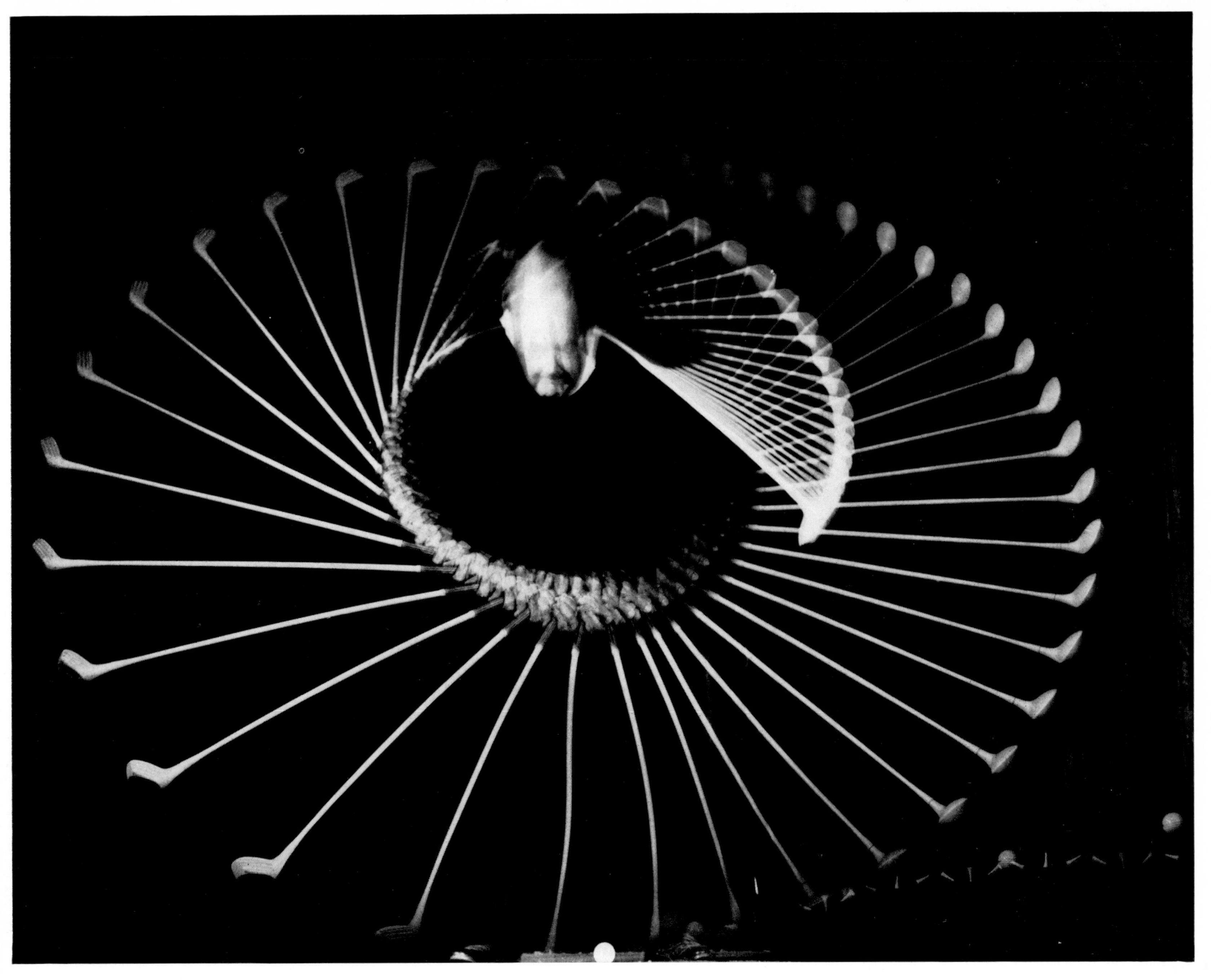

A multiple exposure fans out a golfer's stroke into fifty positions. The interval between exposures is 1/100 of a second. The duration of each exposure, 1/100,000 second.

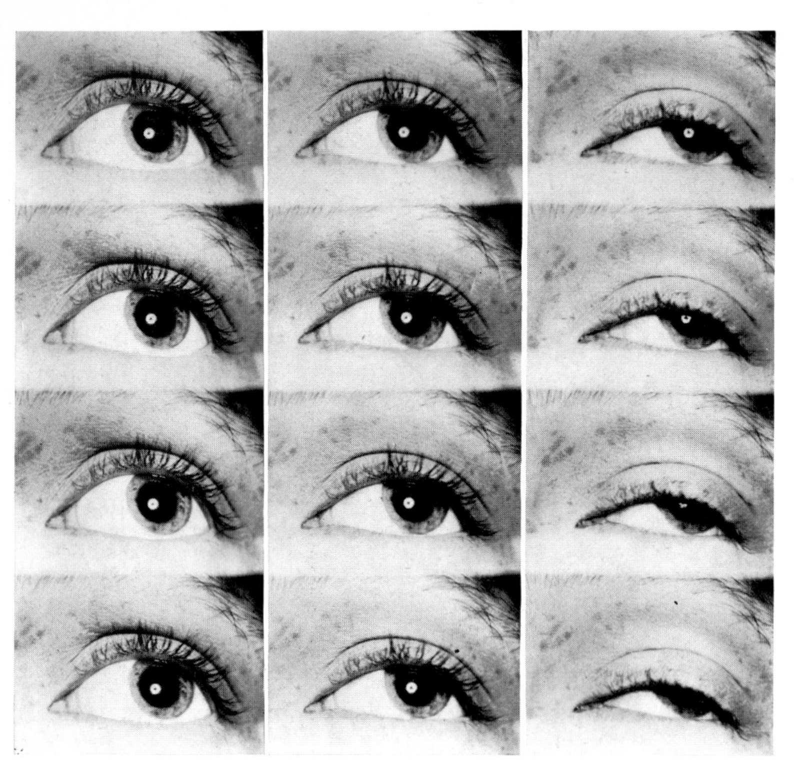

Professor Harold E. Edgerton, MIT

An eye shuts under the glare of a flash. The high-speed camera records the lowering of the eyelid as a drowsily slow motion. The film on which this blink of an eye was caught, raced through the camera at a speed of eighty feet per second. Nevertheless, the quality of the pictures is excellent.

The shot cracks! The playing card is torn in two, and Edgerton has photographed this very event that happened too fast for the eye to see. The flash, one half of a millionth of a second, was perfectly timed! The bullet's impact on the playing card is a masterpiece of high-speed photography. Of course, making bullets in flight visible is not the ultimate goal of ultra-high-speed photography. Modern technology works with dizzying speeds that far exceed the speed of a bullet in flight. Extreme slow-motion catches up with the ever faster events with correspondingly higher film speeds. ▶

Walter Thorwart, Hamburg

Scientist and technicians never tire of photographing the falling drop. There is no institute active in the field that has not experimented with this attractive theme. Four phases have been recorded here with flashes at intervals of five thousandth of a second.

Walter Thorwart, Hamburg

◄ The falling drop can be combined with another enchanting structure, the opalescent soap bubble. Generally, the drops fall right through the soap bubble without damaging it. Strangely enough, it takes quite a while before the tenuous soap skin of only a few molecules' thickness falls victim to the hail of drops. Three phases can be observed in this picture: the impact of the drop, the rising crown, the tearing soap bubble.

Walter Thorwart, Hamburg

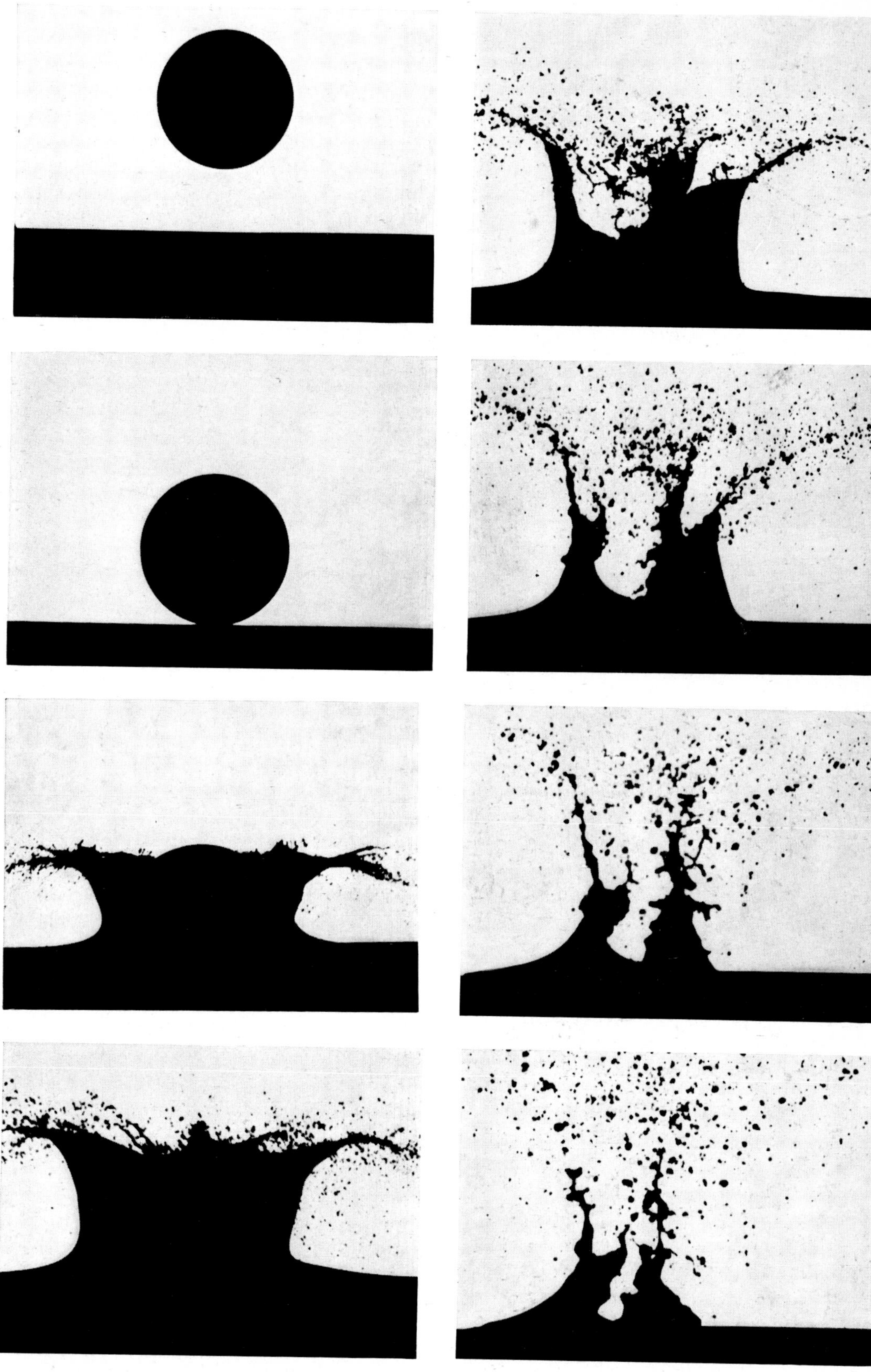

Among fast-moving events, the falling drop is a comparatively slow one, and one that the eye can grasp as a whole but not in its consecutive phases. Here the camera records, with 100 pictures per second, how a crown wall forms, how it collapses. In this sequence taken by the schlieren method, the water looks opaque because of light refraction. A silhouette, similar to an X-ray picture, is the result.

Walter Thorwart, Hamburg

◄ A shot through a lightbulb. The exact split second to "capture" a bullet in flight can be timed by making the bullet pass a photocell sensitive to light and shadow that will trigger the flash almost instantly. Or a microphone, picking up the sound of the shot can be used to trigger the flash.

A bullet hits the edge of a safety-glass plate. Picture frequency is 20,000 photos per second. From the point of impact the zone of breakage travels radially at the high speed of 0.6214 miles per second. The broken segments are of comparatively small and even size and appear to stay in place at first before gravity causes them to cascade downward.
▼

Walter Thorwart, Hamburg

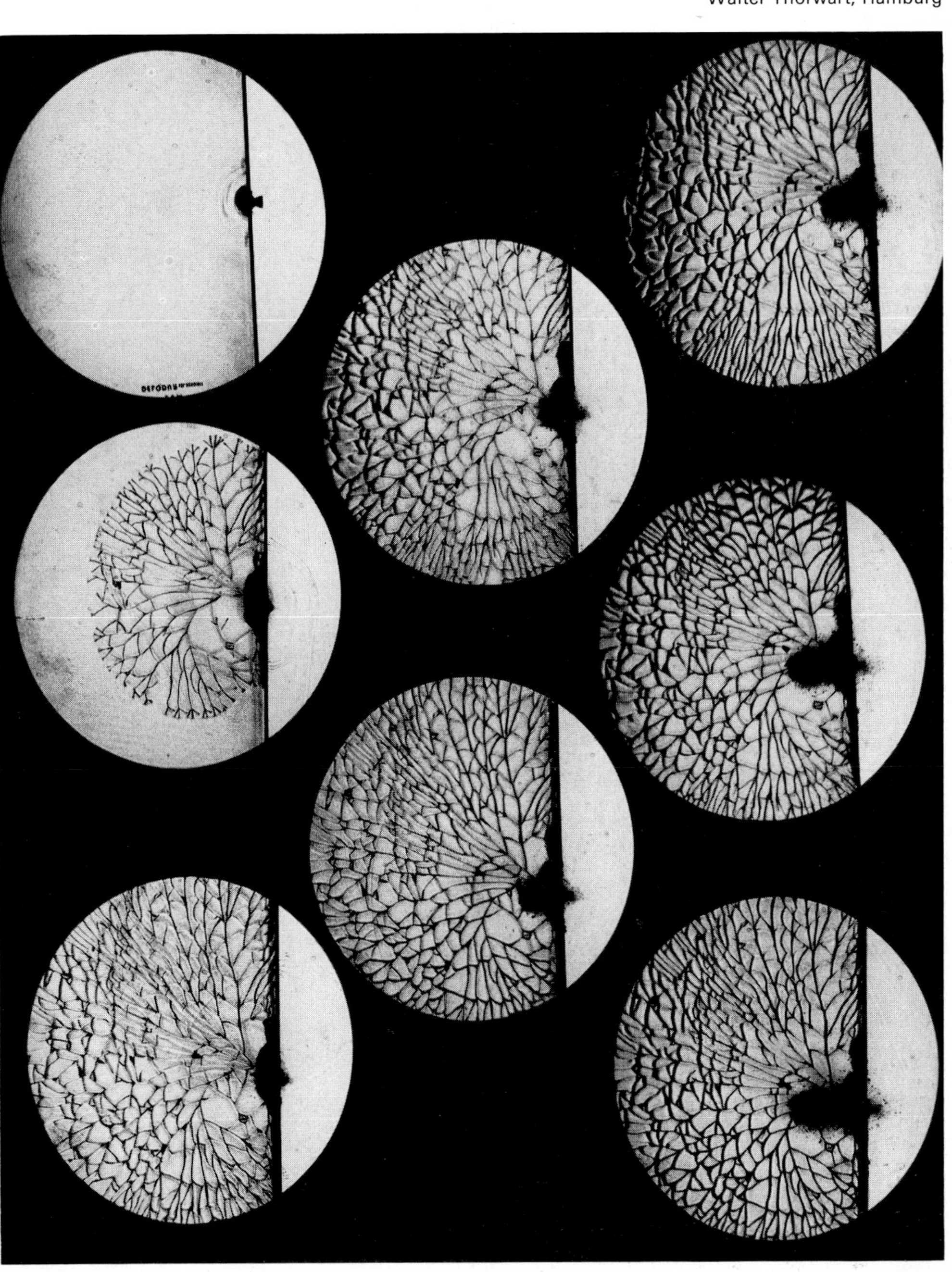

▲
The flight of a bullet "dissected" by technology. The bullet is way ahead of the puff and the gunsmoke.

The bullet transists the warm air current of the candles. Vortex and headwave are altered by the temperature variations above the flames. ►

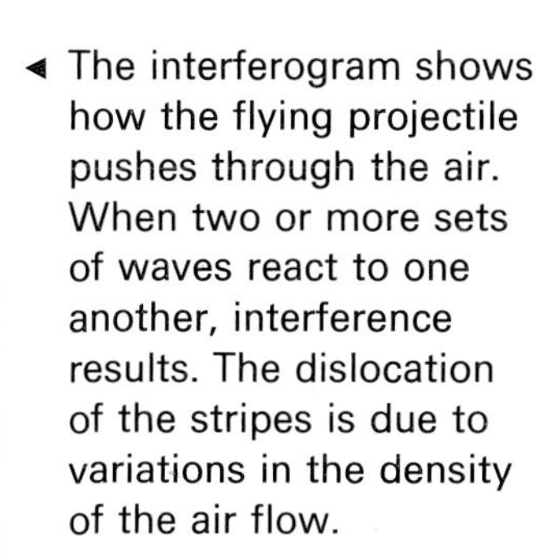
◄ The interferogram shows how the flying projectile pushes through the air. When two or more sets of waves react to one another, interference results. The dislocation of the stripes is due to variations in the density of the air flow.

Prof. Dr. Hubert Schardin

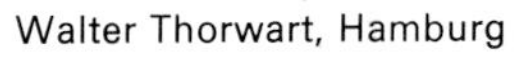
Walter Thorwart, Hamburg

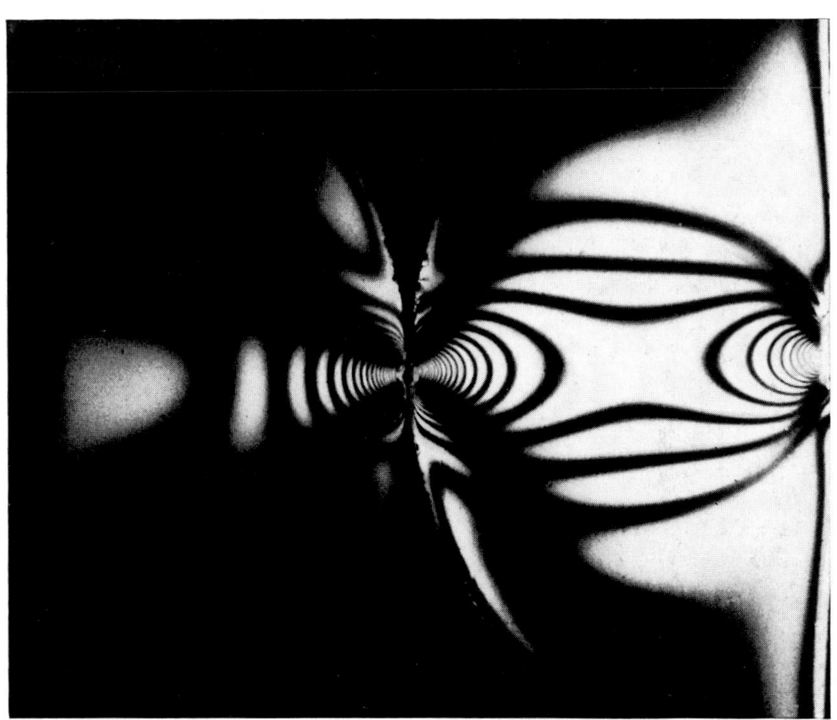

Shockwaves are made visible. A model for a "photoelastic stress analysis" is being transilluminated with polarized light. The slow-motion photography shows shifting patterns. Two phases of the shockwave are shown here: 60/1,000 second and 100/1,000 second after impact of the shock.

Walter Thorwart, Hamburg

Moder technology dea with fast flow

High-tension sparks le between two electrod The spark-trails a being swept along by supersonic air flow in wind channel. Th provide true-to-sca pictures of the invisib air curren

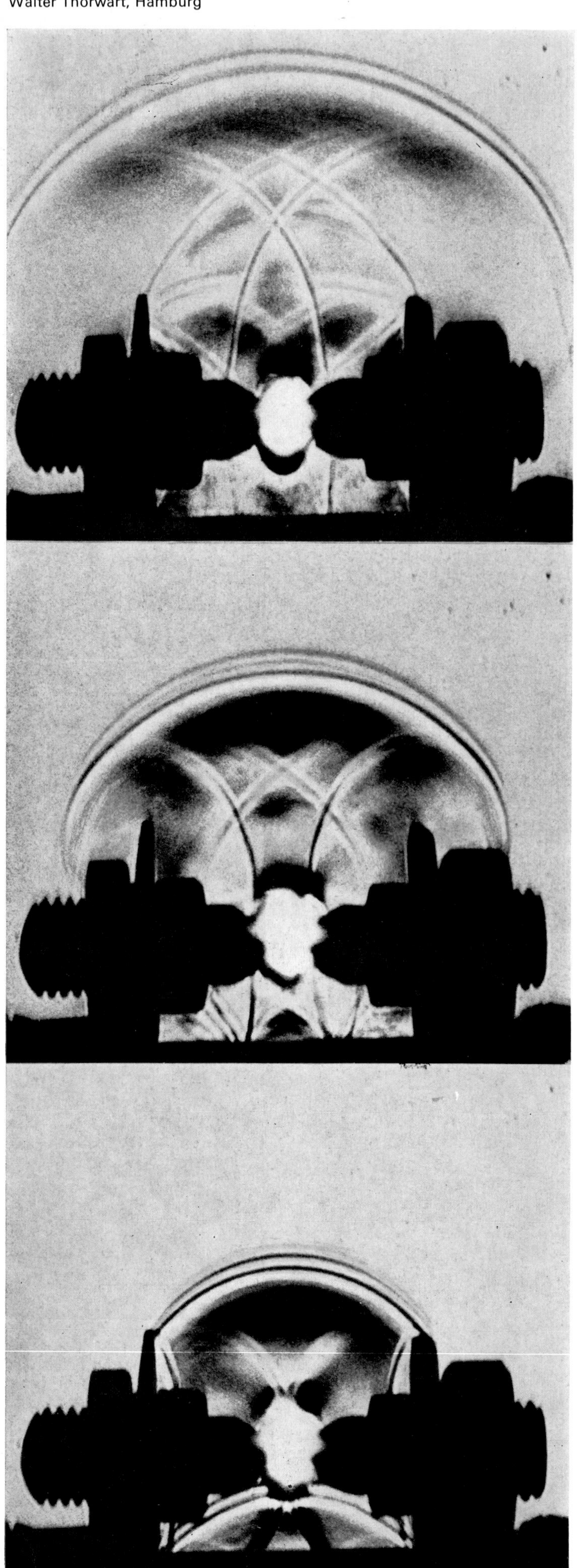

The discharge of an electric spark sends forth sound waves. Compression of air becomes visible in a schlieren photograph. Places of different air density change the light refraction and show up as darker or lighter areas. Exposure time was fifty billionth of a second; interval between pictures, twenty millionth of a second.

Fritz Brill, Laboratory for Optical Photoanalysis, Hofgeismar

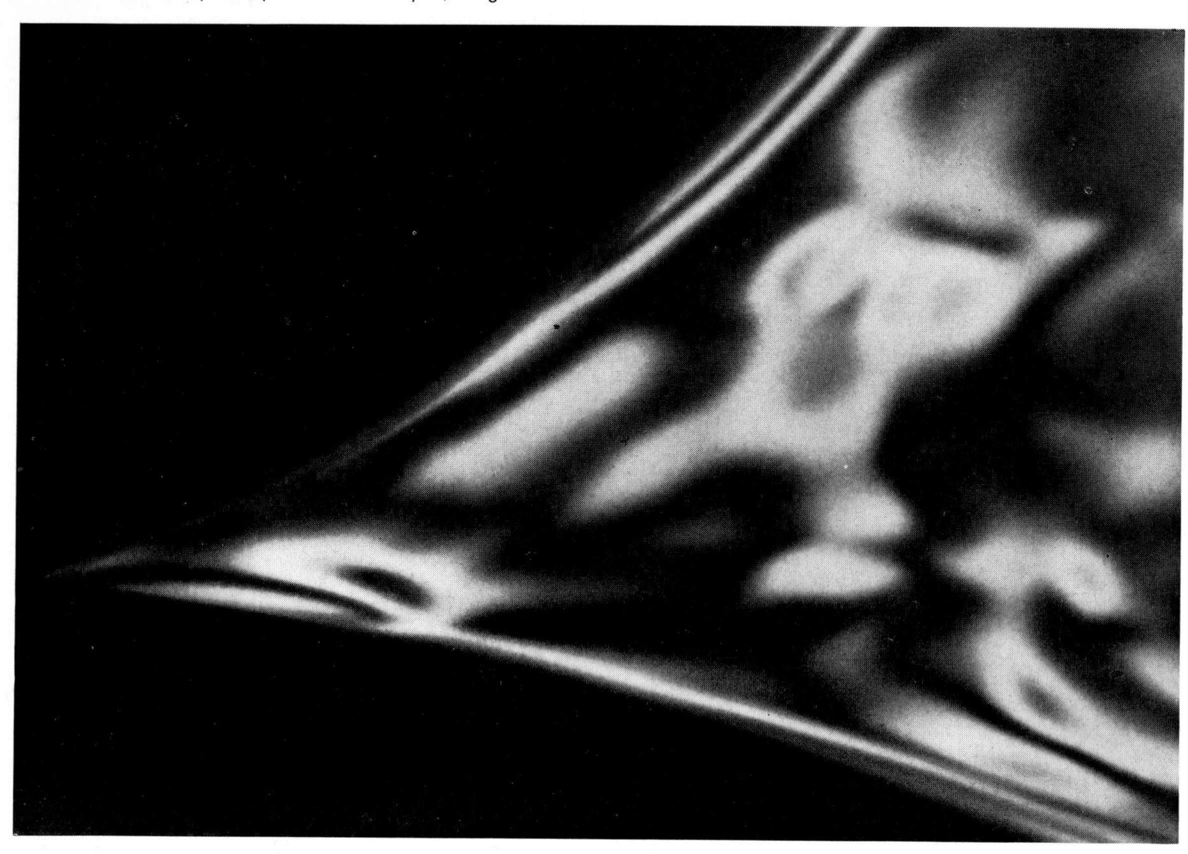

◄ The air flow between the cylinder of a rotary press and the roll of paper. A spark discharger flashes for half a millionth of a second. Electronics and photography are able to "freeze" ultra-high-speed motions.

The shockwave of an underwater explosion traveling upward through a fissure was photographed in a shock tube with a very fast succession of sparks that made shadowgraphs. Exposure time was one ten-millionth second! The result is patterns drawn by nature that scientists can read like formulas. ►

The "true" nature of our world becomes apparent through fantastic pictures.

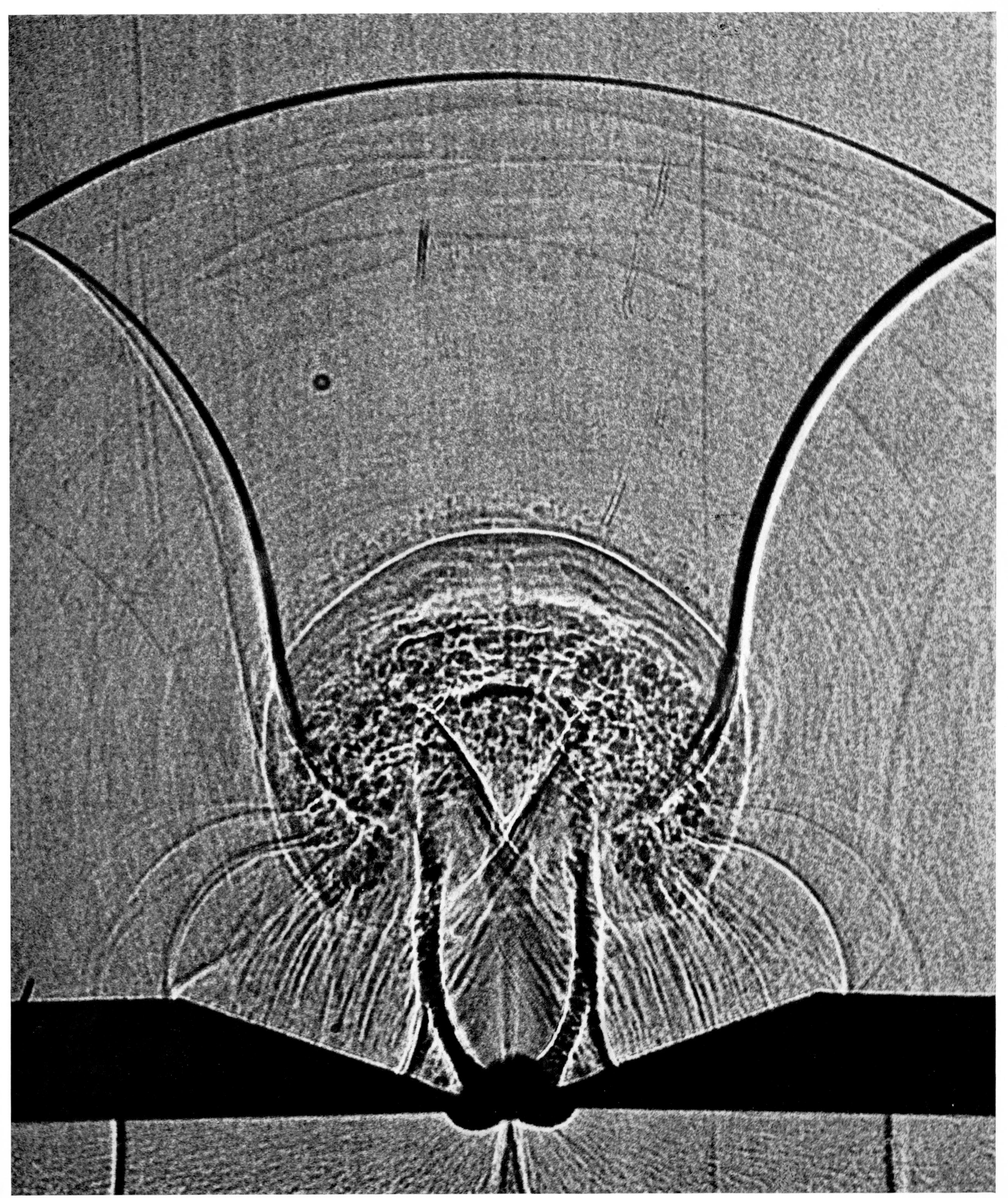

Water and detergent do their job: a grease spot is being removed. Or more scientifically expressed: a dirt particle mixed with grease is dissolved and floated away through the action of surface-tension removing substances. The microscope and time-lapse photography clarify the compelling interplay of forces. One is reminded of pictures of comets with tails of evaporating matter.

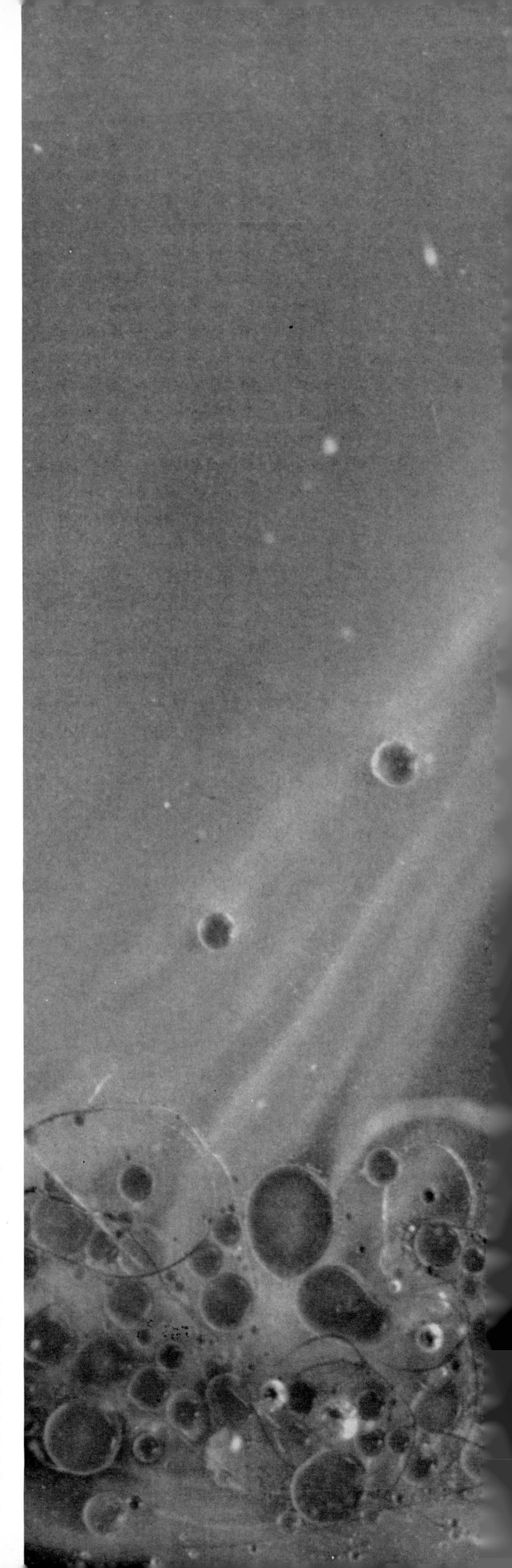

The effect of a laundering agent: oil droplets detach themselves from a fiber; they drift upward. A milky stream of suspended grease "wafts" like a flame over the dancing globules. Time-lapse photography turns the commonplace event into a captivating spectacle. The scientific researcher, the practical technician gain valuable insights, the layman receives a gift of fascinating pictures.

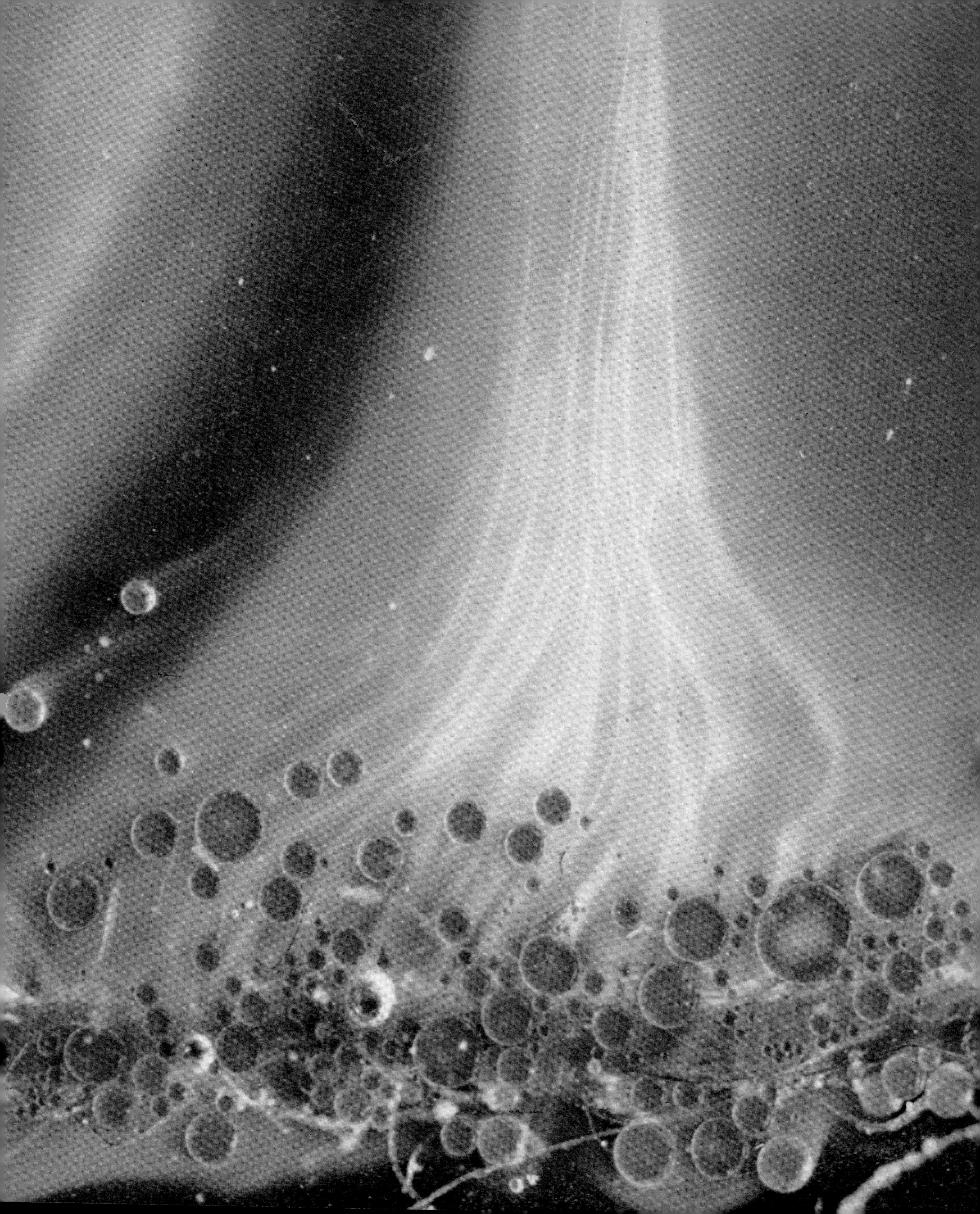

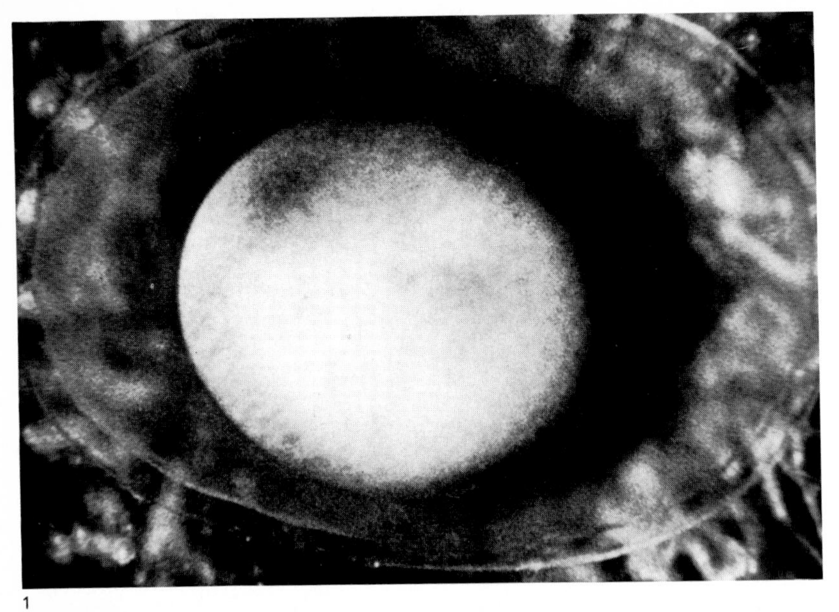

1

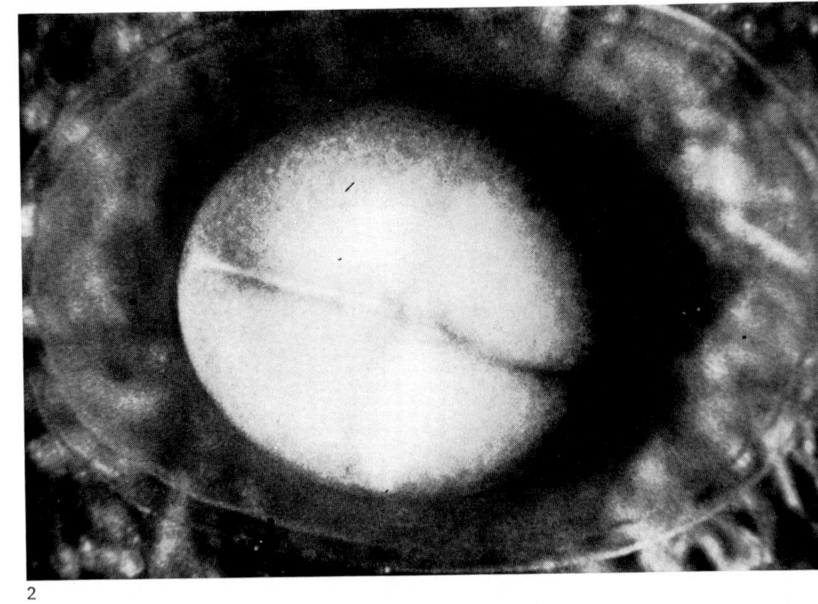

2

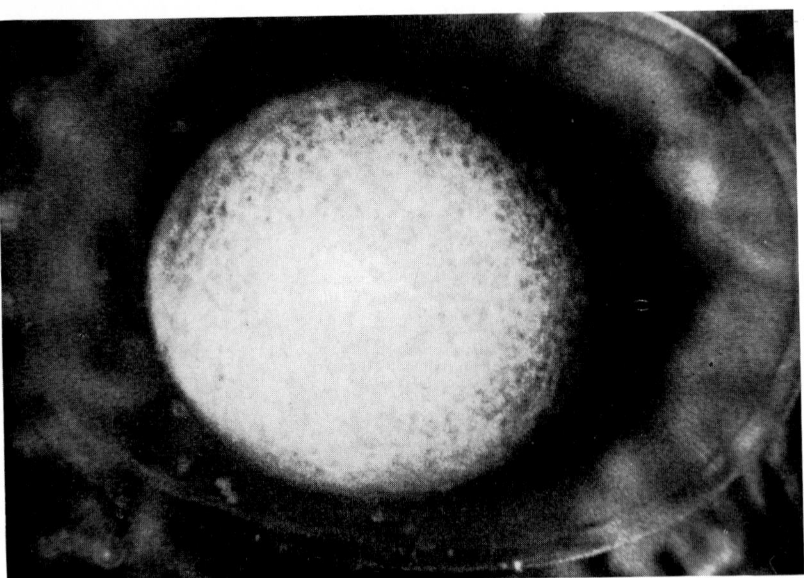

5

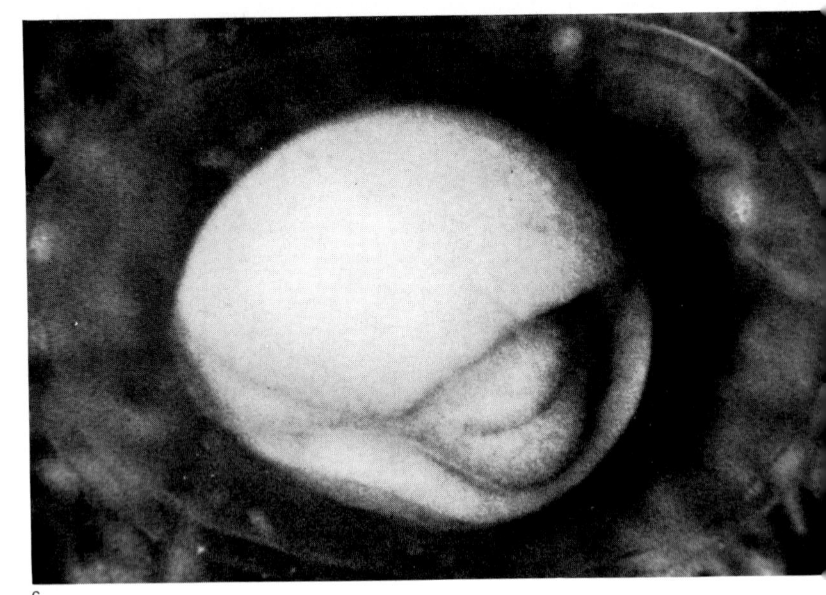

6

Seen under the microscope: the development, over a period of six days, of a fertilized egg cell of an alpine salamander. Pictures 2–5 show how the cell first divides into 2, 4, 8, 16, 32, 64 cells, and so forth. In picture number 6 can be seen how the central nervous system begins to form in the fold. In pictures 7 and 8 the shape of the larva is already gradually appearing. The head with the place for the eye and the beginning of a tail are recognizable. Only a salamander. But salamanders belong to the vertebrates, as do the mammals which include man.

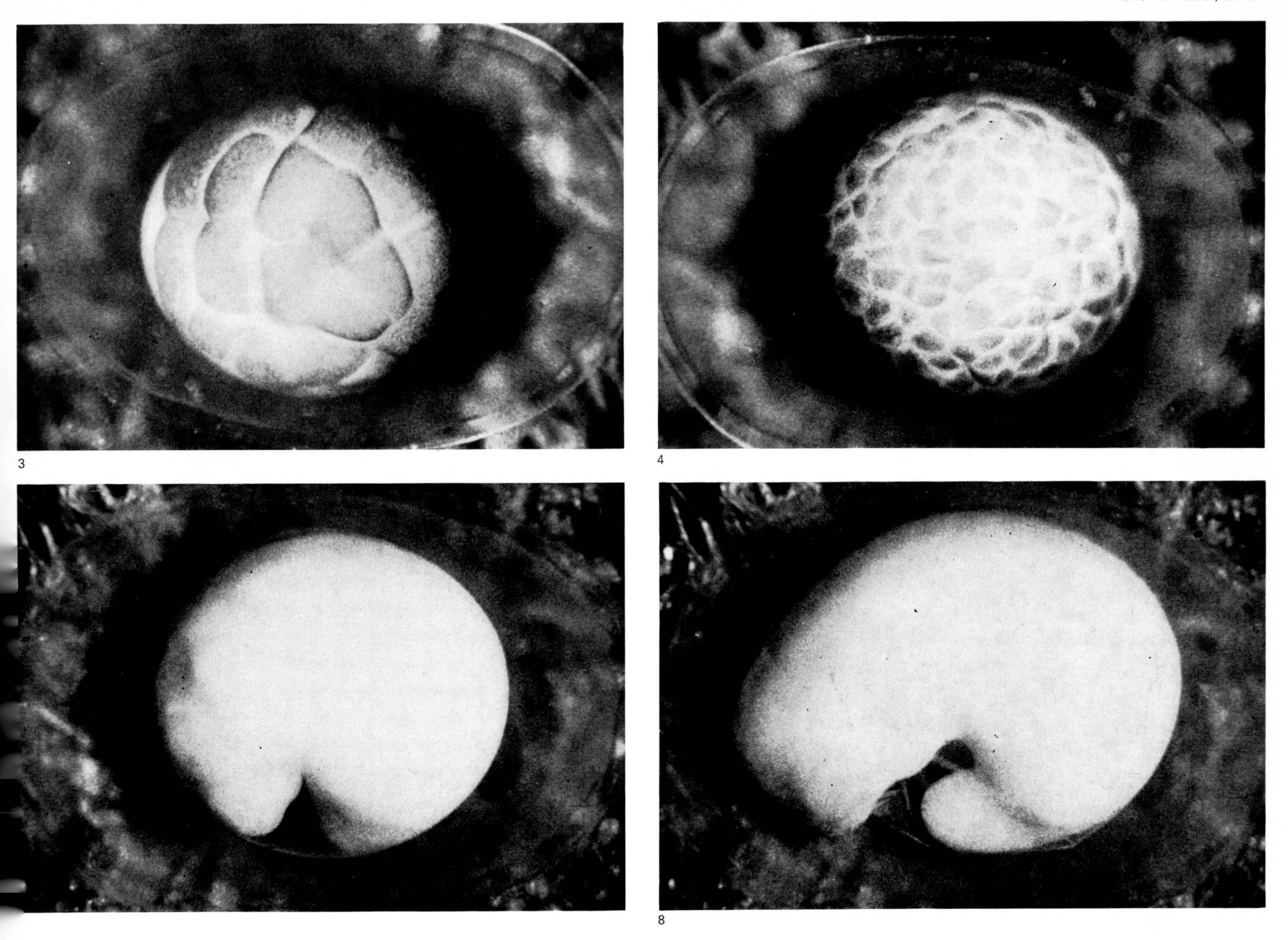

3 4 8

Time-lapse films lead us
to the threshold of the ultimate secrets of life.

Hans A. Traber, Zurich

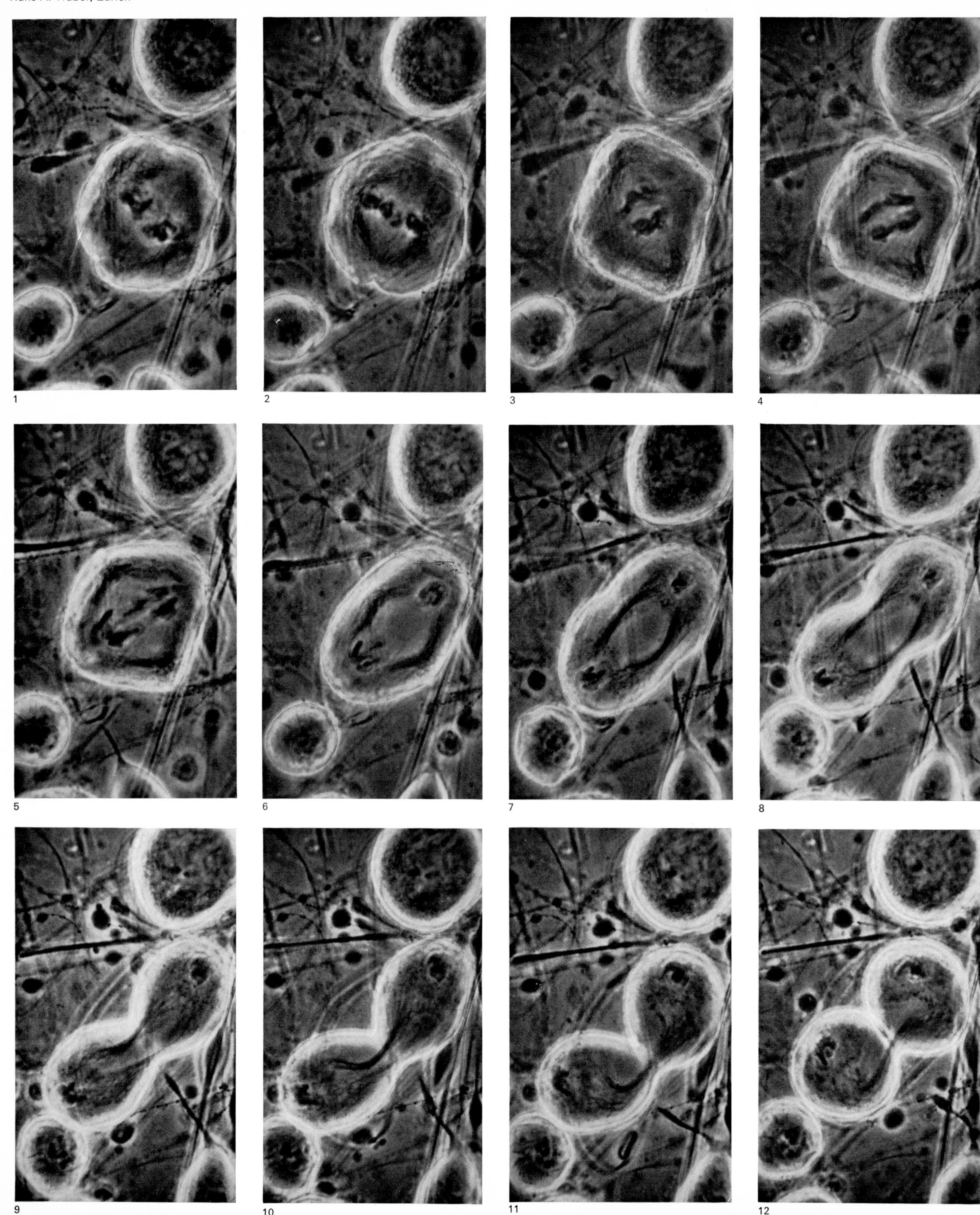

The hereditary information of all of humanity could be fitted into a globule not much larger than a pinhead.

The classical laws of heredity that apply to all living creatures were to a large extent discovered by studying insects. The microscope and time-lapse photography have disclosed a comprehensive view of the amazing mechanisms of life. The maturing of sperm cells in the testicles of a grasshopper, the behavior of the chromosomes that are vital to the still not fully explained workings of heredity, can now be filmed. It can be seen how the chromosomes, small filaments, align themselves in the center of the cell; how they suddenly, as if on command, move apart; how the cell becomes pinched in at the middle; how it divides. The pictures were taken under a light microscope by the phase-contrast method. Magnification is not quite one thousandfold. Twelve hours condensed into half a minute!

INDEX